The Inequality of COVID-19

Immediate Health Communication, Governance and Response in Four Indigenous Regions

Eric E. Otenyo
Department of Politics & International Affairs,
Northern Arizona University,
Flagstaff, AZ, United States

Lisa J. Hardy
Department of Anthropology,
Northern Arizona University,
Flagstaff, AZ, United States

ELSEVIER

ACADEMIC PRESS
An imprint of Elsevier

Academic Press is an imprint of Elsevier
125 London Wall, London EC2Y 5AS, United Kingdom
525 B Street, Suite 1650, San Diego, CA 92101, United States
50 Hampshire Street, 5th Floor, Cambridge, MA 02139, United States
The Boulevard, Langford Lane, Kidlington, Oxford OX5 1GB, United Kingdom

Notices

Knowledge and best practice in this field are constantly changing. As new research and experience broaden our understanding, changes in research methods, professional practices, or medical treatment may become necessary.

Practitioners and researchers must always rely on their own experience and knowledge in evaluating and using any information, methods, compounds, or experiments described herein. In using such information or methods they should be mindful of their own safety and the safety of others, including parties for whom they have a professional responsibility.

To the fullest extent of the law, neither the Publisher nor the authors, contributors, or editors, assume any liability for any injury and/or damage to persons or property as a matter of products liability, negligence or otherwise, or from any use or operation of any methods, products, instructions, or ideas contained in the material herein.

Library of Congress Cataloging-in-Publication Data
A catalog record for this book is available from the Library of Congress

British Library Cataloguing-in-Publication Data
A catalogue record for this book is available from the British Library

ISBN: 978-0-323-99867-3

For information on all Academic Press publications visit our website at https://www.elsevier.com/books-and-journals

Publisher: Stacy Masucci
Acquisitions Editor: Elizabeth Brown
Editorial Project Manager: Sam W. Young
Production Project Manager: Kiruthika Govindaraju
Cover Designer: Mark Rogers
Cover Credits: Jack Ole Lekishon, Maasai Community leader

Typeset by TNQ Technologies

Contents

Acknowledgments

A great deal of attention has been paid to the response efforts carried out by different governments for preventing the spread of COVID-19. The immediate response disclosed, however, that in predominantly rural Indigenous communities, stories of disciplined and selfless mitigation practices and resilience were not given enough scholarly attention. Moreover, as national governments in large and small democracies struggled to fight back against the COVID-19 pandemic, a large body of scholarship and commentary revealed the enduring inequality in health-care services within these countries. More significant, the inequality health care in Indigenous societies remains a matter of global concern.

Logically governance and communication about the pandemic will be subjects of academic and popular inquiry for the rest of our lives. And to the extent that existing unequal COVID-19 mitigation efforts affect Indigenous communities, interdisciplinary debates will continue.

Just having to be a part of the discussions about Indigenous communities and responses to COVID-19 prompted us to write this book. And it has been a joy to think this through with scholarship and personal conversations with many Indigenous people. With this book, we extend humbled gratitude to our Indigenous colleagues and collaborators for teaching us how to be resilient during a devastating pandemic. We acknowledge that we live on lands sacred to local tribes and conducted this work in debt to those who lived and continue to live here. We hope that this book will contribute to shifting the conversation while acknowledging the importance that we are not the only or the most important voices here.

This contribution is also intended to be only one slice of a much larger conversation. From Indigenous leaders and authors we will continue to listen and learn. We can't thank you enough for your support and kindness and we hope to elevate your work as the conversation continues.

Many of the difficulties in writing any book were eased through collaborative efforts by so many of our friends and colleagues. We owe debts of gratitude to scores of friends and colleagues from Indigenous societies who shared their expertise with us. Eric Otenyo is grateful to Pamela Otenyo, Nancy Otenyo-Siage, Anthony Otenyo, and Timothy Otenyo for their moral support. Lisa J Hardy wishes her sincere thanks to Fiona Hardy-Wybranski, Jane and Steve Davis, Gwendolyn Saul, Kerry Thompson, Jennifer Guyer, and Karen Renner. In addition, to members of our respective families, who cheered us on, we will forever be grateful to the excellent reviewers of the manuscript project, Elsevier staff and editorial team including the amazing Senior Acquisition editor Elizabeth (Liz) A. Brown, Sam W. Young, Romulo Viola, and Mani Indhumathi. They are not responsible for any errors in this book. Judgments and analyses made in this book do not represent the views of our employer Northern Arizona University.

The Inequality of COVID-19: Immediate Health Communication, Governance and Response in Four Indigenous Regions, is a beginning of a long nontechnical conversation designed to bring together scholars and practitioners who cherish interdisciplinary discourses on governance and contagion. Our hope is that, in due course, other scholars will bring to light Indigenous knowledge of combatting pandemics that may continue to trouble our planet Earth. And there are painful concerns. Scarce health-care resources must be shared equitably and Indigenous knowledge and governance must be supported. Only then will governments all over the world be lauded for responding to the needs of all citizens. External governments will only been deemed to be successful if they cease to continue the violence of ongoing settler colonialism by supporting and responding to through the active concept of health as a human right.

We dedicate this book to all who have been adversely impacted by COVID-19.

Eric E. Otenyo, PhD

Lisa J. Hardy, PhD

Flagstaff, Arizona

CHAPTER 1

Introduction

Although there have been many outbreaks of contagious diseases in modern human history, nowhere was the use of modern information and communication technologies (ICTs) more widely used than in the novel coronavirus (COVID-19) pandemic. And governments and communities have responded to this challenging episode of global crisis by employing a bewildering variety of communication approaches to mitigate and organize public health initiatives against COVID-19. Existing governmental and community structures have previously demonstrated that effective communication is central to public health emergencies. The global pandemic provides a natural experiment to investigate where and how ICTs are deployed and how they can be improved in future crises.

The World Health Organization (WHO) considers communication and management of contagious disease as inseparable processes. Effective and efficient communication occupies a central place in the WHO's tool kits of managing or containing the spread of global pandemics. The guidelines are issues for the purpose of strengthening health security by implementing international health regulators (WHO, 2008). Similarly, Chapter 12 of the US Centers for Disease Control and Prevention's (CDC) *Epidemiology Field Manual* has set out an elaborate template on how officials can communicate with clarity during disease outbreaks (Tumpey et al., 2018). This is echoed in leading public health journals (Holmes, 2008; Taylor et al., 2009; Vaughn & Tinker, 2009).

The CDC *Field Manual* describes issues of risk management and advice on managing people's expectations depending on severity of the harm from a pandemic. Communication with members of the public must be credible for both social and political reasons. The CDC recognizes that in today's 24-hour news cycles and the proliferation of information sources including digital media devices, television (TV), and social media communications within communities affected must be timely and accurate. There are huge public safety dangers associated with harmful information peddled through the numerous outlets. Government action on communication is crucial for health communication and should be given scholarly attention. The extensiveness and scope of the spread of COVID-19 offers a rare opportunity to examine the essence of communication in various cultural systems.

Given the devastation of the pandemic in Indigenous and economically marginalized communities, we explore the force of communication in defining the magnitude of the virus, the structure and culture of the state organizations communicating to their

The Inequality of COVID-19. https://doi.org/10.1016/B978-0-323-99867-3.00009-0

constituencies, and the tensions between Indigenous responses and externally imposed health and control directives. The COVID-19 experience is arguably the defining issue of public interest in this decade and beyond. Fueled by massive coverage on social media and TV stations, COVID-19, is the story of this time. And since broadcast media is influential in mainstreaming ideas that get onto the policy agenda, one notes the powerful images projected, especially on TV stations across the globe. In fact, during the first half of 2020, COVID-19 stories became the major news items. Such stories contained a number of ingredients for increased viewership and, probably, revenue streams for the various companies. COVID-19 stories had both emotion and drama. Attention-attracting visuals of deaths initially in China, Italy, Britain, and Spain magnified the tragedy of the vulnerability of the global health-care infrastructures. News feeds were dominated by images of Italian coffins lined up next to one another and refrigerator trucks ready to receive bodies. The images also spoke aloud to our common humanity. And as the virus hit the United States, the story of loss of human lives multiplied hundreds of times over. First, it was Washington State and then, the New York, the business capital of the United States became the hot spot, ground zero. Media outlets scrambled to hold the spotlight through numbers of clicks and attention in rapidly changing and crowded global stage.

The coverage of COVID-19 followed an unchartered script. First, it was about body counts and the pain of the suffering families, the deaths in isolation, and the lack of sufficient personal protection equipment by health-care workers toiling to literally save the world. Second, it was the element of political failure on the part of public officials who seemingly were at loss of ideas on how to respond to a novel coronavirus. Third, the script became a part of the "who done it" mind-frame, which is typically an aspect of stereotyping persons and placing blame on certain races or ethnicities. In this case, the disruption caused by COVID-19 was blamed on Chinese people and their communist government and soon some leaders used the term "the Wuhan Coronavirus."

During the coverage of COVID-19, political pundits, especially, on conservative anti-China TV programming shows and on Fox TV stations did exactly that. A number of news anchors used this opportunity to continue existing racialization in the American political discourse by offering opinions intertwined with China's perceived economic threat to the United States. This eventuality is not atypical but is borne out in the vast literature on media—race relations. For example, Campbell (1995), Entman (1990), Hurwitz and Peffley (1997), Gilliam et al. (1996), Peffley et al. (1996) point out that the media coverage of "suspects" and "victims" often embraces racial and ethnic imagery. Epidemics have long been the scene of xenophobic blame (Onoma, 2020).

The eruption of COVID-19 cases became the perfect story to spew criticisms of distant people who could be blamed for contagion and to stereotype populations for their culinary and other cultural practices, which allegedly led to hundreds of COVID-19 deaths. And, when body counts exposed that, in the United States and elsewhere, mortality rates among Indigenous communities and African Americans were highest, the racial imagery of inequalities in health-care policies took center stage of the discourse. As such, it has become a key formal policy agenda item in all governments across the world. We assume that there will be both positive and negative feedback to government actions and announcements. We are aware that positive feedbacks can be shaped by resource flows and by the interaction between specific policy characteristics and the strategic political goals of stakeholders and groups within political systems.

We consider policy feedback concepts as being relevant to the dynamic public communication during the pandemic. In this case, policy feedback is about perceptions of actions taken by governments to mitigate the devastation caused by the spread of the virus in Indigenous communities. The feedback concept allows us to understand whether the initial policy trajectories undermined the existing management practices.

Given the relative instability in global approaches to managing COVID-19, as evident in the varied interpretations of the pandemic's outcomes and the global nature of the governments' policies, one may argue that the countries or regions that achieved lowest infection rates gave positive feedback to their governments. Positive feedback manifested itself in high approval rating for those leaders who communicate defectively, were transparent and relatively successful at instilling confidence within their publics. To be clear, positive feedbacks are only likely to be impactful in democratic societies. Still, in the age of social media and citizen journalism, voices of disapproval are commonplace and offer reflections and assessments of what governments did correctly or incorrectly during the year-long pandemic. Moreover, it is entirely possible that among members of Indigenous communities, the insurmountable odds placed among leadership corps render a sympathetic assessment of COVID-19 management strategies. These processes are complicated because they can and do serve as a means toward ongoing settler colonialism and the refusal of external bodies to recognize sovereignty (Estes, 2020). The scientific communities, and experts, in various governments and at WHO and CDC, more or less provided a set of recommendations that were broadcast through modern ICTs. Federal and state leaders continued with this dissemination along with rapidly changing guidelines sent with rapid force and in some cases to the detriment of tribal nations.

Despite some limitations, we recognize that advances in ICTs include a wired global network of which Indigenous societies are part. This research highlights limitations to full participation due to colonial histories and presents of lack of resources, and where external directives did not contribute to harmonious and effective communication of health directives. We also highlight several projects within nations through which leaders sought to develop internally driven crisis response despite external controls.

Certainly, the outcomes of the communication strategies are a function of various variables including leadership, culture, global institutions, and other powerful socioeconomic influences. Leadership in Indigenous societies is shaped by complex relations among different jurisdictions, different positions within or outside of sovereign nations, and bureaucracies. A vast majority of these bureaucracies are governed through structures which have direct connections to heads of states where Indigenous people are domiciled. In that case, such leaders must communicate to share specific visions and to build trust within the polity. Such trust enables tribal leaders to cultivate good working relationships necessary for goal attainment. The areas between leadership responsibilities of national and tribal leaders are also a site for ongoing tension and disagreement.

Since our attention is focused on various levels of leaders, and influencers, we highlight the importance of communication in sharing responsibilities. Communication, whether through ICTs or the old-fashioned paper and pen ways, is fundamental to management and problem-solving programs. Leadership is about motivation of members of the polity to work together for the common good. How society communicated its reactions within existing structures, in turn, has significant influence on government reactions. In particular, government communications on rules of behavior during COVID-19, availability of resources, and functioning of commerce occupied much of the headlines in popular and official media. Governments also sent messages to citizens on tasks expected and sanctions to be meted in case of noncompliance with mandates. In all instances, governing administrative frameworks were hierarchical, top-down command systems, akin to military communicative approaches.

Images of war

Policy theorists have long understood that policy images help in issue definition and redefinition (Baumgartner & Jones, 1991, 1993, 25–28; Jones, 1994). From a practitioner perspective, the key to successful management of any policy issue is to find and use the appropriate problem image that mobilizes citizens and nationals who would otherwise be disengaged from the policy process. Such images have the ability to provide a mixture of information and emotive appeals. Thus, the image can become widely acceptable to millions of people who may then support any policy pronouncements by concerned authorities.

Indeed, political leaders and journalists framed the management of COVID-19 as a war, which, by intuition, is an existential threat to all citizens. Writing in *The Standard Newspaper*, Kepher Odhiambo, an orthopedics expert, trauma surgeon, and former military personnel in Kenya, eloquently offers one of the best perspectives on this subject:

> *By definition, the situation the world finds in is known as biological warfare. The enemy, COVID-19 is as formidable an enemy as can be found. It destroys the fabric of society. It disables and kills, disrupts the economy, restricts freedom of movement and worship,*

> *disrupts education, and helps undermine authority. It is a mistake to think of coronavirus as only a disaster. A disaster has a timescale, after which the aftermath sets in. But this current crisis self-perpetuates; the disaster worsens daily and the aftermath escalates, critically we have closed our borders. This is a move associated with a declaration of war, this type of enemy is routinely studied in military institutions (including our National Defense College) under the subject of nuclear and biological warfare, and models on how to tackle it have been designed accordingly.*
>
> *The most critical battle front for this disease is the community. Once it is lost, victory is almost impossible, and we will suffer great casualties. We have three weapons at our disposal; awareness, attitude change and actions (3As). Because they do not appear impressive, most countries have taken them for granted. A missing element in this war is a comprehensive command and control structure. Associated with this is a sometimes discordant, inadequate or absent communications and logistics (human and materials) support system (Odhiambo, 2020, p.8).*

From general historical data, wars or all kinds rely on "superior firepower, transportation, and communication" (Parenti, 2011, p.28). The considered final "victory" in wars is theoretically defined as a function of minimizing loss of human lives. Subsequently, it became commonplace in 2020 for analysts to compare numbers of COVID-19 deaths to those killed in conflicts such as the Korean War (New York Times, 2020). As numbers neared 3000 in the United States comparisons to 911 were commonplace as well. And there are military ways of evaluating performance of both commanders and soldiers. These are varied but easy to tease out. In the war against COVID-19, Odhiambo (2020) wonders just "who are the commanders in this war? Is there a hierarchy/apart from the national response team and country teams?" And how low does the command structure go? How much do the lower echelons know? (Odhiambo, 2020).

Military personnel are familiar with these concerns. The Army is highly centralized and uses science and contemporary knowledge of science of human development and suitable management systems (Hajjar & Ender, 2005, p. 522). Army leaders also "need crucial politico-military skills, knowledge, and abilities to most effectively lead at the interface of the professional and political realms" (Hajjar & Ender, 2005, p. 523). These issues bring out the significance of leadership as an important variable in successful conduct of war efforts.

Hence, the seriousness of war in any country population demands astute leadership, especially by the nation's commander in chief. And we must note that the words of the commander in chief matter. As presidential scholar Jeffrey Tulis (1987) points out, political rhetoric is a tool of presidential governance. Tulis (1987, p. 4) states that presidents regularly go directly to the public to appeal for a policy position or to defend an action because they recognize that rhetoric is an essential function in leadership. It may well be the most important action that a president or leader can make publicly during crisis and pandemic situations. The signals, including nonverbal ones that leaders send to administrative levels below them, can aid or negatively affect mitigation processes in any jurisdiction. And indeed, on the ground during the pandemic sociopolitical notions of contagion began to mirror political tropes of contagion.

While lessons from a major crisis like war do not necessarily compare to a global pandemic, as was during 1918 when 50 million people lost their lives to a flu, the principles of managing the messages and mobilizing efforts to combat the disease are strikingly similar. External directives that appear to be helpful or directly useful to people are also imbued with ongoing colonialism and, in some cases, continue to trammel sovereignty. Administrative and political leaders require meeting the needs of the grave pandemic situation. This amplifies the need for harnessing communication skills and to provide combat the COVID-19 pandemic. Yet, the long history of unequal access and rural living for many Indigenous nations create the backdrop for the continuation of oppressive actions and reactions in crisis response. A profile of how several world leaders mobilized through communication efforts to combat COVID-19 exists. While there are variations in style, the vast majority of national responses to the virus framed the pandemic as a war that involved all citizens. Who, however, are these citizens and how are these communications distributed and handled in times of crisis?

Witness these illustrations from across the nations covered in this study. While addressing members of his Bharatiya Janata Party on its 40th anniversary, Modi described the pandemic as a "long war" (Withnall, 2020). And speaking to the media inside parliament, in New Delhi, India's Prime Minister Narendra Modi said: "the country was in the midst of a war." ..." We should not be trapped into over-confidence and nurse the belief that coronavirus has not reached our city, our village, our streets, our office, and so will not reach them ... Physical distancing, covering the face with a cover or a mask and washing our hands again and again will be the biggest medicine to fight this disease in the days to come" (Arora, 2020).

Australian Prime Minister Scott Morrison's initial approach to handling the virus was similar to that of a commander in chief facing a war. He was among the first world leaders to warn of the pandemic and with consent from Australia's National Security Committee, operationalized disease control plans in militaristic terms. By February 1, 2020, his administration had banned the entry of foreigners visiting Australia from China, where large-scale outbreak had first emerged (Packham & Pandey, 2020). His characterization of the coronavirus was dramatic. Speaking in Brisbane, he stated that, "This is a once-in-100-year type event, we haven't seen this sort of thing in Australia since the end of the first World War, but together we are of course up to this challenge" (Westcott & Whiteman, 2020).

And in an address to the people of Kenya, President Uhuru M. Kenyatta stated: that the country was at war. He constituted a National Emergency Response Committee on the

Coronavirus Pandemic and consulted with the National Security Council on ways to manage the crisis. Kenyatta emphasized in stark terms.

> *The entire world is going through an extra ordinary emergency, indeed the last time the world faced a disease of such rapid and dire impact as coronavirus was over a century ago n when an influenza epidemic claimed tens of millions of lives. Indeed, a few decades prior to that a terrible sleeping sickness epidemic rode havoc in our own eastern part of Africa, as such the government will do everything in our power to ensure that we as an action shall not suffer terrible outcomes as a result of the Covid-19." … Fellow Kenyans, make no mistake, we are at war, and we must be together (Kenyatta, 2020).*

He also noted that it is an "enemy that must be arrested." Kenyatta described the situation as an extraordinary emergency situation.

In neighboring Tanzania, President John Pombe Magufuli reported that his own son had recovered from COVID-19 (Magufuli, 2020). Magufuli refused to place Tanzania under lockdown and excluded mosques and churches from lockdown measures. He maintained a skeptical view about COVID-19 and instead problematized it in religious tones. According to him, the virus was a part of a spiritual war against the devil (*shetani*) which required prayers. He also exposed fraudulent COVID-19 test kits that had been approved by the WHO. Magufuli also characterized the scourge of disease as a part of the war against imperialists (*mabeberu*) (Paget & Kwayu, 2020).

Although various opinion columnists and the US Embassy officials in Dar es Salaam regarded his actions as irresponsible, Magufuli's COVID-19 strategy was still very much couched in national security terms. He issued threats to the media and health-care officials about information sharing (Kabwe, 2020).

General Michael McGuire, director of the Arizona Department of Emergency and Military Affairs, told reporters that; "This is a war on this virus …. We plan on doing more and more for the Navajo Nation. We need to make sure we have provided them the essential resources" (Hillyard & Ortiz, 2020).

At the federal level, in the United States, on National Nurses Day, and in other speeches to the press, President Donald Trump referred to himself as being a "war time president" and called for Americans to "sacrifice together, because we are all in this together, and we will come through together" (Oprysko & Luthi, 2020). And he compared it to the attack on Pearl Harbor and September 11, 2001 (Trump, 2020a). Then, and later he called it the "horrible" and "Invisible Enemy" (Boot, 2020). During the May 25, 2020, Memorial Day speech, he further stated:

> *In recent months, our nation and the world have been engaged in a new form of battle against an invisible enemy. Once more, the men and women of the United States military have answered the call to duty and raced into danger. Tens of thousands of service*

> *members and National Guardsmen are on the front lines of our war against this terrible virus, caring for patients, delivering critical supplies, and working night and day to safeguard our citizens ... As one nation, we mourn alongside every single family that has lost loved ones, including the families of our great veterans ... Together we will vanquish the virus, and America will rise from this crisis to new and even greater heights (Trump, 2020b).*

Hence, globally, although the quality of content differed, the official narratives and messaging were intended to rally public support for government efforts to "combat the scourge." This management system was instrumental, and goal directed. The government required citizens to help fight the virus that caused COVID-19. And there were instances when the governments and community leaders emitted consummatory communication styles, which involved fear and an appeal to various emotions.

The literature of managing crises provides practical lessons for what works best under different types of crisis including hurricanes, tornados, earthquakes, and other disasters. Disasters, as Winston Churchill observed, may turn out to be a government's finest hour (Churchill, 1940). Even though governments must work collaboratively, there is an expectation for it to take lead role to return societies to normalcy. This requires collection of necessary data and to provide real-time information to citizens. During disasters, which include pandemics, government leaders rely on a high degree of situational awareness and hence, the importance of technologies such as GPS, smartphones, modems, routers, tablets, and other machine-to-machine devices and network connectivity. Without flexible efficient and scalable networks, the level of response to a disaster or crisis may be compromised. Thus, all communication tools must always be prepared to provide a lifeline for stakeholders engaged in the crisis management.

Much of the recent literature in this genre details how technologies have been leveraged in under disaster situation. For example, Garnett & Kouzmin (2007) gave much prominence to the idea that interpersonal influences and media relationships are important variables in attaining good outcomes during crisis communication efforts. They also acknowledge that communication technologies have certain vulnerabilities. Such a claim is even more important in coordination of the organizations within the networks providing mitigation and preventive services. The different sectors should preferably develop a systemic approach to managing communications during the crisis. Basing their work on Hurricane Katrina, they emphasize the importance of adequate response based on available ICTs. During Katrina, the communication channels among key leaders in different operational organizations suffered due to weak social capital. The various agencies talked past each other and lacked communication strategies resulting in disastrous outcomes. Crisis communication is also about media affairs.

As Kapucu & Ozerdem (2013, p. 181-82) note, officials risk failure when they do not work toward releasing accurate and timely information. Without enabling optimal information flows and failure by officials to inspire compassion and empathy, there is a huge chance that any efforts to prevent crisis escalation and to assist in mitigation may be challenged and frustrated. These ideas from scholars who closely study communications in crisis situations are applicable to how we understand the communication practices during a major crisis such as COVID-19 pandemic.

Indigenous communities and COVID-19

It is no surprise that COVID-19 entered the global ecosystem and immediately resulted in differences in life and death. The framework of syndemics allows for an investigation of how overlapping inequalities play out on morbidity and mortality throughout the world. The concept of syndemics allows for a complex understanding of how different positions of people living in unequal societies result in differences in life and death (Singer & Rylko-Bauer, 2020). However, it is crucial to understand why people live and die differently in different places and among different populations. Indigenous people throughout the world have, for centuries, faced violence, genocide, and ongoing displacement and at the same time have higher rates of disease. To understand why and how Indigenous people face such immense hardships it is useful to engage the theoretical foundations of critical Indigenous theory and the emerging work of scholars who connect current crises with ongoing settler colonialism. This work allows for an investigation of how external and internal governments handled COVID-19 and includes a critique of government functions that trammeled tribal sovereignty and caused additional burdens for the interruption of the highly contagious pandemic.

Indigenous people live across the entire world on tribally owned lands and in diaspora. A common experience of many Indigenous people is a shared history of colonial violence and centuries of oppressive rule. We recognize as well that Indigenous people have varying levels of shared histories and presents of intertwining economic, political, and cultural lifeways that can be specific to tribes and regions. We are non-Indigenous scholars who bring our social science disciplinary backgrounds to this analysis with a strong caveat that we are writing from outside nations. We draw on the work of Indigenous scholars where possible and we do not make general claims about Indigenous societies. Instead, we focus on the interactions between different governments and policies in pandemic response. In future work on COVID-19 we are hopeful that Indigenous authors will continue to analyze and contribute to the global discussion of government response and the use of ICTs from within.

Critical indigenous theory

First Nations people also must navigate shifting and sometimes divisive borders and boundaries. Many have endured violent relocation and multiple forms of oppressive violence through what Jodi Byrd calls the "transit of empire" (Byrd, 2011). It is within First Nations as well that people exhibit survivance by continuing to exist even when all historical and present oppressive forces move to the contrary (Vizenor, 1999). Indigenous survivance remains an active sense of presence in Native stories and cultures. This term can be useful in shifting the script from one that has an exclusive focus on people as victims of circumstance to one that contains the complexities of oppression and survival. Resilience as a concept, however, originated in the natural sciences to describe an ecosystem process of bouncing back after natural disasters. The use of a natural science ecosystem theoretical model to explore a human population can be problematic, especially in working with Indigenous communities. For this reason, we borrow Vizner's framework of survivance to shift the focus of this work to both include analysis of the political and economic forces that continue to oppress First Nations people and at the same time include within that picture the very real layers to survival and negotiation and life that Indigenous people employ to become and stay well. In this way our work contributes to the understanding of inequality and health disparities by analyzing processes that continue to contribute to oppressive forces while also incorporating a perspective that Indigenous people and nations are active participants in legal, political, cultural, and actions at all times.

Other Indigenous scholars writing about dispossession, settler colonialism, environmental racism (Hoover, 2019), food sovereignty, and government oppression and even critiques of government externally designed programs designed for "healing" (Million, 2013) form the necessary background for understanding the moment when COVID-19 arrived.

Unequal impacts

Around the world, the COVID-19 pandemic caused devastation and death. Reporting for National Public Radio (NPR), journalist Philip Reeves (2020) commented that Indigenous communities of the Amazon were among the thousands of those killed in Brazil, one of the countries with the highest death rates. The first case of the coronavirus was first reported in February. Understanding this devastation requires a multileveled lens that includes already existing burdens of unequal health related to multiple threads of interlocking oppressions such as lack of access to care and long-term impacts of racism.

By all indications, Brazil's handling of the pandemic was unsatisfactory. There was no nationwide rule or rules issued for measure such as social distancing and, instead, local officials did as they saw fit. Brazil's President Jair Bolsonaro, a right-wing populist, made the situation worse by firing a health minister who had demanded an aggressive approach to fighting COVID-19. Bolsonaro, at one point, limited the "disclosure of epidemiological data saying the numbers there would no longer be a story on the evening news" (Leitej et al., 2020). This order was reversed by the country's Supreme Court. Because public hospitals in Brazil were already overcrowded and ill equipped, the poor across the nation suffered most.

Over one million Indigenous people live in some 400 communities of the Amazon rainforest. Many of these regions have no information and communications connectivity whatsoever (Wallace, 2011). Human-made catastrophe also unequally impacts tribal areas. Since 2018, the Amazon region wildfires, begun by ranchers supported by top government officials, devastated forest cover and worsened respiratory diseases among Indigenous communities. People living in the regions already experienced high incidence of malaria and other tropical diseases, which exacerbated management of the COVID-19 pandemic.

The first cases of COVID-19 in the small river town of Sao Gabriel da Cachoeira were announced in late April and within a month, there was much panic. Within a month, at least 600 people had COVID-19 and at least 17 villagers had died. The poor health-care systems in tribal lands meant that sick people had to be transported 500 miles away to the nearest city of Manaus where the hospitals were overwhelmed and leading to collapse of the health-care system. The story was repeated in several tribal villages of Amazonia. Michael Fox, reporting for *The World*, an NPR show, noted that many of the health-care facilities in Indigenous communities had already exhausted their capacities to handle the large numbers of COVID-19 cases especially around Novo Progresso region in the state of Parã, Northern Brazil (Werman, 2020).

The tragic death of Paulinho Payakan, a 66-year-old Kayapò chief of one of Brazil's Indigenous people, brought further attention to the devastation caused by COVID-19 (Boadle, 2020). Chief Payakan was well known for his role in fighting for Indigenous rights and environmental conservation. In the 1980s, he led protests against the construction of the Belo Monte hydroelectric dam on the Xingu River in the Amazon forest. Chief Payakan was not the only elder who died of coronavirus. Many more Indigenous tribal elders of the Amazon rainforest lost their lives during the months of May and June 2020 (Boadle, 2020).

In neighboring Colombia, Indigenous communities of the Amazon suffered equal fate. Journalist Jorge Valencia (2020) reported that individuals with COVID-19 were spreading it as they moved along the Amazon River, through Brazil and Peru. Reports indicated more than 1500 members of Indigenous communities had COVID-19 by end of May with high numbers of deaths related to lack of medical personnel and infrastructure.

For example, in the town of Leticia, the regional capital, patients have overwhelmed the two small hospitals.

As a whole, Mexico, which has an Indigenous population of over 12,250,947 or 10.1% of total population, experienced massive COVID-19 casualties. In the months of April and May 2020, the public cemetery in Iztapalapa, Mexico City, witnessed a hive of activities. By end of May over 90,000 Mexicans had been infected by coronavirus (Averbuch, 2020). The National Institute of Indigenous Peoples (INPI) reported that by end of July 2020, 4140 members had contracted COVID-19, of these, 719 had passed on (INPI, 2020). Because the country was not testing much and it hesitated to have a lockdown as strict as in other Latin American countries, the numbers were probably higher. Regionally, only Brazil had more COVID-19 cases than Mexico (Averbuch, 2020). And as in the case of Brazil, the cases spread to Indigenous communities across the nation. They too were most vulnerable, therefore making it difficult for these communities to achieve the regions Sustainable Development Goals (UNESCO, 2020).

As for the Maya people, some of whom live in Belize and others in Mexico, in the Yucatan Peninsula, the coronavirus was brought about to the communities via the United States. Although the area is modern, central and southern parts of the Yucatan Peninsula have generally retained a huge component of agrarian subsistence practices shielding the region from the more urbanized and tourist-based economy at Cancun, Mexico. In July, the government in Mexico indicated that there were 1010 active cases of COVID-19 throughout the state of Quintana Roo (Royal Resorts, 2020). The state government's report suggested that "in Cancun and Playa del Carmen," the zones with the highest incidence of cases "were on the outskirts of the city, not around the tourist areas." The state government constituted a COVID-19 Reaction Police unit that policed the high-contagion neighborhoods such as Villas Otoch, Villas del Mar, in Region 102 (Riviera Maya Times, 2020a).

COVID-19 hit the region when there was considerable strain on the region's ecological system. Because of the exclusion of the Yucatec Maya, their culture, and knowledge in the management of forest resources, commercial logging and agriculture has ruined the ability of the Indigenous people to tap into the rich medicinal wealth they possess. While the government blamed the Mayan culture for lingering poverty, contrary views support the thesis that neocolonial institutions have destroyed the areas' biodiversity and complicated community resources management practices (Barrera-Bassols & Toledo, 2005; Faust, 2001; Gómez-Pompa & Bainbridge, 1995).

There is a reason to bring about this issue of the ecosystem. A WHO conference in 2018 identified that world governments were not adequately prepared to manage the possibility of a disease like COVID-19 emerging across the nations (McKay & Dvorak, 2020). This is in spite of the fact that scientists had long known that humans today are "exposed to deadly new pathogens than ever" (McKay & Dvorak, 2020). Many of the pathogens are found in animals and livestock which now live closer to humans than in previous generations. Forces of global travel, economic development, trade, meat production, and deforestation are known to contribute to this exposure to dangerous pathogens.

In the case of the Maya, observers of the COVID-19 situation in the Yucaten directly linked the impact of the pandemic to "environmental malpractices." That was reported in Riviera Maya Times (2020a), which quoted Vicente Ferreyra, CEO of SustenTur, a foundation with interests in ecotourism. The consensus was that the health crisis was also a crisis of the uncertain future caused by not doing enough to protect the wetlands, seagrasses, coral reefs, and forests (Riviera Maya Times, 2020b). This association between environmental justice and negative externalities associated with climate change is echoed in other narratives. Take, for example, Ed Yong's article in *The Atlantic* on "How the Pandemic Defeated America" (2020). Yong notes that he spoke to more than 100 experts in a variety of fields including conservationists who offer a scientific explanation of the link between bats and COVID-19. And quoting Colin Carlson, an ecologist based at Georgetown University, he wrote that epidemics are spillovers from "land -use change and climate change, both of which are hard to control" (Yong, 2020). He states further:

> *Our species has relentlessly expanded into previously wild spaces. Through intensive agriculture, habitat destruction, and rising temperatures, we have uprooted the planet's animals, forcing them into new and narrower ranges that are on our own doorsteps. Humanity has squeezed the world's wildlife in a crushing grip—and viruses have come bursting out (Yong, 2020).*

The Maya and other Indigenous communities shared the same attributes—negligence from respective governments more prominently through wanton destruction of ecosystems for greedy investors. The resulting displacement and poverty are critical variables in understanding the spread of diseases among these communities.

With centuries of exploitation, these communities faced underdevelopment conditions and as the pandemic reaped through their lands, the decades of negligence exposed the lack of water, lack of adequate health services, limited health literacy, and food shortages. The communities also had, according to UNESCO (2020), "less coverage and accessibility to the internet and an open television signal, and when combined with a lack of equipment such as computers, tablets and mobile phones"; the communities would not benefit from distance learning, telemedicine, and information on health care and hygiene. Many Indigenous communities had cultures that emphasized collective actions and performed ceremonies in which large gathering and lack of social distancing were the norm.

Turning our attention to Africa, there are far too many untold stories of marginalized communities combatting COVID-19 amid multiple-level crises. These crises include political strife and climate change occasioned by massive destruction by logging and poaching of wildlife. The emergence of the coronavirus presents a new front on Africa's search for holistic wellness and health care (Schulman, 2016). Indigenous communities in Africa were already deprived of basic health services. And the United Nations (2013) published reports which highlighted that most Indigenous people live in regions that are marginalized economically and politically and socially isolated. These infrastructural weaknesses and other factors like lack of professional health providers influence Indigenous people's quality of life and rights to health care (Migiro, 2013, p. 15).

Known as the Aka or Bayaka people, these linguistically Bantu nomadic Africans are known for their relatively short heights. The Bayaka too are considered as outcasts and have been subjected to slave condition by the Bilo, another Bantu group based in the Democratic Republic of Congo (AFP, 2020). They live in forested areas of Central Africa and share many of the aspirations of Indigenous people all over the world. They too struggle to "live in their traditional ways," but are caught between worlds (Schulman, 2016). The Pygmies suffer from many underlying conditions (including tuberculosis, malaria, and hepatitis B) and have among the lowest life expectancy rates in the world, estimated at 49 years of age (Schulman, 2016). Other estimates are even more grim, placing men's life expectancy to around 35, and women a paltry 38 years (AFP, 2020).

As the coronavirus reached Bangui, in the CAR, the Bayaka people retreated to the forested areas—Dzanga-Sangha areas. This "strategic retreat" to villages and reserves where the main activities are hunting was vigorously pursued after the disease infected more than 1000 people in the country. By May, at least four deaths were linked to COVID-19. The Bayaka's retreat to areas generally reserved for Wildlife Protection was supported by authorities in charge of the forests run through funding from the World Wildlife Fund (WWF) and other nongovernmental organizations (NGOs). WWF workers moved swiftly to provide food and medicines to the Bayaka. Officials also urged Bayaka people to stay away from towns (AFP, 2020). This confluence of political and social responses that resulted in seclusion away from crowded towns for Bakaya people seemed to be successful. The government also used traditional communication channels and relied on NGOs and religious groups to warn the villagers of the pending dangers.

In North Africa, the Amazigh ethnic group, which is Indigenous to Morocco and Algeria, has sought greater recognition and has now witnessed greater self-governance in Morocco's Constitution of 2011.

This is significant because for a long time, North African states' political structures were organized monarchies. In the case of Morocco, in 1994, although King Hassan II, an Alaouite, acknowledged the importance of contributions by the Amazigh, progress had been slow until King Mohammed VI and his administration yielded to the regional pressures referred to as the Arab Spring opening up democratic spaces in 2011. Still, the monarchy retained the key portfolios of health services and initially acted swiftly to control the spread of COVID-19 (Abouzzohour, 2020). Specific challenges in Indigenous areas were similar to those in other Indigenous areas. For example, communication systems and Internet information channels that broadcast pandemic issues were inadequate. Furthermore, the national TV media was unreliable in Tamazigh areas. There were also complaints from the Amazigh World Congress about inadequate information in Indigenous Tamazigh language (IWGIA, 2020). However, satellite capabilities linking these regions to the diaspora and word of mouth facilitated a high level of awareness and exceptionally competent Indigenous mitigation responses.

The Amazigh World News (AWN) reported that by March 29, 2020, more than 463 COVID-19 cases and 26 deaths were reported in Morocco. For Algeria, the figures were 511 cases and 31 deaths, respectively. The general consensus is that the Amazigh community too lacked sufficient hospitals and health-care infrastructure. The Human Rights Watch and Amnesty International have continuously reported cases of oppression of marginalized communities in parts of northeastern Morocco such as Rif (Riffi), where some Amazigh reside (Az, 2020). These are also areas of economic neglect and unemployment, which translates into lower capacity to pay for suitable health-care amenities. Aware that the state health structures were dilapidated, villages, such as in Kabylia, relied on ancient assemblies inspired by the Tajmaat, maintained vigilance to successfully lower COVID-19 cases in Tamzigh communities (IWGIA, 2020).

Building on the logic of oppression and marginalization, one of the greatest challenges to managing COVID-19 among the Indigenous Amazigh community is that the government of Morocco did not publish sufficient awareness campaign materials against the disease. AWN stated:

> *Despite the latest recommendations released by the High Council for Audiovisual Communication, HAKA, calling for an urgent necessity to include the Amazigh language in all Coronavirus awareness campaigns, along with Arabic and sign languages, Moroccan public radios and television stations continue to ignore public interest during this national emergency declared by the state (Amazigh World News, 2020).*

In spite of calls from human rights activists and correspondence from HAKA, the government declined to act quickly to publish awareness material in Indigenous languages. The government chose to publish COVID-19 awareness pamphlets, flyers, and other educational materials in French and Arabic, even though an estimated 30% of people aged 15 and above cannot read or write in those languages. The case of the Amazigh in Morocco offered a typical example of failed communication and health-care policies among Indigenous communities globally. In Algeria, the community too was affected by the lack of adequate preparation of the central authorities in Algiers. There was widespread public skepticism after the first case of COVID-19 was announced. The "lack of communication undermined the real weight of the outbreak," leading to spread of the disease countrywide (Ghebouli, 2020).

We feel compelled to shed light from Canada's First Nations. While these are hundreds of communities scattered across Canada, they all share similar historical struggles and challenges brought about by colonial rule and the larger socioeconomic logic of capitalism. Through the Indian Act of 1876, colonial powers abolished any form of self-government for First Nation people. Colonialism also introduced capitalist land ownership systems (location ticket system), leading to violence and dispossession. Powerful and intentional efforts to destroy collective identity can create oppressive barriers. While internally First Nations people have resisted and continued with identities in many ways, the pressure of genocide and ongoing settler colonialism complicates the management of contagious diseases, especially, if tools like contact tracing are to be successfully deployed. Many of these units lack the ability to provide jobs, health care, and other amenities. Still, there are multiple tribally led initiatives for health and survival.

The aforementioned forces have, to different degrees, decimated First Nations of Canada. COVID-19 exacerbated the already oppressive economic systems in these communities. And because much of the health-care system in Canada is government funded, managing COVID-19 meant increased the urge to encroach on Indigenous sovereignty. Using the argument that there is need to contain the spread of COVID-19, central governments found themselves at odds with upholding fundamental rights of indigenous communities. The net results are an affront on key tenets of Indigenous sovereignty.

In Canada, the pandemic occurred at a time when Indigenous communities had staged protests against the construction of gas pipeline through First Nation territories (Noakes, 2020). The Wet'suwet'en nation in northern British Columbia had persistently rejected construction of pipelines through its homelands but the government in Canada put negotiations on hold as the pandemic broke out. And to the disappointment of the First Nation, work on the pipeline continued.

The Royal Canadian Mounted Police simultaneously grabbed the right to deny the Indigenous community the ability to self-police and self-determination. There were reports that police officers from the RCMP's Quick Response Team moved freely throughout the Wet'suwet'en clan lands. Since the police are associated with riot control, such free movement especially during the pandemic was threatening. There were indications that members of the Wet'suwet'en nation would not convene meetings to prepare for negotiations with the government of Canada because of the COVID-19-related social distancing guidelines, which also prohibit assemblies of more than 50 persons. However, the workers on the pipeline project did not comply with the guidelines (Noakes, 2020).

Noakes (2020) made another observation similar to the case involving the Sioux in the United States whereby federal powers used the pandemic to violently intrude upon sovereignty. Noakes wrote:

> *Beyond the immediate threat, the pandemic has raised important questions about Indigenous sovereignty in Canada, including whether Indigenous nations have the right to close their territory to outsiders in an effort to contain the spread of the virus. That appears to depend on whether the territory is the site of a major infrastructure project or not (Noakes, 2020).*

There are remarkable similarities to the manner in which the US government disrespected the sovereignty of Sioux Tribes in South Dakota, as we discuss below. While the overall national guidelines for controlling the spread of the coronavirus gave much emphasis to lockdowns and shelter-in-place policies, there was an appearance that state governments could overrule tribal sovereignty.

Witness this, Republican Governor Kristi Noem of South Dakota threatened the Cheyenne River Sioux and the Oglala Sioux Tribes with legal action and use of force, if they did not take down checkpoints put in place to protect Indigenous community members from people who might otherwise transmit the virus to tribal citizens (Lakhani, 2020). Her tweet of May 8, 2020 read: “Today I sent letters to two South Dakota tribes asking them to immediately cease interfering with or regulating traffic on US and State Highways and remove all travel checkpoints” (Governor Kristi Noem, @govkristinoem).

Members of the Great Sioux Nation, who fall within the boundaries of South Dakota, were within their sovereign rights to require those passing through their territory to be subjected to questioning concerning public health. The Indigenous government took these steps to monitor who comes in and out of their territory for purpose of supporting their contact tracing initiatives. In terms of constitutional legality, the Sioux Tribes and their nation predate the state of South Dakota’s statehood. And through treaty obligations, with the US Federal government, the state did not possess authority over the Sioux; yet, the federal government did not challenge the governor’s threat of legal action. Even after several deaths at Smithfield Foods, located in Sioux Falls, the state governor continued to demonstrate disregard for the health and life of people in the region.

Julian Bear Runner, the president of the Oglala Sioux Tribe, wrote to Governor Noem requesting the imposition of a stay-at-home order and a request for closures of nonessential businesses from operations. The governor, however, demonstrated more concern with continuation of existing economic activities than issuing protection for tribal health and life. There was indication that the governor sent mixed messages, on the one hand, acknowledging that the state did not have jurisdiction but later contradicting herself (Hopkins, 2020). The Bureau of Indian Affairs issued a memorandum (April 8, 2020) expressing the opinion that "tribes and native communities were within their right to issue temporary orders banning or curtailing travel into or within tribal lands amid the COVID-19 pandemic." The state, however, insisted that the Cheyenne River Sioux and Ogala Sioux tribes should have consulted with the state (Matzen, 2020).

These cases in Canada and the United States illustrate the imposition of federal guidelines over sovereignty by using the pandemic as an excuse for ongoing settler colonial imposition of guidelines and policies independent of the health needs of people living on tribal lands. The blatant abuse of sovereignty clauses embedded in treaties contributed to already poor development policy outcome in these territories. Compounded by hundreds of years of environmental injustice and ongoing oppression from external forces, managing COVID-19 in Indigenous communities all over the world was not going to be an easy endeavor, even for the best administrations in the world. The structural impediments brought to bear by economic deprivation and social injustice add to the complexity of communication challenges in management of the pandemic. And these images of inadequate socioeconomic development create circumstances in which the COVID-19 management narrative is articulated and explored in detailed selected cases.

The research and its methodologies

This research maintains a focus on all these variables with an understanding that communication is an important component in management of global pandemics. This is because communication is not just about transmitting information, but it also influences disease management and societal wellness. Accurate information promotes better management of disease containment. On the other hand, "fake news," conspiracies, and intentional misinformation and disinformation are dysfunctions that lead to failures in disease program management (Albright, 2017). Unfortunately, the COVID-19 outbreak emerged during an era in which "fake news" has proliferated. In fact, the accusation of fake news is used regularly to discredit science communication and health directives. This reality is a common threat in our ability to go through massive data available through various data sources.

This work is preliminary and relies on government and public documentary, observational, and ethnographic data including informal discussions with informed adult members of Indigenous communities on four continents: People of the Navajo Nation, Maasai people in East Africa, Siddi people living in a Scheduled group in India and Pakistan, and Indigenous people living in what is now Australia. These are geographically delineated units within the community of nations requiring much more awareness than is provided.

It is not lost to us that the selected cases shed unique perspectives of communications to and within Indigenous societies. This, in itself, is unique and extraordinarily interesting.

From an instrumental perspective, shedding light on Indigenous societies provides opportunities to policymakers need to improve material conditions within those societies. The importance of using these case studies is that we are offering insights and even hypotheses that may be pursued in subsequent studies. This is the point that Bruce Berg (1989, p. 217) made while appraising the scientific benefit of case studies. Where data were derived from interviews and observations, interpretive designs permitted participants to be engaged in flowing open dialogue and expand in narratives they are comfortable in articulating. As we all recognize, observations are an important element in qualitative research (Marshall & Rossman, 2016, p. 143). Focused observations help us in describing problems and they unfold. It is also through observations that we can refine and capture key nuances involving government communication strategies. For this research, many of the documentary sources were corroborated with observations and from descriptive electronic resources including video clips from a variety of social media, Twitter, Facebook, YouTube, Tik Tok (a China-owned platform), WhatsApp, and Instagram. These resources mirror mobile ethnographic data which was derived from following subject activities on the numerous open access public online handles and accounts.

Thankfully, a variety of observational online transcripts and actual verifiable recordings of comments from leaders offer limitless opportunities to listen and watch attentively policy pronouncements about COVID-19. And we may also learn a lot from digital storytelling associated with COVID-19. Digital storytelling that is presented in images, sound and videos captured on social media enriched our ability to interpret the feedback processes following official communication on COVID-19. These digital stories offer an insider's view of an era in history (Chilisa, 2011; Marshall & Rossman, 2016, p. 155). And methodologists acknowledge their relevance and wide usage in public health settings (Marshall & Rossman, 2016, p. 158). Besides, stories and voices of Indigenous people are a recognized way through which the contours of our conceptual frameworks are established (Lambert, 2014).

Thus, these resources enable us to quality capture content and meanings of communicative actions which provide credible evidence of implementation styles. When combined with content analysis of speech given by influential community and political leaders, a solid profile of communication behavior patterns can be carefully studied and discerned. Researchers have long recognized that examination of artifacts of social communication including written documents and transcripts of recorded verbal communications is an objective way of making inferences on data so gathered (Berg, 1989, p. 224).

The individuals studied whether from government or society leaders tell us about life and death communications in different Indigenous cultural and political settings.

Their public facing dialogue is what drives the analysis. Unique knowledge systems matter more than simple academic gatekeeping exercises that come from formal training in non-Indigenous academies. As two non-Indigenous scholars we draw from our training in social sciences to analyze the public policy aspect of external communications and the rich always growing literature in Critical Indigenous Theory. In addition, this work is based on our reflection following previous projects designed to tease out the experiences of marginalized communities dealing with COVID-19. The current design aims to help inform debates about official communication during pandemics.

More significantly, messaging about disease and existential threats is a disturbing and fascinating idea because of the emotions that come with the content of the messages. In nonpandemic situations, people are often susceptible to media coverage of issues, and are frequently made to see policy issues in perspectives that are meant to influence their voting behaviors. Messaging during health-care crisis is not so much about fancy ads and persuasive outreach efforts. It is about saving lives and mobilization for caring about others.

Studying communications during a COVID-19 pandemic in Indigenous societies is central to understanding reactions in vulnerable societies. A tweet or broadcast message from a national leader can have very different consequences on these communities depending on existing relations between Indigenous societies and their home countries. Such messages may also be delivered in culturally offensive ways to reinforce negative stereotyping and demonization. Understanding what works and is, therefore, an important step in collective action against a global disease pandemic.

At the core of this type of research is a belief in the value of interdisciplinary perspectives, which certainly enrich the policy processes. While we recognize the drawbacks of the approaches taken, notably, the narrow selection of cases and communities studied, but given the input derived from our data sources, we have hope that future work will uncover other aspects of the issues covered.

Foci and location

By focusing on Indigenous societies, we hopefully add value to awareness of social science issues that are more frequently than not glossed over. This does not imply a lack of academic studies. Quite the contrary, the high quality of Indigenous studies is self-evident. For example, the field of American tribal governance is interdisciplinary and offers opportunities for exploration of social justice, health policy, environmental degradation, communication, and leadership, among other topics (Aufrecht, 1999; Aufrecht & Case, 2005; Barker, 1998; Fouberg, 2000; Ortiz, 2002; Smith, 2000; Taylor, 1972). While the bulk of these studies and commentaries focus on governance issues, few works highlight governance in the context of a pandemic, like COVID-19.

We seek to expand the significance of managing a public health policy situation in a period of great national anxiety. The overall premise of this book is simple: official communications about the COVID-19 pandemic in Indigenous societies are unassembled and not systematically recognized. By drawing on a few examples, from different global communities, we illustrate that social sciences mesh together to provide knowledge on communicating public policy challenges.

Four cases

We chose to share the above examples to begin a conversation about external and existential threats to Indigenous societies: places and people who have survived the use of epidemics and other violence to bodies and lives for decades. Since this is not an exhaustive study of communications and management of COVID-19 in all Indigenous communities, we only scratch the surface of cases to highlight the plight of these societies. Also of note we do not make any claims here to either speak for people living these communities or to claim any sort of pan-Indian experience. The experiences of these communities, in the global effort to contain COVID-19, are our way of being immersed in these parts of the world that enrich our humanity. We do this through a critical investigation of the political processes that already exist and become enlivened and used heavily during times of crises. This approach affords opportunities to add a vantage point of political processes within public health paradigms in ways that are often underrepresented in mainstream discourses. In building an understanding of global health concerns, we, of course, utilize concepts and ideas that we are familiar with, especially the American perspectives. This is for redundancy purposes. Arguably, pursuing such an analysis is in clarification of facts and finding patterns and of complexity, complementarity, and diversity.

The readers would certainly like to know why four specific cases are featured here. The architecture of this book is based on taking examples from different continents because the notion of indigeneity is universal. Our goal is to highlight and explore nuances of a major problem of public health organizations and external governance. We seek to expand collaborations that enrich policy discourses in Indigenous communities. By doing so, we attempt to encourage inclusiveness in studying communication aspects of COVID-19. We also enter this conversation with the deep knowledge and honor of the lives lost due to mismanagement and inability to control COVID-19, inequality, and long-standing injustice. In this way we engage a long view approach to policy while also remembering in a present way that policy impacts everyday lives and everyday deaths and that people are suffering throughout the world from illness and death in ways that do not impact us equally. We do this in solidarity and in the hopes of pushing policy and health discussions in directions that could lead to better crisis response.

This volume centers on the Navajo Nation, Maasai people in Kenya and Tanzania, the Siddi tribe of India and Pakistan, and Australian Indigenous people because of the context disease density in relation to the rates of economic development of these societies. These four brief cases are not selected to rigorously test hypotheses. These help us draw together empirical findings which are valuable in understanding communications about COVID-19.

While these societies may not all be sovereign nations, they all feature prominently in matters of economic marginalization and poverty. It is worth pointing out that Indigenous populations all over the world suffer from similar issues of governance (Ronquillo, 2011, p. 286). The governance leads directly to the increased poverty, social marginalization, general, deprivation, chronic poor housing, unemployment, lower rates of educational attainment, and higher degrees of adverse health security that also characterizes these populations globally (Adams et al., 2007; Harvard Project, 2008). Yet, Indigenous and First Nations people continue to survive through impacts of globally oppressive empires in ways that are flexible, dynamic, and unpredictable; ways that resist and refuse compliance in external and multigenerational genocide (Byrd, 2011; Simpson, 2011).

Because of the nature of the disease, Indigenous communities in Australia and the United States have exercised their sovereignty in a manner that they have never done before. Those in East Africa and India have demonstrated greater degrees of autonomy and empowerment in community control of their fragile health sectors. All the groups discussed in this text have faced inordinate disadvantages in life experiences. These Indigenous societies must negotiate externally transposed physical barriers causing lesser access to health and life than for their nonnative counterparts. Just to give example, more than half of their populations lack broadband internet, which denies them access to important informational resources. Many also have challenges receiving national government identification documents and hence, are cutoff from some services.

Housing conditions too cause severe threats on tribal nations from lack of available resources for upkeep and extreme poverty preventing safe conditions. Even when the federal government in the United States provides housing on tribal lands, these homes are often poor quality and present fire hazards and unsafe indoor air quality and fall into extreme disrepair with little to no follow-through or support once they have been delivered. Since housing is a basic human need and housing shortage is a function of poverty, these situations present abuses to human rights. Many Indigenous people also live without shelter or with shelter that lacks running water, electricity, refrigeration, and transportation. Poor housing also means lack of adequate heating, sanitation, and ventilation which cause numerous diseases. This is in addition to the observed isolation that is exacerbated by families living many miles away from government centers. Despite these outcomes of policies of destruction, Indigenous people continue to survive.

The common theme of this book is the government and public communication process and system with COVID-19 in mind. We start by discussing the appropriateness of various lenses which are also building blocks to laying out a composite picture of communications within a pandemic environment. The standard communication practices between leaders, other stakeholders, and citizens during pandemic situations are relatively easy to discern. Our task is to figure out which formal and informal roles various organizations and actors performed during the initial outbreak of COVID-19.

We recognize that disasters can be opportunities for profiteering and government failures. We recognize, where those occurred as forms of governance failures, albeit tangentially. Naomi Klein's (2008) *The Shock Doctrine: The Rise of Disaster Capitalism* demonstrates how powerful political entities can use disasters and crises to remove people from decision-making and take action toward increasing power imbalances to the detriment of people who live in areas where disasters hit. Distinguished professor of Anthropology Harvey (2012), in the same vein, used the example of the financial and economic takeover of Manhattan to illustrate how and where powerful people and entities work together to intentionally remove and disempower people and gain control. These were legitimate concerns during the entire COVID-19 pandemic experience in marginalized communities. In the case of Federal and tribal control negotiations there have historically been continued efforts to remove and harm Indigenous people through powerful federal means. As power becomes dispersed as state levels tribal and state relationships in the United States, for example, continue to serve as a tense place of negotiations in an ongoing way. However, given the shock and immediacy of the COVID-19 crisis, these points of contention are currently enlivened. Watching how people push and pull policies for survival during this time affords a window into historical tension and present power in a way that simply cannot be disentangled from public health.

As scholars, we also seek to acquire knowledge on modes of government structures and how these impact the management of COVID-19 in the identified Indigenous polities. The tendency, in this case is to recognize that governance in Indigenous societies is complex and the parameters of sovereignty are constrained by colonially induced governance frameworks. This idea is explained as a preamble to recognizing the salience of new ICTs in the organization of public health efforts to mitigate COVID-19 pandemic and its disorganization of "normal" societal functions. After unpacking the digital divide argument, we move on to establish a generic model of functions of communications within a pandemic situation. So conceived, we look at leadership styles and the production of messages about COVID-19. This approach tells us how things transpired in the first few months of the COVID-19 crisis.

In this overview and Chapter 2, the stage is then set for us to examine the communication dynamics within the context of networks. COVID-19 was managed through intergovernmental and interlocal organizations so we must understand governance. This observation may be trite, but it has far-reaching consequences for the delivery of content about the COVID-19 pandemic and what ought to be done to prevent its spread across the nations. In particular, the importance of organizations such as WHO and CDC and nation-specific health ministries is central, as policy venues responsible for communications and message control systems that managed the COVID-19 outbreak. By laying out the core principles of national strategies, we capture the policy instruments that might help us contextualize the ensuing management public health regimes. And, last, the question of roles is presented as a preamble to learning about the political motivations and obligations of various actors who played a part in mitigation against the pandemic.

There may well be other ideas that can help us evaluate the communication problem in various Indigenous societies. Those that come to mind are the lessons learned about the role of communication strategies in management of pandemics in Indigenous societies. It will also be crucial to attend to writing and reflection from Indigenous scholars and government/health personnel working within tribal nations to develop rapid health plans under duress of disaster and the added challenge of negotiations with external state and federal leaders who refuse to adhere to boundaries of sovereignty. Without that on-the-ground understanding of the nuances of these negotiations this work cannot be considered a full exhaustive record of what transpired as the disease spread through various societies. We mean for this to be both a continuation of existing work and an additional chapter in the evaluation of how leaders implement and negotiate policies for health and life during times of crises.

Chapter 3 is a review of the status of government communication and COVID-19 mitigation actions within the Navajo Nation. We provide a narrative on the enabling ICT infrastructure to illustrate the preparation of the community in working within a pandemic environment. The enduring effect of the digital divide in this region is part of the context for crisis response. We first describe the search for digital equality and then outline upward and downward communications strategies and challenges encountered.

Chapter 4 lays out the story of the Maasai and the impact of two different approaches to managing COVID-19 occasioned through contrasting political environments in Kenya and Tanzania. While acknowledging the significance of digital divide issues, we also offer a glimpse of unique culturally relevant messaging strategies put in place to fight COVID-19. The context of Kenya's bureaucratic approach to solving public health policies is addressed and contrasted with Tanzania's elite denial and lack of transparency in containing the coronavirus pandemic.

Chapter 5 is about the Siddi regions of India and how they fit in the country's COVID-19 mitigation policies. With respect to India's COVID-19 organization policies, the Siddi and other Scheduled Tribes faced considerable communications and public policy challenges. These are highlighted to illustrate how poverty and discrimination can compound the public health crisis management function.

In Chapter 6, we focus on well-studied Indigenous societies of Australia. While the fundamental assumption is that these communities would be most affected by COVID-19, the evidence suggests that Indigenous sovereignty served the communities well and the relative loses were not worse than in the larger society. Paradoxically, the pandemic became the shining moment for ethnic minority Indigenous groups to use ICTs to fight for their rights and also to protect their citizens from the COVID-19 pandemic. Members of Indigenous Australian communities pride themselves in leveraging social media for mobilizing for promoting wellness. These issues are highlighted within the context of communications and mitigation against impact of COVID-19.

In Chapter 7 are meant to tease out the problems of communications and how these applied to Indigenous societies. In the end, the discussion sums of the lessons learned from the experience of Indigenous societies without violating the cannons of qualitative research. We recognize the importance of playing a small part to increase representation of Indigenous communities in the historical narrative of the COVID-19 crisis. By shedding light on only a few cases, we continue a conversation about societies whose voices are more often than not silenced by powerful forces around the world. And we expect future studies to expand on our discussions and expand knowledge on modern communications in the service of combatting pandemics in different cultural settings. Armed with this hope, more studies that focus on each Indigenous community should, in greater detail, identify specific challenges emerging from the COVID-19 human experiences. For each community, we look forward to understanding the detailed interventions agreed upon and communicated by respective governments. In our final chapter we detail a brief outline of takeaway messages for future pandemic response.

References

Abouzzohour, Y. (July 29, 2020). *Progress and Missed Opportunities: Morocco enters its third decade under King Mohammed VI. Brookings Report.* https://www.brookings.edu/research/progress-and-missed-opportunities-morocco-enters-its-third-decade-under-king-mohammed-vi/.

Adams, A. S., Lee, J. A., & Lipsky, M. (2007). Governmental Services and Programs: Meeting Citizens' Needs. In Miriam Jorgensen (Ed.), *Rebuilding Native Nations: Strategies for Governance and Development* (pp. 223–245). The University of Arizona Press.

AFP. (June 6, 2020). Last redoubt: Pygmies return to forest to isolate against virus. Daily Nation. *Saturday.*, 36.

Albright, J. (2017). Welcome to the era of fake news. *Media and Communications, 5*(2), 87–89.

Amazigh World News. (March 19, 2020). *Morocco's Public Media Continues to Ignore Languages in its Awareness Campaigns Against Coronavirus*. htps://amazighworldnews.com/.

Arora, N. (2020, April 26). *India's Modi urges citizens to follow lockdown as coronavirus spreads*. Reuters. World News. https://www.reuters.com.

Aufrecht, S. E. (1999). Missing: Native American Governance in American Public Administration Literature. *American Review of Public Administration, 29*(4), 370–390.

Aufrecht, S. E., & Case, D. S. (2005). Indians 78, Washington State 0: Stories about Indians and the Law. *Public Administration Review, 29*(4), 450–461.

Averbuch, M. (June 1, 2020). *Mourning in Mexico. As the coronavirus death mounts, interrupted mourning rituals leave families unmoored*. Foreign Policy, FP. https://foreignpolicy.com/2020/06/01/mexico-coronavirus-pandemic-response-mourning.

Az, Z. (March 22, 2020). *Calls for Morocco to Release Political Detainees Amid COVID-19 Pandemic*. AMAZIGH World News. https://www.amazighworldnews.com.

Barker, M. L. (1998). *Policing in Indian Country*. Monsey: Willow Tree Press.

Barrera-Bassols, N., & Toledo, V. M. (2005). Ethnoecology of the Yucatec Maya: symbolism, knowledge and management of natural resources. *Journal of Latin American Geography, 4*, 9–41.

Baumgartner, F. R., & Jones, B. D. (1991). Agenda dynamics and policy sub-systems. *Journal of Politics, 53*, 1044–1074.

Baumgartner, F. R., & Jones, B. D. (1993). Agendas and Instability in American *Politics*. University of Chicago Press.

Berg, B. L. (1989). *Qualitative Research Methods for the Social Sciences*. Allyn & Bacon.

Boadle, A. (2020). *Kayapó chief dies from COVID-19 in Brazil, led protest against Amazon dam*. Thomson Reuters (June 17) Retrieved from https://www.reuters.com.

Boot, M. (April 21, 2020). *Trump Can't Defeat the coronavirus. But he can wage war on immigrants. (OPINION) Washington Post Washington Post.com.*

Byrd, J. A. (2011). *The Transit of Empire: Indige—nous Critiques of Colonialism*. First Peoples. University of Minnesota Press.

Campbell, A. (1995). *Race, Myth and the News*. Russell Sage.

Centers for Disease Control and Prevention (CDC). (2020). *The CDC Field Epidemiology Manual*. CDC. https://www.cdc.gov/eis/field-epi-manual/chapters.html.

Chilisa, B. (2011). *Indigenous Research Methodologies*. Sage Publications.

Churchill, W. (1940). *Their Finest Hour. Speeches. BBC*. International Churchill Society. June 18, 1940 House of Commons. Retrieved from https://winstonchurchill.org/resources/speeches/1940-the-finest-hour/their-finest-hour/.

Entman, R. M. (1990). Modern racism and the images of blacks in local television news. *Critical Studies in Mass Communication, 7*, 332–346.

Estes, N. (July, 2020). *The empire of all maladies: Colonial Contagions and Indigenous Resistance. No. 52*. The Baffler. https://thebaffler.com/salvos/the-empire-of-all-maladies-estes.

Faust, B. (2001). Maya environmental success and failures in the Yucatan Peninsula. *Environmental Science and Policy, 4*, 153–169.

Fouberg, E. H. (2000). *Tribal Territory, Sovereignty, and Governance: A Study of the Cheyenne River and Lake Traverse Indian Reservations. Garland.*

Garnett, J. L., & Kouzmin, A. (2007). Communicating throughout Katrina: competing and complementary conceptual lenses on crisis communication. *Public Administration Review, 67*(S1), 171–188.

Gilliam, F. D., Jr., Iyengar, S., & Wright, O. (1996). Crime in black and white: the violent, scary world of local news. *Harvard International Journal of Press/Politics, 1*, 6–23.

Ghebouli, Z. L. (April 9 ,2020). *Coronavirus Will Change Both Algeria's Political System and Its Opposition*. FIKRA FORUM. Washington Institute for Near East Policy. http://www.washingtoninstitute.org/fikraforum/.

Goldberg, M. (May 11, 2010). *We're All casualties of Trump's War on Science: The Administration kneecaps experts as the corona virus rages*. New York Time: Opinion. http://www.nytimes.com/2020/05/11.

Gómez-Pompa, A., & Bainbridge, D. A. (1995). Tropical forestry as if people mattered. (408-422). In A. Lugo, & C. Lowe (Eds.), *Tropical Forests: Management and Ecology*. Springer-Verlag.

Hajjar, R., & Ender, G. M. (2005). McDonaldization in the U.S. Army: A Threat to the Profession. In D. M. Snider, & L. J. Matthews (Eds.), *The Future of the Army Profession* (2nd edition, pp. 515–530). McGraw Hill Publishing.

Harvard Project on American Indian Economic Development. (2008). *The State of the Native American Nations: Conditions under U.S. Policies of Self–Determination*. Oxford University Press.

Harvey, D. (2012). *Rebel Cities: From the Right to the Urban Revolution*. Verso.

Hillyard, V., & Ortiz, E. (April 2, 2020). *As Navajo Nation enforces coronavirus curfew, Arizona sends in the National Guard*. http://www.nbcnews.com.

Holmes, B. J. (2008). Communicating about emerging infectious disease: the importance of research. *Health, Risk & Society, 10*(4), 349–360.

Hoover, E. (2017). *The River is in us: Fighting toxins in a Mohawk Community*. University of Minnesota Press.

Hopkins, R. (May 11, 2020). *South Dakota Governor doubles down on her anti-native reputation by targeting tribes' COVID-19 Checkpoints*. The Appeal. https://theappeal.corg/south-dakota-governor-doubles-down-on-her-anti-native-reputation.

Hurwitz, J., & Peffley, M. (1997). Public perceptions of crime and race: the role of racial stereotypes. *American Journal of Political Science, 41*, 375–401.

IWGIA. (2020). *Some aspects of Amazigh life in the context of COVID-19*. https://www.iwgia.org/en/news-alerts/news-covid-19/3698-some-aspects-of-amazigh-life-in-the-context-of-covid-19.html.

Jones, B. D. (1994). *Reconceiving Decision Making in Democratic Politics: Attention, Choice, and Public Policy*. University of Chicago Press.

Kabwe, Z. (May 21, 2020). *Bumbling, irrational, deadly: The Tanzanian president is getting the Covid-19 response all wrong*. Mail & Guardian. https://mg.co.za/africa/2020-05-25-tanzania-magafuli-risk-coronavirus/.

Kapucu, N., & Ozerdem, A. (2013). *Managing Emergencies and Crises*. Jones & Bartlett Learning.

Kenyatta, U. (April 6, 2020). *We are at war*. Nairobi Address. NTV Kenya: State House. Retrieved from httpw://www.youtube.com/watch?v=3zShBp015N4.

Klein, N. (2008). *The Shock Doctrine: The rise of disaster capitalism*. Picador.

Lakhani, N. (2020). *South Dakota governor threatens to sue over Sioux coronavirus roadblocks (May 14)*. https://www.theguardian.cm.

Lambert, L. (2014). *Research for Indigenous Survival: Indigenous-Research Methodologies in the Behavioral Sciences*. Salish Kootenai College.

Leitej, J., Iglesias, P. S., Beck, V. M., & Bronner, E. (June 24, 2020). *The Pandemic's Worst–Case Scenario is unfolding in Brazil*. Bloomberg Business Week. https://www.bloomberg.com/businessweek.

Magufuli, J. P. (May 20, 2020). *God will help us, Tanzania president rejects virus lockdown*. YouTube. https://www.youtube.com/watch?v=nTiKh-ugOvI.

Marshall, C., & Rossman, G. B. (2016). *Designing Qualitative Research*. Sage.

Matzen, M. (2020). Update: Tribes respond to Neom's demands to remove highway checkpoints. *Rapid City Journal*. May 8 https://rapidcityjournal.com/news/local/update-tribes-respond-to-noems-demands-to-remove-highway-checkpoints/article_7ad8c4d3-994d-57c6-bf76-71a35a7e9689.html.

McKay, B., & Dvorak, P. (August 13, 2020). A Deadly Coronavirus Was Inevitable. Why was no one ready? *The Wall Street Journal*. https://www.wsj.com/articles/a-deadly-coronavirus-was-inevitable-why-was-no-one-ready-for-covid11597325213?mod=e2fb&fbclid=IwAR1n_w3INlColz1Fg6fkmJUwpK1GAdIWUueDUrfuYlZqkvwrsfyvdkT0g8o.

Migiro, P. S. (2013). Access to Health Services by Indigenous Peoples in the Africa Region. In *United Nations. State of the World's Indigenous People: Indigenous People's Access to Health Services* (pp. 12–32). New York: Department of Economic and Social Affairs. (DESA).

Million, D. (2013). *Therapeutic Nations: Healing in an Age of Indigenous Human Rights*. University of Arizona Press.

National Institute of Indigenous Peoples (INPI). (2020). *Reporte de casos diarios de COVID-19 a nivel nacional y cobertura de los CCPI del INNPI*. http://www.inpi.gob.mx/codicemexico.

New York Times. (May 27, 2020). *The tally of coronavirus in the U.S. has surpassed 100,000*. New York Times. https://www.nytimes.com/2020/05/27/us/coronavirus-live-news-updates.html.

Noakes, T. (April 20, 2020). *The Pandemic Is Slicing Away Indigenous Sovereignty in Canada*. Foreign Policy. https://foreignpolicy.com/2020/04/20/gas-pipeline-wetsuweten-coronavirus-pandemic-indigenous-sovereignty-canada-oil/.

Odhiambo, K. (April 26, 2020). *We need precision to fight deadly invisible enemy* (p. 8). The Standard.

Onoma, A. K. (2020). Epidemics, Xenophobia and Narratives of propitiousness. *Medical Anthropology, 39*(5), 382–397. https://doi.org/10.1080/01459740.2020.1753047

Oprysko, C., & Luthi, S. (March 18, 2020). *Trump labels himself 'a wartime president' combating coronavirus*. Politico. https://www.poltico.com/news/2020/03/18.

Ortiz, J. (2002). Tribal governance and public administration. *Administration & Society, 34*(5), 459–481.

Packham, C., & Pandey, S. (February 26, 2020). *The Thomson Reuters Trust Principles*. https://www.reuters.com.

Paget, D., & Kwayu, A. C. (May 13, 2020). *The covid-19 pandemic is casting Tanzania's Magufuli in the worst light, in an election year*. Quartz Africa. https://qz.com/africa/1857369/covid-19-puts-tanzania-president-magufuli-in-bad-light-for-votes/.

Parenti, C. (2011). *Tro*pic of Chaos: Climate Change and the New Geography of Violence. Nation Books.

Peffley, M., Shields, T., & Williams, B.((1996). The intersection of race and crime in television news stories: an experimental study. *Political Communication, 13*, 309–328.

Reeves, P. (May 24, 2020). *COVID-19 Threatens Indigenous People of the Amazon* (Weekend Edition). NPR, Transcripts. https://www.npr.org.

Riviera Maya Times. (August 14, 2020a). *Cancun's COVID Reaction Police sanctions 600 for failing to comply with sanitary measures*. https://www.therivieramayatimes.com.

Riviera Maya Times. (July 29, 2020b). *Quintana Roo must invest in the conservation and regeneration of its natural assets: SUSTENTUR*. https://www.therivieramayatimes.com.

Ronquillo, J. C. (2011). American Indian tribal governance and management: public administration promise or pretense? *Public Administration Review, 71*(1), 285–292.

Royal Resorts. (July 24, 2020). *Un update on COVID-19 in the State of Quintana Roo*. http://www.royalresorts.com/covid-19/update-covid-19-state-quintana-roo.

Schulman, S. (May 4, 2016). *Life for the Baka Pygmies of Central African Republic*. Guardian. Global Development. https://www.theguardian.com/global-development/2016.

Simpson, A. (2011). "Settlement's Secret. *Cultural Anthropology, 26*(2), 205–217. https://doi.org/10.1111/j.1548-1360.2011.01095.x

Singer, M., & Rylko-Bauer, B. (2020). The syndemics and structural violence of the COVID pandemic: anthropological insights on a crisis. *Open Anthropological Research, 1*, 7–32. https://doi.org/10.1515/opan-2020-0100

Smith, D. H. (2000). *Modern Tribal Development: Paths Self-sufficiency and Cultural Integrity*. Rowman & Littlefield Publishers, Inc.

Taylor, J. (1972). *The States and Their Indian Citizens*. U.S. Government Printing Office.

Taylor, M., Raphael, B., Barr, M., Agho, K., Stevens, G., & Jorm, L. (2009). Public health measures during an anticipated influenza pandemic: factors influencing willingness to comply. *Risk Management and Healthcare Policy, 2*, 9.

Trump, D. J. (March 22, 2020a). *Coronavirus: Trump calls himself "wartime president" in battle with "invisible enemy" The White House*. https://www.youtube.com/watch?v=taAHtUDo18Q.

Trump, D. J. (May 25, 2020b). *Donald Trump Memorial Day Speech Transcript*. https://www.rev.com/blog/transcripts/donald-trump-memorial-day-speech-transcript-2020.

Tulis, J. K. (1987). *The Rhetorical Presidency*. Princeton University Press.

Tumpey, A. J., Daigle, D., & Nowak, G. D. (December 13, 2018). *The CDC Field Epidemiology Manual. Communicating during an outbreak or public health investigation*. CDC. https://www.cdc.gov/eis/field-epi-manual/chapters/Communicating-Investigation.html.

United Nations. (2013). *State of the World's Indigenous People: Indigenous People's Access to Health Services*. New York: Department of Economic and Social Affairs. (DESA). https://www.un.org/development/desa/indigenouspeoples/wp-content/uploads/sites/19/2018/03/The-State-of-The-Worlds-Indigenous-Peoples-WEB.pdf.

UNESCO. (August 4, 2020). *Indigenous Peoples and COVID -19: The view from Mexico*. https://en.unesco.org/news/indigenous-peoples-and-covid-19-view-mexico.

Valencia, J. (May 29, 2020). *Coronavirus spread threatens Colombia's Amazonian Indigenous communities*. NPR. For The World. https://www.npr.org.

Vaughn, E., & Tinker, T. (2009). Effective health risk communication about pandemic influenza for vulnerable populations. *American Journal of Public Health, 99*(S2), S324–S332.

Vizenor, G. (1999). *Manifest Manners: Narratives on Postindian Survivance*. University of Nebraska Press.

Wallace, S. (2011). *The Unconquered: In Search of the Amazon's Last Uncontacted Tribes*. Crown Publishing Group.

Werman, M. (July 2, 2020). *Fauci: We've got to do something about the coronavirus. Podcast. The World*. National Public Radio (NPR). https://www.npr.org/podcasts/381444246/pri-s-the-world.

Westcott, B., & Whiteman, H. (March 18, 2020). *CNN*. Retrieved from https://www.cnn.com.

Withnall, A. (April 6, 2020). *(Coronavirus: Modi says India in for "long war" as he leads national candle -lit vigil. The Independent*. https://www.Independent.co.uk.

World Health Organization (WHO). (May 21, 2008). *WHO Outbreak Communication Guide*. WHO. https://www.who.int/ihr/publications/outbreak-communication-guide/en/.

Yong, E. (September, 2020). *How the Pandemic Defeated America*. The Atlantic. https://www.theatlantic.com/magazine/archive/2020/09/coronavirus-american-failure/614191/.

CHAPTER 2

Understanding layered dimensions of COVID-19

Understanding the management of COVID-19 means evaluating governance and communication at a global level and within geographically and culturally specific cases. A general public understanding of the virus changes across and within contexts and remains in some areas to be highly intertwined with ideas of politics, economics, and health and illness. The various spaces inform unequal impacts of the virus. Our analysis draws on the global context to situate four geographic and cultural settings for the management and impact of the pandemic with an eye toward inequality and health and illness in each. By focusing on communication strategies, we develop a better understanding of governments at work and the links between different jurisdictions designing and implementing pandemic mitigation policies. Perhaps to put it mildly, managing global pandemics could be understood as a formal process of intergovernmental and community collaborations.

Public Health Emergency

Perhaps, globally Coronavirus is the most used word for the year 2020 and possibly 2021. And rightfully so, it may well be one of the most searched terms of our generation. This means that we have an overload of definitions and explanations to work from. Also, because of the existential threat of the coronavirus, governments across the globe devoted considerable informational resources trying to explain in simple terms its meaning and effects on the socioeconomic well-being of society.

Coronavirus, the disease of COVID-19 (C19), or the virus of SARS-CoV-2, was first reported in Wuhan City, China, in 2019. The first confirmed case in the United States at the time was documented in an individual residing in the state of Washington who had returned from Wuhan on January 25, 2020; however, it has also come to light that there are probable cases from the months prior to that time. There were alternative accounts from the Chinese government–controlled newspaper, the *Global Times*, which cast doubt on the origin of the virus. Journalists suggested the presence of novel coronavirus in Italy and other European regions before a health official identified it in Wuhan (Global Times, 2020). Li Wenliang, the medical doctor who identified the virus, was hailed as a whistleblower and a martyr after he later died of the COVID-19. His death sparked a wave

The Inequality of COVID-19. https://doi.org/10.1016/B978-0-323-99867-3.00007-7

of social media sympathies. China's Weibo, the Twitter-like platform, recorded over 10 billion mentions in his name (Leung, 2020).

Samples collected in Milan and Turin in December 2019 indicated presence of virus in sewage. There were also claims that Barcelona, Spain, had positive samples much earlier than that in March 2019 which predates the December 2019 announcements of the presence of the virus in Wuhan (Global Times, 2020). Later accounts suggested that the virus may have arrived in the United States much earlier. On January 31, 2020, the World Health Organization (WHO) declared it a Public Health Emergency of International Concern. The world entered February coming to terms with the global crisis of an out-of-control pandemic.

There is consensus that the virus, SARS-CoV-19, is not a living organism like bacteria, but a protein molecule (DNA) which enters human cells and mutates. Its origin is disputed but scientists agree that it is a virus that manifests through various symptoms including coughs, sneezing, shortness of breath, and loss of smell among a series of other unpredictable and sometimes unusual symptoms. In simple terms, it is associated with flu-like symptoms and is mostly spread through touching the face, mouth, and from respiratory droplets transmitted from one person to the next. Like most viruses, there is not yet an effective cure. The global death rate from the virus in early March 2021 stood at 2,597,052 with over 117,045,316 officially reported cases (Johns Hopkins University, 2021).

Inequality has been part of epidemics and pandemics throughout modern history and beyond. COVID-19 is no different with widespread xenophobia and local cultural ideologies around contagion and risk. Medical anthropologist and author Lisa Hardy (2020) posited that the spread of COVID-19 invoked numerous sets of emotions among human populations at its inception. Among these are fears of people perceived as "vectors of disease," concern for survival, "inability to do regular activities," and reduced movements in spaces such as grocery stores (Hardy, 2020). News outlets widely covered the virus throughout the world. Fear was and continues to be widespread as Nyla Seweingyawma, a resident of the Navajo Nation, stated: "You hear about [Coronavirus] in the news of it being far from here and you never see it hitting home … and all of a sudden it's just down the hall from your workplace. It's scary because you don't know who's been exposed" (Becenti, 2020). Early interviews on the pandemic revealed inquisitiveness and even pleasant observations about how people were connecting with one another. This later turned to despair as people became ill and lost loved ones to the global crisis.

Globally, although the WHO had declared coronavirus a pandemic, by the end of May 2020, the numbers of those directly affected by the coronavirus was already staggering. Scientists, basing their research on the genetic sequence of earlier strains, also identified a different more lethal strain of the novel coronavirus (Vartabedian, 2020). There was fear that the new strain had appeared in Europe from where it entered the United States through the East Coast (Vartabedian, 2020). By the end of May, close to 1.6 million Americans had been infected with the virus and over 100,000 lost their lives.

Were these outcomes inevitable and if so, what role did communication play? This is the question we have to grapple with. And before we do that, there is need to lay out the structural arrangements through which government communications operate. This means we must identify the literatures and concepts upon which the principles that inform the structures are built. These principles may be distilled into four main areas. First are the national government command and governance structures which shape and confine the communications during crisis situations. Second, we look at the concepts of information and communications technologies (ICTs) and the impact of the digital divide on Indigenous societies. Third, we consider the idea of roles in network governance to show the importance of national and global organizations in managing the COVID-19 pandemic. Fourth, we consider the roles and challenges of national and Indigenous leaders who set the tone of policy implementation. Fifth, we reflect on the perceived and real problems of communication in the current era of fake news and disinformation. Briefly, these broad concepts guide the manner in which we study the selected four cases.

National governments and Indigenous communities: duplicity and layered communication channels

During war situations, commanders choose their messages after serious thought. They send codes that outline desired goals and they use codes to convey strategies that are risky if misunderstood. With abundance of ICT tools, policymakers and health professionals have more than enough avenues to make their cases known and acted upon. Some policymakers appear to be more trustworthy than others. This is important because appeals to the public to follow guidelines for safety and health are effective only when the messages are delivered with clarity and there is some level of public trust. Although difficult to achieve, clarity in communications is an important principle in management. The design of reports, whether written or presented in oral forms, should convey intended meaning and use numbers in a "way that emphasizes the important relationships" (Padovani & Young, 2012, pp. 297–98). Thus, for example, providing numbers of those who have recovered from COVID-19 should be associated with the policy efforts to mitigate against the disease.

The ability for the messenger to display personal qualities like compassion, competence, honesty, and trustworthiness is crucial. It helps if governments working within e-governance networks leverage ICTs for public health policy management. Foundations of trustworthy messaging are more closely associated with the strength of democratic institutions and governance structures. In wartimes, centralized command structures and systems are important in messaging and cohesion in action meant for mitigation against negative outcomes. In this research, we observe that the lack of communication clarity or effective messaging can be attributed to complications arising from the inter- and intragovernmental situations and geographic distance. Each situation is different, though analyzing secondary data on each provides insights into federal public health deployment successes and failures beginning with the Navajo Nation, Maasai people, Siddi Indians, and closing with Australian Aboriginal people and Torres Strait Islanders.

Indigenous societies occupy the world, and all share the common feature of sustaining traditions and customs that distinguish them from politically and economically dominant societies in which they share geographical spaces. Indigenous societies also share histories and presents of survivance despite violent conquest by colonial powers. While a group like Siddi people in India is not Indigenous to that subcontinent, their self-identification as a community continues precolonial cultural practices and they have resolved to maintain their ancestral environments as distinctive communities.

Continuations of colonial oppression

Colonialism is at the root of ongoing settler colonial violence. The intent of colonialism was conducted on genocide and the "logic of elimination" (Wolfe, 2006, p. 387). While colonial practices are often framed as a single event or series of events confined to a timeline, Indigenous scholars demonstrate in writing how colonialism is a process that continues on far beyond primary colonial violence. The outcomes of settler colonialism and other types of oppression include political and economic aggressive disenfranchisement designed to result in underdevelopment of capacity to sustain wellness and health within these communities.

Indigenous and First Nations people have relied upon and developed immensely creative and sustained means for living and obtaining well-being despite oppressive forces. Still, the challenges the federal and state governments continue to employ can be burdensome and threatening to well-being and survival. Even when attempting to institute human rights strategies federal entities tend to inscribe narratives on tribal communities rather than disrupting a dominant narrative in ways that could lead to change (Millon, 2013). Commenting with reference to the Navajo Nation, Dr. Manley Bigey, board chair of the Naat'aannii Development Corp. (NCD), remarked that "effects of colonialism still linger in health care policy in Indian country and that the current Medicaid offerings were not tailored to the specific needs of the Navajo people" (Quoted in Hamway, 2020). Other studies continue to document poor planning and structure on the part of health-care entities and low levels of knowledge of structural barriers to health on the part of health-care providers who serve Indigenous people who need care. Despite these relentless forces of oppression Indigenous people and nations continue to survive. "Survivance" is a term that allows space for the reality of oppression and violence and also values and "an active sense of presence over absence, deracination, and oblivion; survivance is the continuance of stories, not a mere reaction, however pertinent" (Vizenor, 1999: 1). This framework is crucial for understanding the powerful moves Indigenous nations and peoples took during the first year of the pandemic and beyond which are designed for survivance not solely in response to external state and federal laws but as central to survivance. This is true even when it requires navigating the additional layers of the burdens of ongoing sociopolitical racism and regular attacks on sovereignty from states and countries.

Even in light of multiple strategies for dealing with the pandemic, often more successful and well-planned than those in neighboring towns, Indigenous people endured illness and lost at high numbers relative to non-Indigenous neighbors (Akee, 2020; Mineo, 2020; Powers et al., 2020). Public tropes that surrounded this unequal loss failed to consider a layered perspective on why health outcomes were already unequal and continue to be so. Instead, public figures responded through invocations of racialized stereotypes that placed blame on individual or so-called "cultural" behaviors in damaging ways. Anne Nuorgam, chair of the United Nations Permanent Forum on Indigenous Issues (UNPFII), posited that Indigenous communities were more vulnerable to COVID-19 because they "already experienced poor access to health care" and have significantly higher rates of communicable and noncommunicable diseases (Nuorgam, 2020). While it is true that global health data indicate higher levels of poverty, respiratory illnesses, and other health risks for Indigenous people than non-Indigenous people the reasons for this are complicated and rest in histories and current practices of power and dispossession. Moreover, leaders continue to create policies and enact racism in ways that contributed to ongoing stigma and discrimination. A more informed discussion of the agency of Indigenous peoples and tribal governments is necessary to understand the mechanisms through which people access health and wellness and the ways in which settler colonialisms present concrete and repeating challenges to health and life.

Still, external ideas of cultural barriers quickly became part of a public narrative of inequality related to the virus. Some relied on cultural essentialism and stereotypes by positing that high rates of COVID-19 in Indigenous societies might be related to traditional lifestyles, which, some wrote, "pose a threat" because "most Indigenous communities regularly organize large traditional gatherings to mark special events, for instance, harvests, coming of age ceremonies, etc. Some Indigenous communities also live in multi-generational housing, which puts Indigenous peoples and their families, especially the Elders, at risk" (Nuorgam, 2020). These narratives first erase any discussion of structural oppression and second rely on misrepresentations of people who must navigate multiple social and political systems regularly through complex strategies. The single idea of Indigenous people in need is based on the notion that Indigenous people exist in a vacuum of the past whereby knowledge and practice of virus precaution cannot exist.

In fact, in countries like the United States it was White Republican people who continued to gather without care and without adherence to global and local health directives while leaders of Indigenous nations called for stricter enforcement to protect elders.

African Americans share in the same structural limitations to attaining good wellness outcomes. During his June 23, 2020, testimony at a US Senate Health, Education, Labor, and Pensions Committee hearing to examine COVID-19, Dr. Anthony Fauci, Director of the National Institute for Allergy and Infectious Diseases, said African Americans, through no fault of their own, faced a double whammy because they were more likely to be exposed to the coronavirus on the job and many have preexisting conditions that make the virus more dangerous (Brueck, 2020). We premise the discussions that follow with the assertion that the reported large numbers of positive cases of COVID-19 in marginalized communities are not a result of cultural inability to stay safe, but instead they are part of the ongoing oppressive forces rooted in historical and present oppression. Further and perhaps most crucial is the fear that loss of life for Indigenous people also means loss of collective memory, culture, and language. Perley (2014) writes about cultural loss in this way:

> *I am a member of Tobique First Nation, and I experienced and continue to experience the trauma associated with Maliseet language, cultural, and spiritual loss. I feel that trauma on a daily basis as personal alienation from Maliseet living traditions and disempowerment in contemporary regimes of colonial power. My experience is shared by members of Tobique First Nation; all Aboriginal First Nations, Métis, and Inuit in Canada; as well as Indigenous Peoples across the globe.*

These rapidity and intensity of loss associated with COVID-19 introduced a different magnitude of what many already experienced. Indigenous peoples across the globe were well aware of what was at stake and issued calls for successful and sustained efforts to curb the virus.

Situating COVID-19 in the context of new technologies

In the following sections we present limited background for selected regions and peoples to set the stage for a discussion of COVID-19 and ICTs. It is important to note that these descriptions are far from conclusive or thorough and we suggest reading the work of Indigenous scholars to more fully explore the histories and presents for each group. Beyond our rudimentary description of context here the ways in which people in each region and of each identity exercise survivance and negotiate ongoing oppression are not our story to tell.

Tribes of the Southwest United States

The United States rests upon tribal homelands of more than 560 federally recognized tribes. The Southwest region of the country alone is home to more than 20% of Native Americans in the United States. One of the largest tribal nations in this cultural region is the Navajo Nation. The majority of members of the Navajo Nation reside in sections of three different states, northern Arizona, northwestern New Mexico, and southeastern Utah, where a vast majority are in a small area of the southeast. The area is roughly the size of West Virginia. There are also thousands of Diné (Navajo) people in other regions, including Colorado. And although many more are embedded in many urban centers but call the Diné Nation "home." This is a land where people continue to live with and survive the horrors of settler colonialism and genocide (Wolfe, 2006). The Reservation (Rez), as it is known, is where the place of the sacred four mountains and where the spirit of tribal sovereignty reign. The entrance of the Grand Canyon lies on western side of the reservation. It is also the entrance to Lake Powell, and the Little Colorado River. These significant landmarks are key to understanding the Navajo identity and way of life.

The Diné people include diverse ethnicities. Navajo people are dual citizens of their sacred nation and also the United States. As is always the case, there are also Navajo people who have multiple identities and ancestral lineages. Given that Navajo citizens live in spaces that are governed by a layering of administrative units, there is considerable confusion when it comes to establishment of a central command unit in communication of information about life-threatening pandemics such as was presented by COVID-19 pandemic. COVID-19 is not the first health crisis to adversely affect Navajo people. For example, colonial leaders intentionally used epidemics as instruments of genocide through the use of blankets contaminated with Smallpox in the Northeastern parts of the nation and in other occupied native areas (Johnson, 2020).

Technically, the Navajo enclave and territory are a sovereign nation with treaty obligations from the US government. Notwithstanding the role of the Bureau of Indian Affairs and federal responsibility to provide for all its citizens, the sovereignty of the Navajo Nation requires that its leadership negotiates with state governments in New Mexico, Arizona, and Utah. Each of these states had a COVID-19 command system and employed a simultaneously engaged and detached approach to managing the spread of the virus in associated parts of the Navajo Nation. Although the ultimate management and control of the health system in the Navajo Nation was the responsibility of the tribal government led by its president and Council, different sets of obligations complicated the messaging reaching the vast and spread out lands.

One possible explanation for this outcome is that the enabling information and communication infrastructures have been difficult due to the inadequacies and inactions occasioned by the negligence by the various intergovernmental arrangements. The size of the Navajo Nation and the lack of adherence by outside forces to respect and understand how to work with a sovereign nation create its complex relationships with three different states and the federal governments. Such variation in policy venues also affects the quality and coherence of messaging involving emergency situation such as COVID-19. The Navajo

Nation also grappled with conflicting and dangerous directives in border towns where states and towns did not exercise health directives well. Further, then-President Donald Trump openly stated his readiness to assist British Prime Minister Boris Johnson's pandemic malaise but said nothing about his readiness to assist members of the Navajo Nation. Trump's remarks, which centered around providing treatments manufactured by US drug companies, were made after Johnson was moved into an ICU facility in London (Reuters, 2020). Such signaling demonstrates settler colonialism whereby the highest echelons of the US government still invoke greater kinship ties to Britain than Indigenous nations.

Commenting on the internet connectivity situation in the Navajo Nation, Otenyo and Lind (2011, p. 200) observed that "due to poorly executed contracting arrangements and the lack of capacity to successfully monitor e-government infrastructures, ICT projects have often failed to function optimally." They point to reported cases where service providers based in Utah shut down internet access and even double-billed the tribe. This situation is compounded by the inadequacy in provision of broadband services that ensure continuous access of internet services at affordable prices. Thus, the digital divide, characterized as inadequate access to digital services, weakened the ability of community leaders to manage the communication regime during the COVID-19 pandemic.

Maasai people in East Africa

Maasai people live in areas across and between two independent nations of Kenya and Tanzania where they support their livelihoods with products from their animals and land. As nomadic people, they too were and are affected by colonial policies that separated families forcing them to live in different political jurisdictions. Maasai people do not have externally recognized sovereignty. African governments continue to subject Maasai people to human rights abuses resulting in economic marginalization and underdevelopment. Their lifeways and social organizations were interrupted by violent colonialism, yet they have changed and survived. Hodgson (2014) writes,

> *Although they were originally agro-pastoralists, in time certain changes in technology (mastery of iron forging), climate (decades of relatively high rainfall), and social organization (the emergence of male age grades) enabled a group of Maa speakers to specialize almost solely in livestock herding and develop a heightened sense of themselves as pastoralists. In recent years Maasai people have organized non-government organizations and joined Transnational Indigenous Peoples' movements (Hodgson, 2014).*

The national context for Maasai people changed in the 1960s, when Kenya and Tanzania followed very different ideological orientations. Kenya adapted free market capitalism and Tanzania took a socialist path thus subjecting Indigenous Maasai people to two different political processes. The late professor Claude Ake (1984) differentiated the different ideological world views that undergirded policy-making in the two countries. Although the different paths led to different developmental trajectories, later on the two states had remarkably similar outcomes for Indigenous Maasai communities. However, income inequalities are much sharper in Kenya, which is considered a lower middle-income nation than Tanzania. The different income disparities and economic settings have had an indelible mark on Maasai communities. From a policy perspective, the two different states, Tanzania and Kenya, have also established different ICT platforms and infrastructures.

While both Tanzania and Kenya established major international tourism facilities in locations associated with Maasai homelands, the exposure of the community to global capitalism has not necessarily produced advances in material living conditions. Development indicators for Maasai on reservations remain remarkably below national levels, making them among the poorest of the poor. In the age of ICTs, Maasai people have little control over their cultural assets peddled around the globe on social media and other internet outlets (Otenyo, 2017). Maasai people in both countries continue to be disposed and marginalized and both faced the COVID-19 pandemic in the context of different state bureaucratic frameworks. Kenya followed guidelines outlined by the WHO and its Public Health Act, published in Kenya Gazette Supplement 41 and various political proclamations from various political leaders (Otieno, 2020).

On the other hand, Tanzania's handling of COVID-19 was tightly controlled by President John Magufuli's unpredictable proclamations. Although he defended his actions not to take drastic measures such as shutting down the economy, he nevertheless gave his Ministry of Health much autonomy in directing policy actions to combat COVID-19. Tanzania's borders with Kenya were closed for part of the March and April 2020. This closure affected many Maasai families that traverse the region's grasslands. Therefore, Tanzania and Kenya did not have a uniform or coordinated approach to combatting COVID-19. Thus, government communications within the Maasai community reveal different styles and emphasis in management of disease within an Indigenous community spread across different territorial jurisdictions.

Although these communities are domiciled in different territorial jurisdictions, their "tribal citizenship" and moral community affiliations are strong. The separate governments where they pay taxes have not encouraged deeper cross-border nation-building efforts and efforts to combat the pandemic seemed to encourage inward-looking solutions. Here, we suggest that communication processes were overlapping and did not include coherent solutions and political strategies. That mindset can account for some of the abstract and problematic issues in messaging about COVID-19.

We hypothesize that the national governments failed to listen and failed to generate energy required to address the existential health problem that pandemic presented among sections of Indigenous societies and continued a legacy of erasure and interruption of locally derived solutions for health and survival. The health policy strategies reinforced the geographical isolation of Indigenous communities and, seemingly, delinked their efforts to combat the virus from the national and universalistic nature of the scourge.

Scheduled Indian tribes of India and Pakistan: Siddi people

Siddi or Sidi people live in regions that are now India and Pakistan and considered to be descendants of the Bantu people of Africa. Oral histories suggest that these are the original occupants of the Indian land mass. Siddi people are among the scheduled ethnic communities of India, referred to as Scheduled Tribes (STs). Siddi people have different tribal affiliations, identities, languages, and relationships. Identity and race are complex among Siddi people who share ancestry and identity as well as commonalities of living under external rule; however, there are also different identities and ideas of race and ethnicity among different Siddi groups (Khader, 2020). Phenotypically Siddi people often appear to carry traits of African ancestors which socially identify them to others. This means that Siddi people often live as targets of multiple layers of racism. And it is conquest from the Aryans and Dravidians that reconfigured the rich cultural mosaic which characterizes India, a huge subcontinent. Most of the STs have low populations and the dominant groups, those with significant numbers include Gond people of Central India, Bhils people of Western India, and Santal people of Eastern India. Siddi people would count as among those whose numbers are at the lower end of the population count but certainly more populous than the Andamanese.

Siddi people continue to join the global network through their social media connections expanding their narrative presence internationally. As many have now become activists for social justice, their role in politics is beginning to receive some attention (Obeng, 2007). And since they have a rich cultural imprint in Western India, particularly in Goa, Karnataka, Andhra Pradesh, and Gujarat, Siddi art is celebrated in its own right. Without doubt, the significance of their musical art forms and religiosity is well documented (see, for example, Catlin-Jairazbhoy & Alpers, 2004; Catlin-Jairazbhoy & Ali Jairazbhoy, 2002). Hawley (2008) paints a rich picture of Siddi self-identity and their settlement in India as a precursor to modern globalization. Sheth (2014) offers another narrative on sense of place and uncompromising defense of their institutions among Siddi people. Ketaki Sheth and others have also been leaders in producing documentary work on everyday experiences of the Siddi. People belonging to Siddi groups also have recently expanded social networks among different tribes and beyond through "Afro-Indian" and "Afro-Global" networks (Péquignot, 2020). Siddi people also continue to integrate ethnobotanical and other healing practices into their pluralistic models of healing (Solanki, 2020).

It is instructive to look at the lives of Siddi people living in multiple different communities through several documentaries that have been uploaded on different media channels, including YouTube platforms. These documentaries indicate the limits of general press coverage of the condition of Invisible Africans-Indians. The accounts given by Sheth (2014) and Professor Beheroze Shroff (2017) show marginalized condition of Siddi people. They affirm claims that even when Siddi people have sufficient education, they still face discrimination and are denied opportunities for advancement.

And there have been complaints that members of the community are often targeted for kidnapping by those in search for organ transplants. For example, after a member of the Siddi community in Bengaluru was reportedly missing, there was speculation that he had been a victim of the criminal kidnappers. The story was dramatized by the man's relative, "As we belong to the Siddi tribe, originally from Africa, we are discriminated a lot here. Few members of our tribe have been tricked this way and their kidneys have been operated and taken away and that is the reason for our panic" (quoted in Thomas, 2016). Such issues are rare but assume heightened significance when minority groups are disproportionately affected by the criminal gangs preying on the poor.

Although we don't delve into the Siddi populations of Pakistan, people in positions of power and social norms result in racism and abuse and required further attention and scrutiny. However, in Pakistan Sidi people have also entered the national political realm with the election of a Sidi lawmaker to the Pakistan People's Party in 2018 (Sohail, 2018). Siddi people in India have also joined the ranks of lawmakers at the national level. Shantarama Siddi became the first Siddi person to join the legislative council in India during the pandemic (Financial Express, 2020). Press coverage of this accomplishment first included a journalist description of Siddis as a "microscopic community" though they are reported to have a population of 50,000 in the region (Financial Express, 2020). However, another journalist reporting on this victory demonstrated racist violence by stating that Siddi people are "economically and socially backward" (Sayeed, 2020).

According to interviews conducted in villages in Karnataka, Siddi Indians occupy the lowest echelons of the stratified society. By many measures, they are far below the "Untouchables," the lowest social class (Mboya, 2019). Being so low on India's caste system has continued to position this minority group among the world's most disrespected outcaste ethnicities. The Siddi legally hold Indian citizenship and speak local dialects but because of their African heritage suffer outright discrimination and have reportedly lost their land (Shodhan, 2015; Stuart, 2018). Luke Duggleby, a British photographer documented the living conditions of the Siddi people, commented on his outsider view of Siddi people when he said, "because of discrimination, many marry outside the Siddi community and dilute their African appearance. For some that is how they can avoid discrimination, but for many this is seen as the very disappearance of the Siddi people themselves" (Chari, 2015).

One of India's leading media content platforms, Youth Ki Awaaz (YKA) (2020) reported that this inhumane treatment is because of their skin color and traditions by discussing the depths and extent of racism adding that, "the caste system is not the only source of discrimination; the quota reservation for ST, SC, and OBC in universities and jobs fuels such hate" (Youth Ki Awaaz, 2020).

There are also groups of ethnic Siddi in parts of Pakistan, and very few in Sri Lanka (Chari, 2015). Although Siddis don't feature much in Indigenous studies programs, they share colonial pasts and presents of living with continued forceful rule and disenfranchisement experienced by Indigenous communities across the globe. And their indigeneity is not in doubt. In India, Siddi- or the "lost Africans of India" live in villages in three states, Maharashtra, coastal state of Karnataka, and Gujarat. Many also live in the city of Hyderabad and in Andhra Pradesh.

There are various often disjointed accounts on the origin of the Siddi tribe. Some historians trace the origins of the Siddi in Africa (Bhattacharya, 1970; Lodhi, 1992). These accounts trace their origins to Bantu peoples of East Africa and diverse regions of Abyssinia (Ethiopia). It is suggested that the Siddi were moved from East Africa to India as slaves (Habshis) of Arab traders in the 7th century. Habshis (Siddi) is the Arabic word for an Ethiopian or Abyssinian (Amarat, 2006, p. 13). The exact dates are disputed and some state that slavery was not as organized as that which existed in the 16th century (Economic Times, 2008). This suggests that the trans-Indian Ocean slave trade involving sale of Africans predates that of horrific sales across the Atlantic Ocean to the Americas. It has been estimated that between 4.7 million and 11 million Africans were sold by Arab merchants in the region (Amarat, 2006, p. 19).

There are clearly two data sets on the slavery story. First, the European obsession with slavery accounts for the existence of thousands of Siddi villagers in the Goa region of Western India. Thousands, if not millions, of Africans perished during this relocation, in ships nicknamed as navios tumbeiros (Coffin Ships) (Herald, 2020). Portuguese slave traders actively transported Africans from Mozambique as well. The history of Goa is well documented. It is the place where the Portuguese set up camp upon arrival in India in 1510. After arrival, the Portuguese brought with them Africans whom they had captured in Africa. The Portuguese referred to the slaves as Kafris or Khaapri, or Cafres (Blacks), and to themselves as "White" (Paakle) (Herald, 2020). African labor was required for manual work in and out of their houses. Many of the former slaves relocated from Goa to Karnataka. Some still speak languages such as Goa Konkani, which has some Portuguese words.

Groups of Siddi people moved to India on 7th-century Arab slave ships. The Arab slave dealers took Africans from diverse ethnicities, from Central and South Africa. The slave dealers moved through Zanzibar, in southeastern Africa before shipping off to India. The Arabs, through the Mughal (Mogul) Empire which, who were culturally oriented with Persian, controlled much of South Asia in the 16th and 19th centuries, moved Africans to serve as soldiers and unpaid workers. Some of their African slaves were considered royal slaves and rose through the ranks to be important leaders in military and society. However, a vast majority were left behind.

The slave narrative is countered by different historical account. That is, of a good kind. The remarkable resilience of the Habshis or Siddis and the rise of many of them to high positions in ancient Indian society provide a different narrative. Amarat (2006) presents examples of distinguished Sidis, especially those who became high profile leaders within the Mughal Empire, which dominated India in the 16th and 17th centuries. Amarat (2006, p. 19) adds that many of the Africans who came to India were traders.

A genome-wide survey to trace their ancestral origins concluded that they inherited ancestry from Africans, Indians, and Portuguese. Siddi people were in India long before Vasco da Gama arrived in the Indian Ocean port of Mombasa, Kenya, in the 15th century. Additional analyses of their uniparental chromosomal and mitochondrial DNA linked them to Bantu people of Africa Shah et al., 2011. There are also records of Africans transported by Portuguese and British colonialists much later on.

Still, there are also groups of Siddi who moved to India as merchants and sailors outside of the slavery experience. At the end of slavery, some Siddi fled into bushes of India, for example, Jambur village, Gir forest in Gujarat and settled there. Gir forest has the highest number of Siddi tribal people (United Nation, 2017). Although some of the Siddi became influential within the larger Indian society, the vast majority remained poor and marginalized. It has been suggested that the Siddi "scattered across the woodlands of the Indian subcontinent" to escape slavery (Awaaz, 2020).

Any visits to Siddi villages show African indigeneity, even though some people believe they have Yemeni, Omani, and Bahraini origins (Minda, 2004). Siddi folk songs and drumbeats (including dance called Sidi Goma or dhamal) are African and many Siddi demonstrate the complexity of cultural identity by adorning saris and other native Indian attire and garb. Professor Amy Catlin, an ethno-musicologist and her collaborator Nazir Ali Jairazbhoy, found similarities between Siddi musical styles and instruments to those in parts of Africa (Catlin-Jairazbhoy & Jairazbhoy, 2002, 2003).

Many Siddi people who are not fluent in African languages still enjoy African songs (Whitehead, 2000). Because of their triple heritages, they have names from all sources including Portuguese and/or Konkan, for those in Kamataka State clusters. But, outwardly, Siddis look like Africans but are completely assimilated in Indian cultures and languages, hence the reference to them as Afro-Indians. Although a vast number of Siddi are Sufi Muslim, they also revere their African saint, Gori Pir (Gor or Ghor), and Siddi saints like Chishti Sufi. Gori, as it is claimed, can cure diseases such as COVID-19. Still, hundreds are Catholic, thanks to Portuguese influences.

Siddi people share one characteristic with all marginalized Indigenous societies–government indifference and economic neglect. Many serve as farmers and manual contingent workers. Still unemployment is rampant in Siddi villages. Because of government failure to end poverty on Siddi tribal lands, the community lacks adequate health care and education institutions (Ramdev, 2015; Shastri, 2015). Many Siddis also endure human rights abuses including kidnapping for alleged organ harvesting (Thomas, 2016).

Although efforts were made to register all marginalized Indian communities through the Aadhaar Digital ID program, the delivery of welfare services remains a challenge among clusters of Siddi scattered in different states. Due to poverty issues, affordability of various ICT products, including mobile apps, is still a source of concern and the digital divide continues to affect the Siddi and other Indigenous societies (World Bank Report, 2016).

Siddi populations in neighboring Pakistan live in the city of Karachi and in Baluchistan and Sindh provinces (Chari, 2015). In Pakistan, they are also known as the Sheedis. Journalists refer to this community as the African-Pakistanis. The Siddis are among the smallest ethnic community of Pakistan. In terms of culture and religion, many Siddi people in Pakistan are Muslims but maintain a distinct African identity "in the midst of dominating South Asian cultures" (Abbas, 2002). Siddi people hold distinct ceremonies that honor both Muslim and African traditions, which include conducting an annual Mangopir festival. The Siddi ways of worship incorporate African customs and Islamic mysticism. Many Siddi traditional ceremonies and rituals involve use of their language, which is a mixture of Baluchi and Kiswahili dialects.

During times of economic success, India's leaders did not include Siddi people in investments to education. Although the late Prime Minister of India, Indira Gandhi was supportive of the tribes, most state and federal political leaders have made only a few investments in regions inhabited by close to 60,000 members of the Siddi communities. Lack of investments is exacerbated by low land ownership rates occasioned by lack of land title deed and tenure. With some notable exceptions, the vast majority of people have only been able to achieve third or fourth grade levels of education. These externally oppressive forms of disinvestment are forms of global settler colonialism whereby access to resources for economic and political power and advancement continues to be limited by racism and national political power.

Many young Siddi women must find employment as low-wage domestic laborers in neighboring towns, while men often only find employment as underpaid agricultural workers (Mboya, 2019). In terms of health care, the Siddi suffer worst outcomes as well. In fact, during the last major plague of the 18th century, the community survived by moving into the deep forested interior regions of India and depended heavily on religion for survival, often without adequate government support. Unquestionably, then, the Siddi have continued to struggle for greater political and economic freedoms, much like other marginalized societies (Shroff, 2008). That said, like their kin in India, the Siddi of Pakistan also live in abject poverty.

Indigenous people of Australia

Australia's Indigenous communities include different groups with multiple identities and different socioeconomic situations; however, all have endured colonial and settler colonial systematic oppression, economic disadvantage, and exploitation of the community (Pink & Allbon, 2008; Tatz, 2001). Marginalization and settler colonialisms are manifest in other dimensions including lower life expectancy, political underrepresentation, and cultural insecurity. Despite ongoing oppression, however, Indigenous people within the region of Australia and elsewhere also continue to survive and contribute to robust theoretical and artistic ways of knowing.

Like the other cases, the colonial model in Australia resulted in the destruction of Indigenous environments and oppression hinged upon notions of terra nullius. The power relations between people of the First Nations of Australia and the occupiers led to destructive outcomes, especially loss of nature and cultural devastation. Although considerable damage continues to be done in the name of modernization, the Indigenous people of Australia have through community organizing and increased fostering of solidarity continued to reclaim knowledge about self-preservation. As evident from their approaches to combatting COVID-19, which arrived in Australia through China, the return of native control over their destiny is evident from efforts put together to combat the pandemic.

Still, First Nations communities have taken advantage of existing technologies and leveraged various applications in culturally relevant ways. In spite of the existence of digital exclusion and divide, these groups increasingly use mobile phones and other computer-mediated devices and applications.

Following Carlson (2014) and Carlson and Frazer (2015, 2018), we acknowledge the power of social media in managing health and well-being. Among the most widely used applications are Facebook, Twitter, Snapchat, and Instagram. An Application such as Facebook, which is popular for community networking, has proved to be valuable resources for community support networking and mutual assistance. On the other hand, Twitter users are often engaged in political commentary. Many studies report that Aboriginal and Torres Strait Islanders report high incidence of mental health issues. And it is through connections and networks that several have sought and received groups therapies. In addition, social media offers opportunities for grief and support to those mourning. People of the First Nations of Australia have made extensive use of social media to comment about those deceased (Carlson, 2014). Furthermore, Facebook is a very popular platform that reconnects Indigenous people, especially those separated through past government actions. Younger members of the Indigenous communities use social media platforms to connect with elders and also to facilitate knowledge preservation and build community.

Carlson and Dreher (2018) and Edmonds et al. (2012) provide further evidence that First Nations people apply innovative ways to leverage social media in daily living experiences and social networking. It is on social media that members of Indigenous societies shared knowledge about the scope and dangers of COVID-19. It is also in social media that culturally relevant message of hope and strength was shared. Indigenous communities employed social media to explain what COVID-19 meant and the communities' willingness to join in the national efforts to prevent the spread of the disease in their native land. Indigenous communities embraced social media platforms to publish messages on efforts to use ceremonies and community circles to stop the virus.

ICTs and concepts of the digital divide in Indigenous societies

Technology is undoubtedly integrated into changes over time. Perhaps, in recent times, proliferation of ICTs has participated in shaping some aspects of cultures in the industrial world but gone beyond the scope of the transformations brought about by the industrial revolution to reach the most traditional enclaves. And digital technology ubiquitous, AI has multiplied just how much data we can carry around on our fingertips. The outreach of mobile phones makes it possible for a vast majority of both rich and poor societies to be part of the global connectivity system.

To a large degree, the ability of a nation or people to benefit from the advantages of modern ICTs, the by-product of the information revolution, is the availability of these technologies in the first place. The COVID-19 pandemic proliferated at a time when several countries, in the developed world, had already attained maturity in their digital presence and ICT infrastructural development. It was expected that communities that invested heavily in online and digital resources and services would, arguably, have better opportunities for citizens to interact with each other and the government.

However, due to the continued existence of the digital divide and unequal regional investment, internet access remains a challenge in many poor and Indigenous societies. For example, in Appalachian southeast Ohio, studies showed that the region lagged behind the rest of America in terms of availability of quality internet services. The Gallipolis Daily Tribune (2020) stated so eloquently: "Lack of quality Internet access, skills, and motivation to are creating even more challenges as the nation and world transition to conducting vital business and communications to online environments during the Covid-19 pandemic" (Gallipolis Daily Tribune, 2020). The lack of internet resources in Appalachia is well documented as is the case in most Native American and Alaskan Indigenous societies (Vick, 2020). In America, although universal telephone service has been in existence for over a century, that is not the case with internet services. Evidence from the Federal Communications Commission suggests that "less than half of households living under $ 20,000 are connected" (Vick, 2020). These societies continue to lag behind other communities in using internet for critical tasks that promote engagement and wellness.

Indigenous societies often experience lack of connectivity and investment. The technical definitions of the digital divide as offered by David Garson (Garson, 2006, p. 98) are simple: the dichotomous access discloses whether an ethnic group or social group has or does not have access to the internet. Tribal nations in some cases span across geographically dispersed rural areas that, in some cases, do not have electricity or running water let alone effective broadband.

On the other hand, the continuous access argument is another way of understanding the divide and reflects on the availability of the internet through high speeds and at any time one chooses to, seems applicable to tribal societies that don't have stable electric power and connectivity. There is also the dimension of skill access, which speaks to education and training equity, that is enormously disadvantageous to elder members of society in Indigenous communities. The external control of information infrastructure resources continues to compound inequality due to the inability to access the benefits of vital information uses (VIUs) that impact their quality of life. Perhaps, researcher Professor Laeeq Khan, who mapped the digital divide issue in Appalachia, offered one of the best templates for highlighting the importance of digital services in the context of VIUs during the COVID-19 global pandemic. This goal-oriented template provides a clear picture of the benefits of a robust information and communication infrastructure in managing COVID-19. We summarize it in Table 2.1.

Scholarship on the digital divide in Indigenous societies is vast. For example, Willis and Tranter argue that scholarship focuses on income inequalities as a key driver of digital inequalities (Willis &Tranter, 2006). This line of thinking is also expressed in

Table 2.1: General uses of ICT services.

Internet use focus area	Functions and service specifics
Health	Health tracking, appointments, doctor–patient interaction, televisits (telehealth, teletherapy, telemedicine) services. Health information about noncontagious diseases that lower life expectancy.
Education	Formal and informal educational resources. During COVID-19 pandemic, most educational activities shifted to e-learning platforms.
Job searching	The internet provides information on job opportunities during economic downturn.
Social media use	Social media is key player in provision of networks that keep friends and families together for mutual support. Social media are also conveyors of information and news.

Cullen's (2003) work on the Maori of New Zealand. Oppenneer (2009) and Samaras (2005) observe that the availability and use of ICTs directly contribute to improved standards of living. And without access to adequate digital products, a society's development efforts are likely to be stifled (Cullen, 2006, pp. 292–3; Norris, 2000, 2001).

While it cannot be maintained that all societies handle digital divide issues equitably, the cases in India demonstrate inequalities as well. The official Telecom Regulatory Authority of India (TRAI) push for digital equity has recognized the gender gap especially associated with cultural factors (Pandey, 2020). Wireless subscription rates also differed between rural and urban communities. Adding to the intersectionality problem, women have lower access rates to mobile internet connections which weaken their ability to use uninterrupted modern communication devices during pandemics.

These challenges in India, particularly for females, especially the dichotomous lack of computers and broadband digital resources in rural communities where the STs reside are enormous obstacles to health-care equity. Further, those villages lack well-connected roads and electricity which are necessary for pandemic communications. According to UNESCO reports, the general profile of India's school attendance in the public schools remains concerning both at primary and secondary levels, notwithstanding the admirable enactment of Right to Education Acts. Government runs more than 73% of the schools and most of which are underfunded (UNESCO, 2019).

If observations made in India are a tip of the iceberg, then we can posit here that a community without extreme digital divide challenges is best positioned to be effective at managing the technical aspects of communication during pandemics. Additional layers of impact of the divide during the pandemic will become clear as time goes on. Already there are documented inequities associated with education. For instance, American Indian students already face a teacher shortage and lack of access (McGill, 2018).

On the Navajo Nation recent research suggests only 17% of individuals have access to internet (Running Bear et al., 2021). COVID-19 had an immediately high impact on the health and life of people living in the nation and, in response, the nation has continued to have full lockdown measures to prevent the spread of COVID-19. Therefore, lack of connectivity in the region halts the ability to access education and other crucial information for those living outside of the range of connectivity (Running Bear et al., 2021). However, contact tracing through Indian Health Services (IHSs) was effective via phone calls (where possible) and in-person home visits. The modern world is defined by the expansive networks of communication enabled by complex information and communication systems. As of 2017, at least 5 billion people had mobile phones of one kind or another. The notion that we are a global village is evident in how quickly information spreads. In fact, globally, during the pandemic, uses of email, websites, search sites, e-government (including courts), and other applications increased several times over (Cloudflare, 2020; Quaintance, 2020).

In the United States, several city government functions including approval of building plans, business permitting, traffic information management, payment services, and new employee onboarding activities were offered online (Westrope, 2020). Some cities used chatbots to answer citizen questions concerning pending services. Some cities in Maryland, for example, added COVID-19 content to chatbots (Westrope, 2020). These, telemedicine and other telework endeavors continue to define the emerging post-coronavirus environment. There are many reasons why the growth of telework was eminent. First, the interest in teleworking was already visible even before the pandemic. Second, on a global stage, the lowered costs of connectivity were appealing to many stakeholders (Daniels et al., 2000, p. 1).

Many communities, including those in Africa, saw an increased use of mobile apps that tracked COVID-19 hotspots and provided other pertinent information. However, African states had already tested community-based approaches to contact tracing, which were undergirded by a philosophy that villagers know their neighbors and their kin. The June 26, 2020, US COVID-19 Task Force seemed to acknowledge the importance of community-based initiatives in managing Ebola and HIV epidemics in Africa and announced a paradigm shift to deemphasize the prevailing institutional approaches that had been driving the implementation agenda in the United States.

Prior to June 26, much of the tracing and containment had focused on "bending the curve" to allow for hospitals to handle COVID-19 patients. The June 26th meeting with the press also highlighted, albeit in a subtle manner, shift to national goals to recognize individual state efforts while at the same time using the platform to reassure the nation of the need for common sense approaches that would trickle down throughout the country. Most important was the announcements in improved therapeutics, improved capacity of hospitals to handle cases, and the need for surveillance.

Applications in COVID-19 contract tracing

The term contact tracing can be tricky, if not properly conceptualized. As a public health measure, digital contact tracing is much more nuanced. It involves public health staff conducting interviews with people who have received a positive test result. Public health street-level bureaucrats (contact tracers) seek to find out who else was in close proximity with the infected person or persons and then alerting them. Armed with the specific information, the officials then move on to recommend isolation of all those affected. According to people living through COVID-19 in the United States the use of contact tracing during COVID-19 has been spotty and ineffective in some cases and overly bothersome in others.

Ordinarily, telephone contacts are a part of the tool kit that augments manual contact tracing. Some of the difficulties involved in contact tracing are the following. First, people do not necessarily pick up the phones to respond to unknown callers. There is knowledge of phone scams involving people calling and asking for personal information. Contact tracing requires the collection of personal information such as the names and health status of all family household members and activities the recipient has participated in prior to their test result. People have reasonable hesitation about providing these personal details to an unknown person calling them on the phone. Second, there may also be issues with phone numbers that are not in service, which in Africa is a common occurrence. In America, too, there is the added complexity of phone owners blocking robo calls and spamming, which is possible when call blocking applications mislabel calls from tracers. Third, tracers have to grapple with the possibility that people have multiple phone lines. Again, this is a common feature in Kenya, where a mobile phone owner may speak of a "social Safaricom line" and an "official Airtel" line. Tracers have to know whether or not persons being traced have more than a single phone. Tracers have another difficult barrier if they don't know the appropriate time to call individuals being sought after. Even though contact tracing is described as involving information technologies, it is also labor intensive. The use of digital contact tracing devices was a part of the toolkits enabled through ICTs during the beginning of COVID-19. Most of the applications were based on smartphones, which are the conduit of most of the communications across the world. Mobile phones have the advantage of providing real-time decision-making and privacy controls.

In 2020, in an unprecedented move, technology giants Apple and Google collaborated to develop Bluetooth digital contact tracing applications (apps) which allowed public health officials to use logs of people the phone's user came into contact. This is because the "apps mimic traditional tracing by uploading to a central public health server the ID numbers of smart phones that have been close to an infected person's smart phone" (Chakravorti, 2020). In mid-June, Google and Apple added COVID-19 Trackers (COVID-19 Exposure Notification) on mobile phones including those in Kenya. The assumption was that the contacts would point to those most likely infected by COVID-19 (Rodrigo, 2020). In theory, the applications had the potential to infringe on privacy through tracking people's movements and their favorite interactions.

Although the various digital devices were not fully deployed even in developed countries, the absence of the ICT infrastructure in many parts of the developing world raised awareness of the failure of the world economy to eradicate poverty, even when rich countries continued to benefit from extracting resources from areas in which the majority inhabitants are Indigenous people. And the absence of COVID-19 platforms in many of the marginalized communities created additional burdens on any ICTs positive externalities. Contact tracing still occurred through in-person visits and other means; however, this required more human labor, risk to workers, and resources. It is worth noting as well, however, that even with digital assistance and access to full technological resources contact tracing did not engage effectively throughout the world, indicating the lack of a rapid and ready plan to curb the spread of the virus.

The complexities were also compounded because of the expansions in social media, which are part of the mosaic of disruptive technologies that transcend political jurisdictions. Indeed, social media provided new opportunities for communities to learn more about the coronavirus and also to document real experiences of people and communities suffering from the coronavirus. Some large social media groups became beacons for those suffering from COVID-19 to ask one another questions prior to the release of vetted scientific data. The CoRecovered support group on Facebook, for example, gained thousands of members from over 100 countries who continue to share information, comfort one another, and build community. All of these aspects of connection are especially essential during periods of social isolation related to lockdowns and self-isolation.

Roles and networks in governance

Because of the global nature of COVID-19, it behooves us to recognize the networks within which official communications about the disease flowed. The increased role of international organizations and states was, of course, facilitated by advances in communication technologies. One remarkable realization of the importance of the concept of communication in networks was articulated in the announcement that the WHO and International Telecommunications Union would send SMS messages on COVID-19 to users worldwide. This was because these organizations recognized that at least 3.6 billion people were offline and that text messages were the most effective means to reach them.

As expected, weaker states delegated much of their technical knowledge from more developed governmental and international partners. Organizations such as the WHO and Centers for Disease Control and Prevention (CDC), which traditionally are sources of health care and policy research, were placed in the spotlight. In fact, polls showed that the CDC was among the best performing government agencies in the United States (Newell et al., 2008). The expectation was that the two agencies would offer the lead agency services in managing the coronavirus epidemic. Although they offered critical guidelines and advice on setting the communications foundations for local, Indigenous, and national governments, political leaders criticized their efforts. Criticisms leveled at these technical organizations grew as the failure to contain the spread of COVID-19 caused havoc on national economies and causing political discontent.

In theory, the various governments and organizations involved in COVID-19 policy management networks were embedded in complex relationships. To put these relationships in perspective, the COVID-19-related horizontal and vertical communications, we have to understand the place of governance as a part of the problem-solving equation. Managing the pandemic and its attendant problems required focusing on widespread communities, organizational structures, economic and social systems, and political structures. Network governance provided a suitable framework for explaining the underlying communication patterns, community participation, and conflict resolutions. For example, the Maasai community is embedded in two national governing structures and systems. And the governments of Kenya and Tanzania too are beneficiaries of the United Nations' system in which the WHO communication protocols are articulated and promoted for adaption by member states. Within these networks are numerous other relationships that shaped the mechanisms through which management of global pandemics takes place. Native communities in the United States vary in their relationships to state, federal, and tribal governments depending on sovereignty. Health systems are just as complicated with the navigation of IHSs, the CDC, state health departments, local nonprofits, local medicine people, and other overlapping systems.

These and other organizational ties affect the content of policy management and communication about COVID-19. Under conditions of low capacity, WHO became a primary provider of technical information on management of COVID-19. As a specialized body, whose funding was from voluntary contributions from members of the United Nations, states, and private organizations, its decisions reflected the will of those interests (Kapucu & Ozerdem, 2013, p. 235).

In effect, the involvement of Kenya or Tanzania in WHO's COVID-19 management processes is a function of network structures put in place to minimize the spread of infectious diseases on a global scale. Because of the technical expertise available at WHO, there is consensus that its leadership in providing technical answers to the COVID-19 crisis was relied upon more vigorously in developing countries. This led to a trickle-down effect in terms of information flowing from national governments to the counties and Indigenous jurisdictions.

This network also provided opportunities for relationships among governments. Thus, based on WHO recommendations, different nation-states made arrangements to tap into resources available within individual member states of the WHO. The underlying relationships within the global wellness framework are meant to reduce uncertainty in solutions prescribed for containing COVID-19 and other pandemics.

There are two points that we prefer to highlight: first, networks are about intergovernmental relationships. And within governments, they often assume an interorganizational nature. The ability of these organizations to communicate effectively makes a difference in the management of disease and, in extreme cases, pandemics. That is the case for COVID-19. Secondly network governance offers opportunity to discern areas of weakness in articulation of information about contagious diseases. Also important is the place of sovereign nations working to direct tribal health measures even when encumbered by external governments and health agencies with leaders who intentionally and unintentionally place additional challenges of settler colonialism in the way.

This issue of network functionality came into play especially in terms of the disputes concerning the role of the WHO in articulation of an early warning system after COVID-19 was identified in Wuhan, China (Florko, 2020; Hudson et al., 2020). Initially, President Donald J. Trump tweeted a message to the Chinese government praising it for doing a great job. On January 24, 2020, Trump tweeted: "China has been working very hard to contain the Coronavirus. The United States greatly appreciates their efforts and transparency." He also wrote, "It will all work out well. In particular, on behalf of the American People, I want to thank President Xi!" (Serwer, 2020). He repeated those praises often, though this commendation turned out to be short lived. In March, as the virus spread globally and begun to take the toll on Americans, Trump, for political reasons, found a convenient scapegoat.

Buoyed by conservative radio and TV stations, he blamed China as not being transparent, in addition to leveraging the WHO's interventions to its advantage. Now, the administration had found a major talking point for its audiences. The narrative was that the administration was less than satisfied by the role played by the WHO in communicating the scope of COVID-19 and as a remedy, Trump threatened to cut funding the organization. This was followed by a rash of conspiracy theories including an idea among conservative Americans that the WHO was working with powerful elites to depopulate the world and inject survivors with tracking nanoparticles.

The Trump administration, for domestic political mileage, had now firmly shifted blame for the novel coronavirus pandemic to the WHO. Other governments like China quickly pledged to step up and boost WHO finances. China reportedly offered to provide funds to support efforts to fight COVID-19 and boosting public health systems in developing countries (Perper, 2020). On May 28, 2020, Trump terminated the US relationship with the WHO. In his statement, President Trump noted that the WHO had failed to reform changes within the organization's critical operating tasks, especially its failure to take lead on the spread of COVID-19. In Trump's own words, "because they had failed to make the requested and greatly needed reforms, we will be today terminating our relationship with the World Health Organization and redirecting those funds to other worldwide and deserving urgent global public health needs" (Holland & Nichols, 2020). Although the official withdrawal notification to withdraw the United States from the WHO was filed on July 6, 2020, the effective date of leaving was set for a year later, July 6, 2021 (Samuels, 2020). This said, the global network of countries and organizations controlling and managing COVID-19 crisis were also engaged in resource dependence and contingency negotiations. Observers presenting on popular TV Show, Fareed Zakaria's GPS (2020) linked the withdrawal of the United States from the WHO as an additional data point in the growing evidence of the country's decline as a global leader. Parag Khana, author of

The Future is Asian, while reflecting on this eventuality, saw the handling of COVID-19 as an aspect of the delegitimizing of American soft power (Zakaria, 2020). Another commentator, Natalie Nougayréde, a columnist of *The Guardian*, suggested that withdrawal from the WHO indicated Trump administrations affront on a rule-based international order (Zakaria, 2020).

Needless to say, the different authorities commenting on the progression and scope of the disease are all aspects of networks and intergovernmental relations. In the United States, this framework allows us to have deeper awareness of the nature of national capacity and interrogate it against tribal sovereignty, which is our key spot for attention. One could argue that if the US government was dissatisfied with WHO's performance was it also not appropriate for tribal nations to express similar frustration with federal responses to COVID-19? By emphasizing the interconnectedness of organizational networks in management of public health crisis, we discern the process by which institutional processes reinforced each other's concerns about COVID-19.

Writing about the United States, Ortiz (2002) observed that the federal government through the Bureau of Indian Affairs (under the U.S. Department of the Interior) administered grants to support service delivery in areas such as education and health care. This realization does not negate the significant autonomy and self-reliance initiatives demonstrated by tribal governments (Adams et al., 2007). If anything, as we shall note, there was considerable burden placed on Native American communities as they struggled to contain the spread of COVID-19.

Setting the tone: leaders roles and policy venues in COVID-19 communications

The question of roles is important because the efforts to prevent the spread of the COVID-19 were both local and international. This means that communication systems and patterns were intertwined in the structural complexities borne out the existence of multiple centers of power, which meant the development of coordination strategies (Table 2.2). Managing a global pandemic forces us to move beyond client-oriented bureaucracies and consider responses within cross-jurisdictions, cross agencies, and cross-sectors which are thoroughly networked. The ensuing e-governance framework, enabled by digital services, allowed for public health policy consultations to assume a heightened level. And multiple agency organizations enjoined in seamless information exchange virtual networks were at the core of the collaborative interventions that emerged.

Specifically, the role of actors such as the United Nations, WHO, USAID, and powerful nation-states which also have multiple bureaucratic agencies all indicated a complex nature of interactionism in solving challenges brought about by COVID-19. Although the partnerships often appeared troubled by political divisions, at the professional level, a new networked public health system of e-governance was established.

Table 2.2: Network governance: COVID-19 coordination strategies.

Government strategy	Lead agency and style
Oversight and policy direction	Command and control authoritarian styles. Police use of bull horns and robots to force curfew
Resource provision	Distribution of masks, medication, hotels for quarantined individuals, hospital care, ventilators
Liaison and negotiations with other states and international organizations	For example, supplies from China, agreements with other states over border closures and movement of people and goods, resource exchange, and joint actions with neighboring states
Facilitation of efforts to combat COVID-19	Coordination of strategies including opening and closing economic activities, flow of information and incentives, and common standards and norms in business practices; establishment of new ties
Situational awareness and global systemic developments	Understanding of global trends and learning examples of best practices in containment of the spread of the disease

(Table 2.2) Network Governance: COVID-19 Coordination Strategies.

For example, a country like the United States has 50 states with governors and also a national (federal) government, which oversees agencies such as the CDC, and affiliated organizations. The CDC deserves some further mention. As an organization, although CDC has been described as a "work in progress," it is considered a world-class agency in health protection (Goodsell, 2011, p. 123). The centers for previous messaging strategies were highly effective during previous influenza pandemics. For instance, in 2009, H1N1 Flu (Swine flu) virus had been designated by the WHO as a pandemic. At that time, the CDC took lead and had an imprint on all government policy levels. The CDC put in place a nation-wide strategy that was highly successful.

By all early indications, in 2020, the CDC's role in providing a national strategy for managing COVID-19 was vigorously debated amid concern that political interference was jeopardizing its communication strategies. The depth of the debates was vast and beyond the scope of this discussion. However, CDC's strategic direction appeared to be missing. There were no clear forecasts about the trajectory of COVID-19 pandemic. The Trump administration issued directives to limit communications between scientists and the public even prompting a lawsuit in early spring of 2020 (Knight Institute, 2020; Sullivan, 2020). The notion that the scientific community at CDC was in charge of what was happening was largely missing as key members of the CDC were not the face of briefings on the pandemic explosion. And the almost contradictory statements released by the different political leaders made it appear to be invisible. Once President Joseph (Joe) R. Biden appointed Dr. Rochelle Walensky to lead the CDC in 2021, Walensky confirmed that the CDC had been "muzzled" by the Trump administration (The Lancet, 2021).

Earlier in April, however, the public raised questions about the "silence." Leading scientists and advocacy group praised CDC for sounding an alarm about the virus early, after a

January 26 statement by Dr. Nancy Messonnier, the Director of the National Center for Immunization and Respiratory Diseases of the CDC, had indicated that disruptions to daily life "could be severe" if efforts were not put in place to stop COVID-19 from spreading (Herper, 2020). However, preparations proved to be inadequate and disjointed, in part, because of mixed messages from the country's political leadership. Throughout the year, there were inconsistencies in the manner in which different leaders assessed the threat. The lack of clarity and consistence was evident in how different states and locations approached their policies to open up the economy after the initial countrywide lockdowns.

Perhaps, one the most chilling indictments of the Trump's administration of the COVID-19 crisis was George Packer (2020) *The Atlantic*'s staff writer remark that "we are living in a failed state." He did not mince his words: "The crisis demanded a response that was swift, rational, and collective. The United States reacted instead like Pakistan or Belarus—like a country with shoddy infrastructure and a dysfunctional government whose leaders were too corrupt or stupid to head off mass suffering. The administration squandered two irretrievable months to prepare. From the president came willful blindness, scapegoating, boasts, and lies. From his mouthpieces, conspiracy theories and miracle cures. A few senators and corporate executives acted quickly—not to prevent the coming disaster, but to profit from it. When a government doctor tried to warn the public of the danger, the White House took the mic and politicized the message" (Packer, 2020).

The New York Times highlighted the lack of funding and Trump administration's failure to establish a coherent national strategy and structure which proved disastrous and forced the European Union to ban travel from the United States in July 2020 (Frieden, 2020; Shear & Jacobs, 2020; Slotnick, 2020).

President Trump's administration "chipped away at the CDC's capacity to combat infectious diseases" (Lancet, 2020). Significantly, the administration withdrew key CDC staff from China in July 2019 leaving behind an intelligence vacuum which was devastating when COVID-19 began to emerge (Lancet, 2020). The undermined role of the CDC was displayed by the administration burdening it with low-priority actions such as tracking cruise ships and American citizens returning from Europe (Frieden, 2020). A former CDC director, Tom Frieden, who served between 2009 and 2017 defended his former colleagues in the following words, "To blame the C.D.C for failing to deliver in this situation is like blaming a person who was bound and encased in concrete for failing to swim." Thus, the seemingly eroded role of the CDC was significant, but it did not deter the agency from playing a significant role in advising health operatives in state and local governments. Many state and local agencies cited CDC guidelines in articulation of COVID-19 policies.

As we are aware, states have county and local governments, with thousands of health centers, which formulated their own microlevel COVID-19 contingency plans. And for the

Navajo community, the IHS served that operational purpose. In reality, the networks model was reified during the crisis. Amid the challenge of a lack of unified nation-wide strategy, there was much confusion and disappointment in how the United States managed the COVID-19 crisis. Typically, in any network or web of organizations, the need for clarity in roles becomes a critical component of the crisis management endeavor.

In January 2020, when the White House constituted a task force to guide policy on COVID-19, the hierarchical roles of the various players in policy process were evident. The White House Coronavirus Task Force had a diverse array of expertise in areas including worker safety, vaccines, and therapeutics. Members included United States Global AIDS Coordinator, Surgeon General of the United States, Secretary of Health and Human Services, Secretary Housing and Urban Development, Advisor White House Chief of Staff, Secretary of Homeland Security, Office of Science and Technology Policy, Director of National Institute of Allergy and Infectious Diseases, Director of the Domestic Policy, Commissioner of Food and Drugs, Director Office of Management and Budget, Deputy National Security Advisor, Secretary Treasury, Secretary Veterans affairs, Secretary of Agriculture, Secretary of Labor, Director of the National Institutes of Health (NIH), FDA Directors of the Center for Biological Evaluation and Research, Administrator of the Health Resources and Services, Administration, and the CDC.

The vice president, Michael (Mike) R. Pence, was designated chair of the task force (White House, 2020). For publicity and his political communication style, US President Trump seemed to take much of the prefect's role. One would assume that this team of experts would at least apply principles of "Smart collaboration," which according to Professor Heidi Gardner (2016) was about intentional acts of bringing together the right kinds of experts and integration of holistic solutions to tackle complex problems such as COVID-19. As evident from the numerous press briefings, initially, the task force often showed a lack of congruency in policy prescriptions.

As time went by, it became clear that the task force did not prepare adequate policies and undermined the disjointed disease control strategies. Of concern was the fact that President Trump had continuously downplayed the need for testing and even publicly recommended slowing down and removing available testing. There was also schism among task force members, some of whom felt that the CDC had not handled the testing issue well enough. At the end of June 2020, Dr. Fauci, then serving as Trump's Chief Medical Advisor, admitted new strategies for pool testing were necessary and that "something was not working" (Acosta & Fossum, 2020; Edwards, 2020). Later on, after the election of Joseph Biden as president, Dr. Fauci acknowledged that Trump's administration was not transparent enough in how it handled the pandemic (Connolly, 2021).

The evidence supports the lack of transparency. For 40 days, Trump's administration suspended White House–based press briefings and only reemerged June 26 when Vice

President Pence led the team in addressing the rising number of COVID-19 cases in Arizona, California, Texas, and Florida. The briefing session held at the Department of Health and Human Services (DHHS) did not provide ample time for questions. And the task force glossed over key questions on implementation of its strategy to combat the pandemic. Dr. Fauci's June 29 admission to a senate committee on Health, Education, Labor, and Pensions that the country was going in wrong direction was an indirect admission that there was no systemic synergy in efforts to combat COVID-19.

On the ground across the United States this silence and division again contributed to conspiracy theories. We encountered these in our missed-methods research of people living during the pandemic and some of them rose to attention through podcasts and blogs. Conspiracies again suggested that government and global health leaders were working with business leaders with nefarious intent. One such false theory was that Dr. Fauci has a partnership with Bill Gates, George Soros, and Jeffery Epstein and together they planned to remove the cash economy through the use of vaccine injected nanoparticles activated by 5G (Evstatieva, 2020). These theories eroded trust in health institutions and created additional layers of chaos in a world with rapidly changing and emerging data about the behavior of SARS-CoV-19.

The relative dysfunction in attention to key details of the health crisis at the national level not only impacted public reactions to information but it also trickled down to state levels. A prevailing impression was that each state was on its own and hence the disjointed approach to managing the ordinances and regulations guiding the time frame for various locations to open up their businesses to the larger public. Some governors seemed to employ approaches similar to those by the White House. Notably, some governors assumed the roles of official spokespersons for state government response to the coronavirus in their jurisdictions. The overriding reason for the governor's decision to direct operations in their states was the lack of cohesive strategy from the national government. The apparent friction in messaging resulted in Trump taking a quick retreat from his earlier position of having total authority in COVID-19 management (Smith, 2020). Trump's revocation of his unconstitutional tweet and provocative pronouncements concerning his authority paved way for governors to make their own decisions about "when to reopen their states" (Smith, 2020).

New York Governor, Andrew Cuomo pushed back on the White House push for easing restrictions from coronavirus mitigation efforts for purposes of reviving the US economy (Mangan, 2020). At one of his daily news conferences, Governor Cuomo rejected the president's insistence on opening up the economy against the advice of health professionals. For Cuomo, accelerating the economy at the cost of human life was unacceptable. He did not believe Americans would choose the economy over public health (Mangan, 2020). Cuomo distinguished himself by giving daily briefings that lasted more than 3 months. In most of the briefings, widely covered by national and local TV stations, he presented data on Power Point slides earning much praise for his honesty. By mid-June, the daily briefings were halted due to declines in rates of infections in the state of New York (CBS News, 2020).

As the numbers of those dying from COVID-19 increased, the messaging became increasingly disorganized and uncoordinated. In June 2020, Vice President Mike Pence issued an opinion piece in *The Wall Street Journal*, which was intentionally meant to change the narrative to reflect government's position that the war was proceeding on well. Pence contended there was not a "second wave" of the coronavirus and that the government had ramped up testing and the media was sounding wrong alarm bells (Pence, 2020). He stated that national economic activities should be proceeding on as if the virus was nonexistent. This optimistic view of the spread of COVID-19 was not supported by data from the scientific community, including statements issued by the head of the National Institute of Allergy and Infectious Disease, Dr. Fauci, who warned against reckless opening up of economic activities in various states and also attendances in public rallies organized by President Trump's campaign (Gajanan, 2020). Dr. Fauci considered rallies as risky and reiterated that avoiding crowds and wearing masks was necessary in prevention of viral activity (Chiu, 2020). Dr. Fauci also lamented that "Clearly, we have not succeeded in getting the public as a whole uniformly to respond in a way that is a sound scientific, public health and medical situation" (Lovelace, Jr., 2020a).

The Washington Post doubted accuracy of Vice President Pence's numbers he used to make claims that the cases of COVID-19 had stabilized and that the daily average case rates across the United States dropped down from 30,000 in April to 25,000 in May 2020 (Bump, 2020). *The Washington Post*, using data from the COVID-19 Tracking Project, indicated that there were increased positive cases in multiple states including in Alabama, Arkansas, Michigan, Mississippi, South Carolina, Washington, and Wyoming. The records showed that the new cases were not trending downward as Pence claimed (Bump, 2020). As time continued data on rates of infection and deaths followed directives with evidence like the rise in numbers of cases and deaths 2 weeks after the governor of Arizona, Doug Ducey, allowed health guidelines to expire.

There was wide criticism of the manner in which the different agencies and personalities handled the crisis. Former Vice President and soon-to-be President Joseph (Joe) Biden criticized President Trump for his self-declaration as "a war time president" who "surrendered" (Renshaw, 2020). Biden stated: "American people sacrificed so much to fight this virus … we may lose some of the progress we've begun to make. All because he's lost interest … Now he's just flat surrendering the fight" (Renshaw, 2020). At another address from Wilmington, in his home state of Delaware, Biden gave his road map to tackling the COVID-19 pandemic and chided Trump for being a "cheerleader" at a time America needed a president (Wilkie, 2020).

President Trump himself made the messaging worse by tweeting and speaking with a lack of clarity and logic. His incoherent messages were widely distributed through social media where his supporters took on a strong association between adherence to health directives and left-wing conspiracies. In an interview with *The Wall Street Journal*, Trump stated that some Americans might wear face covering and masks not as a way to help reduce the spread of the virus but to "signal disapproval of him" (Lovelace, Jr., 2020b).

President Trump also had added fuel to fire by blaming the rise in coronavirus cases to increased testing and ordered a defunding of federal testing programs (Feuer, 2020; Gabbatt, 2020). According to *The Hill*, Trump said, "if we stop testing right now, we'd have very few cases, if any" (Weixel, 2020). His administration cut off funding to a long-standing research project on coronaviruses that sought to understand the risk of bat coronaviruses causing infections to humans. According to Dr. Fauci's testimony to Congress, the White House had instructed the NIH, which is a part of the DHHS, to cut the funding but it is the DHHS which made the final decision (Mole, 2020a,b).

Trump, also, most of the times refused to wear masks at public meetings, which from a public health standpoint was communicating the wrong message to many Americans especially those who needed presidential signaling to buy in the idea of using protective masks. Not wearing face masks gave a sense of false invincibility, which was not particularly helpful in a pandemic situation, especially because political leaders are key influencers. Trump supporters began to call masks, "muzzles" and complain that demands for masks and shutdowns were liberal moves toward a regime of control. Many began referring to people who wore masks as "sheeple," or people who could not think for themselves.

By late-June 2020, most state governors were literally making more uncoordinated speeches and issuing disjointed proclamations about opening up the economy amid rising COVID-19 cases in states that had not been severely impacted in the early months of February and March. In Florida, for example, Governor Ron DeSantis told the press, including WFOR-TV, that clusters of "overwhelmingly Hispanic" day laborers and agricultural farmworkers were responsible for the spike in coronavirus cases in that state (Woods, 2020). His insensitive remarks delivered in Tallahassee were controversial: "Some of these guys go to work in a school bus, and they are all just packed there like sardines, going across Palm Beach County or some of these other places, and there's all these opportunities to have transmission" (Woods, 2020). This assertion was quickly refused by the Farmworkers Association of Florida and other Industry associations which instead placed blame on poor timing in the state's testing protocols. The associations pointed out that the testing came too late to those communities.

Poor communication strategies at the state levels, especially in signaling best practices for reopening the economy, made an already dire situation worse. New York State which had been the country's hot spot in the month of April was relatively successful at maintaining a consistent science-based approach. Governor Cuomo consistently urged his colleagues to adapt a common strategy. He consulted widely with governors in Connecticut and New Jersey to find collaborative solutions to stopping the spread of coronavirus and reopen the regional economies in a coordinated way. Governors in the southern states including Texas, Alabama, Georgia, and Florida retained, for the most part, disjointed strategies as COVID-19 ravaged through the region. With numbers rising as June 2020 came to a close, it was not surprising that New York was now no longer the hot spot. And in response to actions by mainly Republican governors in the new hot spot southern states and the White House, Governor Cuomo was quick to retort:

> *You played politics with this virus, and you lost. You told the people of this state, you told the people of this country, the White House. Don't worry about it. Go about your business. This is all Democratic hyperbole. … Now you see 27 states with numbers going up. You see the economy going down … It was never politics. It was always science (Quoted in Lovelace, Jr., 2020c).*

Clearly, the roles of who was in charge of managing the "war" against COVID-19 in the country was unclear and buried in the lack of national leadership, which would have streamlined the communication processes and also offered benchmarks of roles to be performed by the various government institutions. Evidence of Trump's disoriented messaging emerged in September after Bob Woodward (2020) of *The Washington Post* and author of book *Rage* released recorded tapes in which Trump admitted to not taking the COVID-19 seriously and downplayed threat of COVID-19 (Parker et al., 2020). Trump had repeatedly told the American public that the virus would disappear and cause no damage. When criticism mounted, he excused his remarks as a means to instill calmness. Immediately, after this disclosure, the Trump administration announced it had contracted a private firm to manage the messaging around the pandemic. The HHS awarded a contract to "inspire hope" (Butler & Gianfortune, 2020).

Role theory, which has roots in structuralist scholarship applied to foreign policy, is an inspirational framework to help unpack the complexities in communication challenges and opportunities during COVID-19. In particular, the behavior of the state in reaching out or addressing its citizens is central to shaping behaviors during pandemic situations. Indeed, social behavior can be difficult to understand and manage if in the first instance the roles of those providing survival information are unclear. Or is the process by itself untrustworthy, especially in situations where regimes are governed by undemocratic political establishments. During the COVID-19 crisis, the various government spokespersons were, for some, placed in boxes of "political policymakers" while others, especially media personnel who cover pandemics, were widely accepted as "medical health experts." The range of spokespersons varied widely and included members of the Department of Defense, especially the Defense Health Agency, Department of Homeland

Security, community leaders, presidents of different tribal governments and universities, governors, and mayors of cities, among others.

In international politics, states are core actors in policy-making situations. Understanding state behavior during a crisis or public health situation provides opportunities to unpack domestically shared views on state conceptions of its own importance. This leads us to recognition of state roles in relation to interests of ruling elites. Ruling elites perceive their roles to be those of custodians of domestic order and tranquility, hence act according to the centrality and salience of a situation's effect on those broad objectives. And as they represent political jurisdictions such as governments, their roles are assumed to constitute visions of a national mission.

Such views are inherently generalizations and do not allow us to account for all manner of communications taking place between states and societies. However, we have to make sense of the roles by those in charge of health policy, especially in communication of information that is prescriptive and in public interest. Such prescriptions are supposed to be culturally constructed and offered within shared norms and expectations. Thus, some of the communications experienced within the international system mirror national elite norms and expectations of themselves and their obligations to the world community.

In other words, expectations of elite roles within the context described by Walker & Simon (1987) as mutually significant to the interconnected system of the global community of nations. This is important because COVID-19 is a disease that spreads from person-to-person and does not know boundaries. Therefore, those playing roles of national spokespersons and providing information about the spread and scope of disease are also playing a role in the international arena. The challenge was, therefore, to recognize the importance of the communication styles and content of all those who conveyed ideas about COVID-19. They are in our context, playing multiple roles.

Communication gatekeepers and problems in the age of "fake news" and misinformation

Communications during the global COVID-19 crisis were conducted in numerous networks. Some communications were among individuals within their different social media settings like WhatsApp family groups, Twitter handles, Facebook platform among thousands of channels. Of course, the directions of communications were downward, upward, external, horizontal, and complex. As students of communication tell us, these communication systems are all governed by specialized role and function assignments. In other words, within the COVID-19 communication systems, we can recognize individual roles of those transmitting messages. In such a theory, roles fit into categories like the "gatekeepers," "opinion leaders," and those who provide a bridge to the various stakeholders. These are also individuals who are important in liaison services (Rogers & Agarwala, 1976).

Rogers and Agarwala (1976, p.133) help us place the various actors in their different roles. We extrapolate from their important work and consider the relevance of organizational communications between government and its citizens within an identifiable pattern or schema. These roles inform us that gatekeepers are expected to control the messages which flow from government to the community. At the same time, within government agencies and networks, there must be liaison officials who act as bridges that link officials to persons in the communication subsystems. It is important to recognize that Rogers and Agarwala's work recognize the role of opinion leaders, who, in theory, are a vital element in management of any crisis. Besides, they identified cosmopolite as an individual who performs the role of boundary spanning and helps strengthen the filtering mechanisms in and out of all the relevant organizations. In most countries, the cosmopolite is a high-level public official with direct links to members of the national intelligence bureaus and the political establishment.

Although these roles and concepts do not necessarily affect all COVID-19 communication patterns or situations, we can easily discern the process of communication about the pandemic in all state apparatuses. For example, we can picture in our minds who was the gatekeeper, the opinion leader, and cosmopolite in any media platform including WhatsApp, Facebook, or government TV COVID-19 Special Reports. It is also possible that some individuals can play more than one role.

As the popularity of social media increased, the roles of managing pandemic-related information became a significant element in mitigation efforts. Because the online platforms as highly decentralized and amenable to unprecedented opportunities for public engagement, the task of creating and distributing accurate COVID-19 information proved to be challenging. Perhaps, to build on the roles described above, we must reflect on specific roles of "trusted messengers" and opinion leaders. This is where a rather stunning and ultimately pessimistic view of the mitigation efforts unfolded. For example, in many instances some people were unable to differentiate the unscientific remedies prescribed by their opinion leaders who also happened to filter information about the COVID-19 outbreak. Thus, the concern about accountability in managing pandemic information is not just theoretical but a matter of practical significance.

Managing infodemics

The COVID-19 experience illustrated some of the best- and worst-case scenarios in management of information systems within government operations, more specifically, the instances of distorted perceptions about COVID-19. With the rise of conspiracy theories and misinformation spreading rapidly online, the PBS lamented that "infodemic" made the management of COVID-19 costly (Yang et al., 2020).

Among the reported bogus claims were that COVID-19 could be blown out by using a hair dryer to a patient's face, or that 5G mobile phones transmit the virus (Yang et al., 2020).

The dangerous global flood of misinformation was harmful and required immediate attention and techniques to identify the falsehoods. Since all human societies have their own perceptions of disease and how it affects humanity, there were motivations to consider the numerous distortions and alternative realities that mushroomed during the COVID-19 crisis. Although there were different interpretations about the pandemic even within members of the same family, the government positions, including among Indigenous societies, were significantly varied.

Many factors contributed to the rise of rumors and "fake news" about COVID-19, which, in itself, can be considered unethical when involving public safety matters. It is an example of lying which Sissela Bok (1999) described in her book *Lying: Moral Choice in Public and Private Life*. In that book she notes that a scientist may, for example, believe that a new discovery to be true, but the available data are yet inconclusive. This reality came to be a prominent feature in the hurried rush to find a vaccine and treatment for COVID-19. Bok wrote, "Confident that future experimentation will bear him out, he may then falsify the data in order to gain support for what he feels sure is true. He lies, in part, in the service of what he takes to be the truth. Sometimes these initial hunches turn out to be correct; often, they do not" (Bok, 1999, p. 85). Globally, many health practitioners offered different and sometimes conflicting solutions to the problem, creating more confusion than solutions.

At the same time, social media and other news outlets in the era of right-wing and authoritarian-leaning governments suffered setbacks due to the proliferation of "fake news" and propaganda meant to distort facts. Fake news aims at many goals, including propaganda and disinformation. These are issues which former editor of *Time Magazine*, Rick Stengel associates with organizational cultures in several agencies such as the State Department (Stengel, 2019). During the cold war and its aftermath, acts of disinformation were quite prevalent and expansive (Corera, 2020). A leading security correspondent, Gordon Corera (2020) considers disinformation as a Russian intelligence strategy against the United States and the West. The strategy includes exploitation of the openness of Western societies and using "cyber illegals" to disrupt social life in the United States and the West. According to him, cyber illegals are spies and other persons who hide or obscure their identity over the internet to influence changes (Corera, 2020, pp. 353—70). Disinformation campaigns against the United States by Iranian government officials were also reported in various media. These we launched as pro-China campaigns aimed at providing conspiracy theories, especially concerning the origins of the coronavirus.

Disinformation over COVID-19 divided societies across the globe. Several politically motivated groups employed disinformation to frustrate government mitigation plans and public health measures that shut down economic activities. In addition, other groups seeking to take advantage of people's fears and COVID-19 panic and anxieties were often financially motivated, and some ransomware hackers created fake health websites to lure unsuspecting and frustrated users. As Chief Executive Officer of the Bond Health Center Dr. Temple Robinson opined, acts of disinformation and bad information are costly and take away resources from tacking the pandemic (Freeland, 2020). In addition, Dr. Robinson stated that resources are used for patient education to squash rumors which were amplified on social media outlets. Using resources in that manner is, therefore, not wise (Freeland, 2020). In the current epidemic, there is evidence that the Trump administration increasingly used overt political content to present inaccuracies for the purpose of political gain.

As head of the COVID-19 Task Force, then-Vice President Mike Pence claimed in *The Wall Street Journal* editorial that America was winning the fight against the virus, "thanks to the leadership of President Trump" (Pence, 2020). This statement ran contrary to existing evidence. Instead, figures showed that in about 100 days, over 120,000 Americans had died from the coronavirus and that most Americans disapproved of the administration's response to the pandemic (Alonso-Zaldivar, 2020; Gittleson, 2020; Netburn, 2020; New York Times, 2020; Pence, 2020). Gallup and West Health survey released mid-June 2020 found that 57% of US adults rated Trump administration's response to COVID-19 as poor, even though the country has the world's most expensive health-care system (Alonso-Zaldivar, 2020). Thus, arguably, for the US presidency to inject aspects of misinformation as a tool to manage the pandemic was unprecedented for governance in an advanced democracy.

Criticisms of President Trump's handling of the coronavirus pandemic hinted at the lack of truthfulness in his handling of his administration's response. In particular, he claimed that internal use of ultraviolet light and disinfectant was treatment for COVID-19. The WHO issued an advisory against spraying disinfectants across broad spaces indoors to try and kill coronavirus in response to this bizarre suggestion (Fox, 2020). The WHO advised that disinfectants had negative effects on those in facilities where these methods applied (Fox, 2020). Still, social media became a sounding board or multiple theories about the use of bleach to kill the virus (Havey, 2020).

In the same vein, Trump publicly disagreed with professional assessments of experts like leading immunologist Dr. Fauci and derided the CDC experts for issuing guidelines that were too expensive and very tough (Bidgood, 2020; Carvajal, 2020). *USA Today* reported that Trump's action revealed a power struggle within the administration that had been lingering on for months (Murphy & Stein, 2020). These utterances, more than anything, reflected lack of cohesion within the task force that he had appointed. This pattern of wishing to impose his ideas on people even if the ideas were meritless was a common feature in Trump's overall approach to governance. He was repeatedly accused of firing whistleblowers and persons who challenged his policy positions (Risen, 2020).

It is also an era of the rise of "fact-checking" organizations including divisions of commercial news organizations, such as Fact Checker, a part of *The Washington Post*.

Indeed, the industry's work complemented journalism which even earned PolitiFact a Pulitzer Prize. These fact-checking organizations which include PolitiFact, Snopes.com, FactCheck.org, and segments of government communication departments are a part of making journalism more accountable (Craft & Davis, 2016, p. 217). Though the public uses these resources to check stories for accuracy, conspiracies on the political right continue to question the legitimacy of all media including these outlets by suggesting they are biased left-wing outlets.

Fact-checking predates the COVID-19 era and has been a critical part of political communications, especially during presidential campaigns when ads scrutinized for truthfulness. In 1988, the presidential campaigns featured ads that were subjected to newspaper scrutiny (West, 2018). Television ads were also reminded whenever corrections were necessary. Professor Darrell West (2018, pp. 124–26) underscores the importance of fact-checking during election seasons. Fact-checking organizations not only evaluate political claims but also elevate the roles of truth squads in political discourse. West (2018, p. 126) references Trump's penchant for deception and his consistent misstatements and his rebuttal of valid criticism by labeling contrary information as "fake news." To Trump, news stories that did not give him the coverage he desired were tantamount to misreporting. For example, Trump's staff desired that the news report that he had the largest audience to ever witness an inauguration (Kettl, 2018, p. 9). This was, of course, not true. However, Trump and his staff wanted alternative facts to show him in a different light. He continued this approach well into his presidency and applied the same misinformation tactics in the management of the COVID-19 pandemic.

Such behavior is troubling from ethical and practical standpoints. Thus, the fact-checking organizations have to continuously debunk outrageous and harmful claims and conspiracy theories spread through social media on Facebook, WhatsApp, Twitter, YouTube, and Instagram, among others. As implied above, narrow and misguided objectivity persisted throughout the COVID-19 crisis, which made it quite difficult to manage the pandemic.

As an element of its social responsibility model, the British Broadcasting Corporation (BBC), for instance, employed fact-checking investigations to debunk the claim that certain medications were a cure of COVID-19. A BBC team tracking coronavirus misinformation found that such falsehoods led directly to acts of violence including mob attacks and individuals mistakenly consuming poisonous substances such as tank cleaners as "preventive medicine" (Spring, 2020). BBC fact-checkers and investigators also published a March 21, Tweet from President Donald Trump which stated:

> *HYDROXYCHLOROQUINE & AZITHROMYCIN, taken together, have a real chance to be one of the biggest game changers in the history of medicine. The FDA has moved mountains-Thank You! Hopefully they will BOTH (H works better with A,* International Journal of Antimicrobial Agents*) (Spring, 2020).*

Trump's tweet on the efficacy of Hydroxychloroquine and Azithromycin caused much debate in global public spheres. Trump went further on his claims of the drug's efficacy and boldly announced that he was taking it himself as a preventive measure. CNN, MSNBC, and other news networks gave much coverage to Trump's apparent hubris and stubbornness in advocating for the consumption of hydroxychloroquine.

As it turned out, the WHO later on found his prescription to be dangerous and recommended against self-medication of antimalarial drugs for treating COVID-19. There were also press reports that more than 100 Americans died after taking hydroxychloroquine for COVID-19 (Rahhal, 2020). Data showed that in the first few months of 2020 over 293 Americans died after taking hydroxychloroquine compared to 75 such deaths in 2019. Trump's endorsement of the drug was therefore counterproductive.

Trump also made other incorrect comments about COVID-19. This was after being criticized for his bizarre opinion that disinfectants should be used to "knock out" the virus in a minute and/or that ultraviolet rays be used to neutralize the virus. When criticism increased, he responded by claiming that his comments were sarcastic (Spring, 2020).

Through its Reality Check, BBC also threw out the claims that 5G technology helps transit coronavirus. There were also claims that 5G can suppress the immune system, thus making it easier for people to succumb to the disease. Thousands of social media messages had been shared suggesting that mobile phone masts were responsible for the spread of coronavirus pandemic. BBC responded, in part, because videos of arsonists attacking mobile network masts in Birmingham and Merseyside were posted on Facebook, YouTube, and Instagram (Schraer & Lawrie, 2020).

Giant new media operations, including CNN, added features such as Coronavirus: Fact versus Fiction-Podcast to their COVID-19 information featuring Dr. Sanjay Gupta, a leading medical personality. CNN's videos served to debunk myths about COVID-19. Many other news companies across the world took similar measures to "set the record straight" on matters of global health.

In fact, the social media companies, including Facebook, also established partnerships with fact-checking organizations to combat untruthfulness. For example, in late May 2020, Twitter highlighted two of President Trump's tweets that were patently false and misleading. Trump had urged people to "get the facts about mail-in-ballots" because mail-in-ballots would lead to widespread voter fraud (Conger & Alba, 2020; Fung, 2020). Twitter linked the "get the facts" message to a curated fact-check page which, in turn, provided further information and news articles debunking the president's untruthful assertions.

Facebook responded by building its own COVID-19 app embedded within its platform to update its community on the ongoing events and efforts to manage the scourge of coronavirus. However, the large number of citizen accounts spewing narratives about

COVID-19 made it nearly impossible to manage the operations of the social network "nervous systems." Still, much of society continues to contend with navigating the world of information overflow.

The many dimensions of irresponsible communication include existence of a culture of questionable political manipulation and distortion of information practices. Studies, for example, West (2018, p. 123) and Snow (2010, p. 71), demonstrate that government officials are known to manipulate and relay biased amplified information, especially during political campaigns, there was similar expectation for the United States, in part because 2020 was also a year of presidential election. It is conceivable that the same logic applies to Tanzania, where the government was not transparent in how it handled the COVID-19 crisis. 2020 was an election year in that country as well.

Subsequently, issues of manipulation and withholding of aspects of the disease were, in theory, addressed by centralizing the communication command system. However, there were also problems of style and evasion by political officials when it came to matters such as "opening up" the economy when scientific data were not supportive of the bureaucratic decisions.

For the United States, population in general, there were significant differences along party lines. Initially, the NPR/PBS NewsHour/Marist Poll of March 12–13, 2020, revealed that 60% of Republicans perceived coronavirus to be of very little concern as a threat to their lives. 50% of Independents believed that coronavirus was a real threat. The numbers were even more startling if we consider the question of how President Trump handled the coronavirus crisis. According to a late May survey by FiveThirtyEight, 82.2% of Republicans approved of Trump's handling of the pandemic compared to a paltry 12.5% of Democrats who did likewise (Skelley, 2020).

Data showed that many Republicans were not enthusiastic about government lockdown and caution on the consequences of not wearing masks in public spaces. And Republicans were doing less social distancing than those who identify as Democrats (Allcott et al., 2020). This was a contentious partisan response to risk reduction measures required by government to slow down spread of COVID-19. Right-leaning media houses echoed similar partisan messages concerning dangers associated with the COVID-19 pandemic (Aleem, 2020; Kantrowitz, 2020). Donald Trump urged his supporters through his tweeter handle to "liberate" Michigan from lockdowns (Liptak, 2020). Many observers saw this as a partisan attack on Michigan's Democratic Governor Gretchen Whitmer who had rejected calls to reopen parts of the state's businesses amid the coronavirus pandemic.

Those affiliated to the Democratic Party were perceived to be more sympathetic to following scientific scripts published by the CDC and other technical departments on managing infectious diseases like COVID-19. In fact, Trump's administration was criticized for sidelining scientists at CDC by providing testing guidelines that were often a misrepresentation of the organization's researchers (Mandavilli, 2020). And perceptions that the president was waging war against "Science" were articulated in opinion pieces in influential news outlets including *The New York Times* (Goldberg, 2020). His actions seemed to be consequential, even for those in the military profession. Writing before the outbreak of COVID-19, James Burk, a professor of Sociology at Texas A & M University, observed that the American publics' "declining confidence in science as a reliable form of practical knowledge is another troubling trend for the military professionalism .... The decline is a problem for all science-based professions, not just the military, because it makes it more difficult for all to establish the legitimacy of their authority" (Burk, 2005, p. 54). As has been articulated, even before COVID-19 crisis, the "Twitterverse" was already lively with arguments against the rising tide of antiscience. Handles on #ScienceMarch, #resist, #ScienceNotSilence, among others, vigorously warned against the attacks on science. Chris Mooney developed an eloquent argument on this subject in his book, *The Republican War on Science*. In this he argues that Republicans, since the Regan era, right through to George W. Bush, the 43rd president, staffed federal bureaucracies with people not in favor of science-based solutions to public policy problems. Mooney (2005) concluded that the attack on science in favor of pseudoscience and religious activists is largely a Republican phenomenon. Thus, the behavior of most Republican leaders during COVID-19 crisis resonates with this anti-science mentality. And this behavior earned President Donald Trump an editorial rebuke from the *Scientific American*. For the first time in 175 years, the October 2020 issue carried a presidential endorsement calling for Americans to elect Joe Biden for president because Biden would offer fact-based plans to protect Americans' health (Bellware, 2020). This followed the medical journal, *The Lancet*'s May 16 editorial on Reviving the US CDC, which criticized President Trump's incoherent handling of the coronavirus (Lancet, 2020).

However, many argued that science was not a partisan issue and that science was the foundation of American society (O'Leary, 2020, p. 34). This mistrust of science is problematic from many fronts, not just from ethical reasoning. First, it provided a path for destructive ignorance to flourish. Second, it fueled the antivaccine culture that had already proved deadly in America. Recall, for example, in October 2018 and July 2019, more than 600 people in New York City contracted measles because a good number of residents had refused to vaccinate their children from the disease (Zucker et al., 2020). The city of New York lost at least US$ 8.4 million in public health and other expenses (Zucker et al., 2020). Most who became ill were members of the Orthodox Jewish community.

The strategy of evasion

Leaders employed evasion through refusal to answer questions about the progressions of COVID-19 and instead, turned their focus on perennial distractions such as reining in on immigrants or linking the pandemic to culture wars. Journalist David Frum (2020) argued that the Trump administration's hesitancy to confront its failure to contain COVID-19 was a factor in its choosing to invoke culture wars and announce support for opening up churches for worshipers, even when the scientific community was emphatic that such a measure would result in increased spread of the virus. The anti-coronavirus lockdowns in effect banned largest gatherings including sports, sales conferences, college graduations, and religious services. Trump's administration moved ahead to issue statements in support of exemptions of religious group gatherings from the state pandemic control regulations (Frum, 2020).

In view of the Trump administration's persistent proclamations that immigrants were a problem to American prosperity, it is fair to conclude that his communication and actions amounted to evasion and xenophobic scapegoating. Following the huge economic devastation caused by COVID-19 lockdowns, the Trump administration attacked immigrants for taking American jobs. Writing for *The Washington Post*, journalist Max Boot (2020) pointed out in his opinion piece that, "Trump having lost against the corona virus shifted his battle toward the visible enemy, the immigrant! He continued his policies against immigrants and while signing an Executive Order to temporarily suspend immigration into the United States attacked immigrants for taking American jobs" (Boot, 2020). This tactic amounts to evasion because, in reality, the data did not support the claims provided in explaining reason for suspension of visas. On the contrary, data from the National Science Foundation indicated that more than half of the United States doctorate holders in computer science, engineering, and mathematics were foreign-born immigrants. And many of the doctors and nurses who provided health care during the COVID-19 emergencies were also immigrants as were drivers distributing food supplies across the nation (Boot, 2020). Yet even when facts provided contradictory evidence, several politicians associated with the Republican Party chose evasion by communicating information tangential to the core threats from the pandemic.

Xenophobia

While the use of evasion can produce devastating public health consequences, the use of intentional xenophobic imagery is even more dire. This has been the case historically with epidemics and pandemics (Barreto, Jr., 2020; Onoma, 2020). And settlers have used contagion as a tool of genocide (Diamond, 1999; Ostler, 2015). This pandemic is no different in that immediately there was international and US-based blame on China for the virus. Some politicians began invoking racist associations by calling it the "Wuhan Coronavirus" and the "Kung Flu." Donald Trump promoted the image of coronavirus being "Chinese" induced. He repeated his characterization in several speeches, often resulting in hate crimes against Asian Americans (Jackson, 2020; Kelley, 2020; Serhan & Mclaughlin, 2020). In one speech delivered in Tulsa, Oklahoma, he described the pandemic as "Kung Flu," a very derogatory and insensitive remark. Weijia Jiang, a CBS journalist, observed that

Trump's slur was racist (Chan, 2020). And several Asian Americans felt the slur was inappropriate. The *Los Angeles Times* quoted University of Texas Professor Madeline Y. Hsu's summary of the reaction to the statement from the president:

> *It reads like a cheap pun used as a low-level jibe from someone who is arrogant and poorly educated, rather than the assessment of an extremely powerful leader who has for months had the fullest array of information about the nature and dangers of Covid-19. It's something that a 6-year-old might say, trying to bully a classmate. Unfortunately, the crudeness of the statement echoes the extreme mismanagement of the coronavirus pandemic and the callous endangering of millions of American lives. Even those who survive the illness suffer terribly, some for the long term. It is even more of a tragedy that this illness arose with such a heartless demagogue in power (Hus, in Chan, 2020).*

By describing the COVID-19 pandemic's alleged origin, the president and others did not want to change their perceptions about China. Those whose reality is shaped by being selectively perceiving certain notions of "others" were most unlikely to register COVID-19 as a human condition that was not spread by a particular ethnicity or race. In official communications, the selective perceptions of COVID-19 cases extended beyond stereotyping China and, in many instances, extending negative narratives to those who are at the lower levels of socioeconomic development.

Adam Serwer (2020) writing for *The Atlantic* offers the explanation that Trump and conservatives defend bigotry as a strategy to show that they are protecting America from illegal aliens. "The strategy works for Trump's base because it considers foreigners as culprits and offers his supports to the emotional satisfaction of venting fury at liberals" (Serwer, 2020). Our mixed methods research also indicated micro- and macrolevels of blame with people stating that they needed to stay away from others who might be contagious while the others were usually people of lower socioeconomic status including, for example, Indigenous people (Hardy, 2020). We can argue further that various government reports continued to "other" the victims of COVID-19 in spite of the clear evidence that even leaders of developed countries such as Great Britain became ill.

The othering continued even as the search for therapeutics and vaccinations accelerated. Initially, comments from government spokespersons and scientific researchers from developed countries were received with much skepticism in developing countries. For example, the announcements that the British and France were contemplating to conduct COVID-19 vaccine trials on Africans was widely criticized in various social media platform in African countries and the Black diasporas. It may well be that these perceptions of injustice and government induced suffering in Indigenous and marginalized communities were evident during much of the COVID-19 crisis.

Similar arguments exist when stereotypes are employed to communicate about a serious communicable disease as is COVID-19. The mere existence of stereotypes in the discharge of public services is a negative attribute with possible negative consequences in outcomes (Zalkind & Costello, 1962, p. 227). Ethnic Chinese and other Asians were stereotypes as being responsible for the spread of the scourge (Otieno, 2020).

Trump's penchant for not being accountable is now well documented. For example, communications Professor Jennifer Mercieca of Texas A&M considers him a dangerous demagogue (Mercieca, 2020). In her analysis of Trump's rhetoric, he found the president's speeches were a strategy to unify his followers against everyone else. Trump made speeches that were meant to distract the nation's attention, and, in many instances, he uttered words that allowed him to have plausible deniability thereby avoiding accountability of the dangers caused by his remarks. So, for example, by amplifying the origin of coronavirus, he would invoke a false sense of patriotism while avoiding responsibility for the deaths of thousands of Americans whose lives would have been saved had the administration taken the right decisions.

In addition, Trump's use of threats, as he did with the WHO, was a part of his pattern of silencing the criticism from the international community of scientists. By taking away funds from the WHO at a time of great need, Trump's ad baculum behavior fits a pattern that Dr. Mercieca associated with demagogues changing the subject of debate. Again, by using words like the WHO had failed leadership, he was simply increasing distrust and polarization in the scientific community. This, as professor Mercieca (2020) suggests, makes it more difficult to solve problems and global challenges like COVID-19.

Another layer of challenges exposed in discussions on communications about COVID-19 among officials' government channels is the observable distortion of concepts occasioned by barriers in language. We learn from the literature on administrative discretion that implementing policies is often a tough task because the enabling laws are written in languages, which are in many instances flawed (MacIntyre, 1986, p. 79). As Cohen (1973) and Fuller (1969, pp. 227—28) observe, the languages we use are always going to be fallible. This was exactly the case in many of the instances of messaging about methods of managing the COVID-19 crisis.

Technically, the potential of sender and receiver of the messages missing out on the meanings of words and ideas was real. The problem is exacerbated when the translations of technical terms are confusing. Communications were an important part of the policy management but the design of the programs to prevent the spread of the virus to already economically marginalized communities was problematic from many fronts. The design in many instances showed insensitivity to the needs of communities. This is beside the fact that there was, in the case of the United States, a lack of congruence in the meanings of the messages projected in official federal communications. The lack of congruence was evident in the nonverbal cues and behavior by sections of those charged with official responsibilities to manage COVID-19.

Perhaps, the problems discussed above were compounded by perceived use of technical and medical jargon and gobbledygook especially in newspaper reporting about COVID-19. Jargon was evident in some of the more urban settings of Indigenous communities. In addition, widespread use of circulars in public bureaucracies, especially in communication affecting the Maasai in Kenyan and Tanzania, exhibited some aspects of poor communication strategies and circumlocution.

In long developed societies in late capitalism, concern over the scale of use of modern technologies prompted experts to advise governments on best practices over and above those implied. During the stay-at-home and social distance orders put in place in all societies, various governments were expected to ensure that employees who worked from home (telecommute) were able to achieve reasonable levels of productivity. However, the magnitude of resources for such work was hampered again by levels of poverty across the globe including in some Indigenous communities. For example, many Native Americans working in state governments and attending or working at universities in Arizona had to relocate to tribal lands where they did not have access to internet and reliable cell service. For those working at the state universities such as Northern Arizona, there was an effort to provide broadband capability through mobile satellite equipment; however, without cell service this was only spotty at best.

KPMG LLP, a consulting company, identified some of the actions required to make government operations viable in the wake of COVID-19 disruptions (KPMG, 2020). Among these are support services for cloud-based conferencing and teleworking, help desk operations, capacity for VPN concentrators, portals, and gateways to handle large number of employees that moved outside their workstations and were required to work from home. In addition, where government offices are located on rural Native American and other Indigenous communities, there was concern about the vulnerability of the security systems. According to KPMG (2020), there was real possibility that "organized crime groups could use the fear of Covid-19 to carry out highly targeted spear phishing campaigns and to set up fake websites, leading to an increased risk of a cybersecurity incident." The need for enhanced cybersecurity was made more urgent as government organizations made enormous transitions to offsite operations. The argument here is that all organizations face the same threats, but chances are that government organizations in less endowed jurisdictions would potentially be at higher risk as well. This is, in part, because of high general Information Technology failure in those communities occasioned by lower levels of specialist support.

Conclusion

In the absence of a unified tradition of studying novel coronavirus global pandemic, we selected lenses that are generally well understood by most social scientists, especially in the realm of contemporary governance. The above concepts are important as we proceed in the next chapters to look at cases of communication about the COVID-19 in different Indigenous societies. Over and over again, we recognize that the ability of nations to leverage ICTs is key to how they successfully reduce cases of the coronavirus in respective geographic jurisdictions. The avowed purpose of highlighting these concepts is to provide organizational context in diagnosis of how the pandemic was managed on a global scale. There are other lenses through which elements of the pandemic can be interrogated but we find these sufficient and realistic for a multidisciplinary study. The point that emerges most evidently from the framework is that national leadership sets the tone of subsequent altering behaviors within the Indigenous societies. In areas where Indigenous sovereignty has been periodically abused, even genuine efforts to implement programs that supported efforts to reduce morbidity were made much harder. ICTs then became another tool in settler colonialism creating additional burdens for Indigenous people working to protect themselves and their communities from the pandemic.

References

Abbas, Z. (March 13, 2002). *Pakistan's Sidi keep heritage alive*. BBC. http://www.bc.com

Acosta, J., & Fossum, S. (June 22, 2020). *Fauci says task force 'seriously considering' new testing strategy*. CNN. Herald & Review. https://herald-review.com/news/national/

Adams, A. S., Lee, A. J., & Lipsky, M. (2007). Governmental services and programs: Meeting citizens' needs. In M. Jorgensen (Ed.), *Rebuilding native nations: Strategies for governance and development* (pp. 223–245). The University of Arizona Press.

Ake, C. (1984). Ideology and objective conditions. In J. Barkan (Ed.), *Politics and public policy in Kenya and Tanzania* (pp. 127–138). Heinemann.

Akee, R. (May 13, 2020). *How COVID-19 is impacting Indigenous peoples in the U.S. Nation*. PBS News Hour. Arizona State University, Arizona PBS https://www.pbs.org

Aleem, Z. (March 15, 2020). *A new poll shows a startling partisan divide on the dangers of the coronavirus*. Vox. https://www.vox.com/2020/3/15/21180506/coronavirus-poll-democrats-republcans-trump

Allcott, H., Boxell, L., Conway, J., Gentzknow, M., Thaler, M., & Yang, D. (May 2020). *Polarization and public health: Partisan differences in social distancing during the coronavirus pandemic*. Stanford University. https://doi.org/10.3386/w26946. https://www.nber.org/papers/w26946

Alonso-Zaldivar, R. (June 18, 2020). *Poll: Americans not buying white house spin on coronavirus*. US News. https://www.usnews.com/news/health-news

Amarat, H. (2006). *African elites in India*. Mapin Publishing.

Awaaz, K. (June 9, 2020). *The invisible tribal lives of India and their experiences with Racism. HT Digital India*. Digital Streams Ltd.HT Live.com.

Barreto, R. C. (2020). The COVID-19 pandemic and the ongoing genocide of black and Indigenous peoples in Brazil. *The International Journal of Learning: Annual Review, 4*, 417–439. https://doi.org/10.1007/s41603-020-00126-y

Becenti, A. (March 26, 2020). *Workers battle coronavirus -and jinni-at epicenter*. Navajo Times. http://www.navajotimes.com

Bellware, K. (September 15, 2020). *Scientific American backs Biden for its first presidential endorsement in 175 years*. The Washington Post. https://www.washingtonpost.com/science/2020/09/15/scientific-american-joe-biden/

Bhattacharya, D. (1970). Indians of African origin. *Cahiers d'Études Africaines, 10*, 579–582.

Bidgood, J. (July 8, 2020). *Trump threatens schools and blasts CDC guidelines in an effort to reopen.* Boston Globe. https://www.bostonglobe.com/2020/07/08/nation/trump-threatens-schools-blasts-cdc-guidelines-an-effort-reopen/

Bok, S. (1999). *Lying: moral choice in public and private life*. Vintage Books.

Boot, M. (April 21, 2020). *Trump can't defeat the coronavirus. But he can wage war on immigrants.* Washington Post Washington Post.com. https://www.washingtonpost.com

Brueck, H. (June 23, 2020). *Fauci says racism is fueling the coronavirus 'double whammy' killing Black Americans at high rates*. Business Insider. https://www.businessinsider.com

Bump, P. (June 16, 2020). *Pence is right; we're not seeing a second coronavirus wave. We're still in the first one*. The Washington Post. https://www.washingtonpost.com

Burk, J. (2005). Expertise, jurisdiction, and legitimacy of the military profession (39–60). In D. M. Snider, & L. J. Matthews (Eds.), *The future of the army profession* (2nd ed.). McGraw Hill.

Butler, A., & Gianfortune, R. (September 16, 2020). *Courtney Bublè joins the podcast to discuss the $250 messaging contract at HHS that democrats want to investigate*. Government Executive. https://www.govexec.com/oversight/2020/09

Carlson, B. L. (September 29, 2014). *Remote Indigenous Australians rely on Facebook to stay in touch*. http://sbs.com.au/news/article/2014/08/26/remoteindigenous-australians-rely-facebook-stay -touch

Carlson, B. L., & Dreher, T. (2018). Introduction: Indigenous innovation in social media. *Media International Australia, 169*(1). https://doi.org/10.1177/1329878X18803798

Carlson, B., & Frazer, R. (2015). It's like going to a cemetery and lighting a candle: Aboriginal peoples, sorry business and social media. *Alternative: An International Journal of Indigenous Peoples, 11*(3).

Carlson, B., & Frazer, R. (April 4, 2018). *Indigenous voices are speaking loudly on social media, but racism endures*. Macquarie University (Wes Mountain).

Carvajal, N. (July 7, 2020). *Trump says he disagrees with Fauci: "I think we are in a good place". I disagree with him*. CNN. https://www.cnn.com/world/live-news/coronavirus-pandemic-07-07-20-intl/h_980c3fd6931a1535a28ac5a56c4b3be2

Catlin-Jairazbhoy, A., & Alpers, E. A. (2004). *Sidis and scholars: Essays on African Indians*. Red Sea Press.

Catlin-Jairazbhoy, A., & Jairazbhoy, N. A. (2002). *Sidi Sufis: African Indian mystics of Gujarat*. Apsara Media.

Catlin-Jairazbhoy, A., & Jairazbhoy, N. A. (2003). *Sidi Sufis: African Indian mystics of Gujarat*. University of Texas Press.

CBS News. (June 19, 2020). *Cuomo hails New Yorkers for doing "the impossible" together.* CBS News. http://www.yahoo.com/news/watch-live-york-governor-cuomo-112229613.html

Chakravorti, B. (July 10, 2020). *Digital contact tracing's mixed record abroad spells trouble for U.S. Efforts to rein in COVID 19*. Government Executive. https://www.govexec.com/technology/2020/07/

Chan, S. (June 22, 2020). *Opinion: The link between anti-black racism and Trump's 'kung flu' comment.* Los Angeles Times. https://www.latimes.com

Chari, M. (June 30, 2015). *Meet the fast-disappearing community of Indians and Pakistanis of African origin.* Quartz India. https://qz.com/india/440978/meet-the-fast-disappearing-community-of-indians-and-pakistanis-of-african-origin/

Chiu, A. (June 17, 2020). *'Of course, not': Fauci says he personally wouldn't attend Trump's Tulsa rally, citing coronavirus*. The Washington Post. https://www.washingtonpost.com

Cloudflare, Inc. (2020). *How was the Internet designed to handle such a large spike in traffic?*. https://www.cloudflare.com/builtforthis/

Cohen, M. L. (1973). Book reviews -legislative history: research for the interpretation of laws by G.B. Folsom. *Harvard Journal on Legislation, 10*, 521–527.

Conger, K., & Alba, D. (May 26, 2020). *Twitter refutes inaccuracies in Trump's tweets for first time.* New York Times. http://www.nytimes.com

Connolly, G. (January 21, 2021). *Fauci says Biden administration will be 'transparent and honest' on Covid unlike Trump regime*. Independent. https://www.independent.co.uk/news/world/americas/us-politics/fauci-press-conference-today-covid-biden-trump-b1790969.html

Corera, G. (2020). *Russians Among Us: Sleeper cells, ghost stories, and the hunt for Putin's agents*. HarperCollins Publishers.

Craft, S., & Davis, C. N. (2016). *Principles of American journalism: An introduction*. Routledge.

Cullen, R. (2003). The digital divide: A global and national call to action. *The Electronic Library, 21*(3), 247−257. https://doi.org/10.1108/02640470310480506

Cullen, R. (2006). E-government and the digital divide. In P. Hernon, R. Cullen, & R. Harold (Eds.), *Comparative perspectives on E-government: Serving today and building for tomorrow* (pp. 289−313). Scarecrow Press.

Daniels, K., Lamond, D., & Standen, P. (2000). Managing telework: An introduction to the issues. In K. Daniels, D. Lamond, & P. Standen (Eds.), *Managing telework* (pp. 1−8). Thomson Learning. Business Press.

Diamond, J. (1999). *Guns, germs, and steel: The fates of human societies*. W. W. Norton & Company.

Economic Times. (March 5, 2008). *Africa's lost tribe, the Siddis face poverty in India*. https://economictimes.indiatimes.com/africas-lost-tribe-the-siddis-face-poverty-in-india/articleshow/2838616.cms?from=mdr

Edmonds, F., Rachinger, C., Waycott, J., Morrissey, P., Kelada, O., & Nordlinger, R. (2012). *Keeping intouchable: A community report on the use of mobile phones and social networking by young Aboriginal people in victoria*. Institute for Broadband −Enabled Society, University of Melbourne.

Edwards, C. (March 21, 2020). *Trump has a problem: The government is failing to get control of coronavirus testing*. The National Interest. https://nationalinterest.org/blog/buzz/trump-has-problem-government-failing-get-control-coronavirus-testing-135677

Evstatieva, M. (July 10, 2020). *Anatomy of a COVID-19 conspiracy theory*. NPR. All Things Considered. https://www.npr.org/2020/07/10/889037310/anatomy-of-a-covid-19-conspiracy-theory

Feuer, W. (June 23, 2020). *Trump blames rise in coronavirus cases on increased testing, despite evidence of more spread*. CNBC. http://www.cnbc.com/2020/06/23

Financial Express, FE. (July23, 2020). *Who is Shantarama Siddi? The first from Siddi community in Karnataka to become a lawmaker*. FE Online. https://www.financialexpress.com/india-news/who-is-shantarama-siddi-the-first-from-siddi-community-in-karnataka-to-become-a-lawmaker/2032967/

Florko, N. (April 14, 2020). *Trump announces a formal freeze on WHO funding, pending an investigation*. STAT. https://www.statnews.com

Fox, M. (May 16, 2020). *Don't spray disinfectants to kill coronavirus, WHO advises*. CNN. https://www.cnn.com

Freeland, C. (2020). *More and more complications: COVID-19 and Youth*. NPR podcast. The 1A. Celeste Heedlee (Host) https://the1a.org/Segments/

Frieden, T. (June 2, 2020). *What the C.D.C. did wrong, and why*. (Opinion) New York Times. https://www.nytimes.com

Frum, D. (May 1, 2020). *Trump brings religion into the coronavirus culture war: Religious services shouldn't be exempted from state pandemic regulations*. The Atlantic. https://theatlantic.com

Fuller, L. L. (1969). *The morality of law*. Yale University Press.

Fung, B. (May 26, 2020). *Twitter labeled Trump tweets with a fact check for the first time*. CNN Business. http://www.cnn.com

Gabbatt, A. (June 25, 2020). *Trump plan to cut federal support for Covid-19 testing sites sparks alarm*. The Guardian. https://www.theguardian.com/us-news/2020

Gajanan, M. (June 17, 2020). *'Of course, not'. Fauci says he wouldn't attend Trump's Oklahoma rally amid pandemic*. Time. https://time.com/5854925/anthony-fauci-coronavirus-risks-trump-rally

Gallipolis Daily Tribune. (June 3, 2020). *Appalachian Ohio study highlights challenges of digital divide*. https://www.govtech.com/network/Appalachian-Ohio-Study-Highlights-Challenges-of-Digital-Divide.html

Gardner, H. (2016). *Smart collaboration: How professionals and their firms succeed in breaking down silos*. Harvard Business Review Press.

Garson, D. (2006). *Public information technology and e-governance: Managing the virtual state*. Jones & Bartlett.

Gittleson, B. (June 16, 2020). *Trump pushes ahead with Tulsa rally, while Pence misleads, and coronavirus cases rise*. ABC News. https://abcnews.go.com/Politics/trump-pushes-ahead-tulsa-rally-pence-misleads-coronavirus/story?/id=71275621

Global Times. (July 9, 2020). *Virus found in sewerage in Italy strong evidence of earlier existence of novel coronavirus: Zhong Nanshan*. https://www.globaltimes.cn/content/1193977.shtml

Goldberg, M. (May 11, 2020). *We're All casualties of Trump's war on science e: The administration kneecaps experts as the corona virus rages*. New York Time. Opinion http://www.nytimes.com/2020/05/11

Goodsell, C. T. (2011). *Mission mystique: Belief systems in public agencies*. CQ Press.

Hamway, S. (February 19, 2020). Navajo -backed Medicaid program progresses. *Albuquerque Journal*. https://abqjournal.com/1422106/navbacked-medicaid-program-progresses-ex-tribal-managed-care-entity-would-be-the-firts-of-its-kind–in-the-nation-html

Hardy, L. J. (2020). Negotiating inequality: Disruption and COVID-19 in the United States. *City and Society, 32*(2), Article CISO12312. https://doi.org/10.111/ciso.12312

Havey, N. F. (2020). Partisan public health: How does political ideology influence support for COVID-19 related misinformation? *Journal Computational Social Science, 3*, 319–342. https://doi.org/10.1007/s42001-020-00089-2

Hawley, J. C. (2008). *India in Africa, Africa in India: Indian ocean cosmopolitanisms*. Indiana University Press.

Heral. (June 17, 2020). *Marginalized lives matter in goan art*. Herald HT digital content services (Goa, India). M/S Herald Publications Pvt. Ltd.

Herper, M. (April 13, 2020). *U.S. is less safe from coronavirus when CDC goes quiet, former director says*. STAT. https://www.statnews.com/2020/0413/

Hodgson, D. L. (2014). Culture claims: being Maasai at the United Nations. In L. R. Graham, & G. H. Penney (Eds.), *Performing Indigeneity: Global Histories and Contemporary Experiences* (pp. 55–83). University of Nebraska Press.

Holland, S., & Nichols, M. (May 29, 2020). *Trump says terminating US relationship with World Health Organization over virus*. Reuters. https://www.reuters.com/article/us-health-coronavirus-trump-who/trump-says-terminating-us-relationship-with-world-health-organization-over-virus-idUSKBN2352YJ

Hudson, J., Dawsey, J., & Mekhennet, S. (April 25, 2020). *Trump expands battle with WHO far beyond aid suspension*. The Washington Post. https://www.washingtonpost.com

Jackson, H. (March 31, 2020). *Rep. Chu: Attacks on Asian-Americans 'very, very alarming'*. MSNBC. http://www.msnbc.com/

Johns Hopkins University. (2021). *COVID-19 Dashboard by the Center for Systems Science and Engineering (CSSE) at Johns Hopkins University (JHU)*. Coronavirus Resource Center. Retrieved at https://coronavirus.jhu.edu/map.html

Johnson, M. K. (2020). COVID and Indian country: Being Indigenous in corona virus world. *Practicing Anthropology, 42*(3), 4–5.

Kantrowitz, A. (March 18, 2020). *Conservative media still isn't sure what to think about the coronavirus*. Buzzfeed News. www.buzzfeednews,com/article/alexkantrowitz/conservative-media-still-isnt-sure-coronavirus

Kapucu, N., & Ozerdem, A. (2013). *Managing emergencies and crisis*. Jones and Bartlett.

Kelley, A. (March 31, 2020). *Attacks on Asian Americans skyrocket to 100 per day during coronavirus pandemic*. The Hill. https://www.thehill.com/changing-america/

Kettl, D. (2018). *Fake news*. Thousand Oaks, CA: Sage.

Khader, K. S. (2020). Translocal notions of belonging and authenticity: Understanding race amongst the Siddis of Gujarat and Hyderabad. *South Asian History and Culture, 11*(4), 433–448. https://doi.org/10.1080/19472498.2020.1827598

Knight Institute. (March 19, 2020). *Knight Institute V. CDC, Knight Institute asks CDC for policies restricting employees' speech.* https://knightcolumbia.org/content/knight-institute-asks-cdc-for-policies-restricting-employees-speech

KPMG. (April, 2020). *COVID-19 technology impacts: Actions government agencies should consider to keep their business operating.* Delaware. kpmg.com.

Lancet, M. (May 16, 2020). *Reviving the US CDC.* Editorial. https://doi.org/10.1016/S0140-6736(20)31140-5

Leung, H. (February 7, 2020). *An Eternal Hero. Whistleblower doctor who sounded alarm on coronavirus dies in China.* Time. https://time.com/5779678/li-wenliang-coronavirus-china-doctor-death/

Liptak, K. (May 1, 2020). *Trump tweets support for Michigan protesters, some of whom were armed, as 2020 stress mounts.* CNN. http://www.cnn.com

Lodhi, A. (1992). African settlements in India. *Nordic Journal of African Studies, 1*, 83–86.

Lovelace, B., Jr. (June 19, 2020a). *Dr. Anthony Fauci says it's 'frustrating' Americans are ignoring science amid coronavirus pandemic.* CNBC. https://www.cnbc.com

Lovelace, B., Jr. (June 18, 2020b). *Trump says some Americans wear coronavirus masks 'to signal disapproval of him'.* CNBC. https://wwwcnbc.com/2020/06/18

Lovelace, B., Jr. (June 25, 2020c). *Cuomo criticized other state leaders' response to the coronavirus pandemic playing politics in deciding to reopen their economies.* CNBC. https://www.cnbc.com/2020/06/25/new-york-gov-cuomo-to-other-states-you-played-politics-with-this-virus-and-you-lost.html

MacIntyre, A. A. (1986). The multiple sources of statutory ambiguity: Tracing the legislative origins to administrative discretion. In D. H. Shumavon, & K. H. Hibbeln (Eds.), *Administrative discretion and public policy implementation.* Praeger.

Mandavilli, A. (September 17, 2020). *C.D.C testing guidelines was published against scientists objections.* New York Times. https://www.nytimes.com/2020/09/17/health/coronavirus-testing-cdc.html

Mangan, D. (March 24, 2020). *Cuomo fires back at Trump on coronavirus: No American will say speed up economy at cost of human life.* CNBC. https://www.cnbc.com/2020/03/24

Mboya, T. (July 24, 2019). *The Siddis: African tribe in India.* Documentary. Nairobi: Citizen TV https://www.youtube.com/watch?v=yKbj0JiKlnQ

McGill, M. H. (February 7, 2018). *The least connected people in America.* Politico. https://www.politico.com/agenda/story/2018/02/07/rural-indian-reservations-broadband-access-000628/

Mercieca, J. (2020). *Demagogue for president: The rhetorical genius of Donald Trump.* Texas A & M Press.

Million, D. (2013). *Therapeutic nations: Healing in an age of Indigenous human rights.* University of Arizona Press.

Minda, A. Y. (2004). *An African Indian community in Hyderabad: Siddi identity, its maintenance and change.* PhD dissertation. Gottingen: Cuvillier Verag Gottingen, Gottingen University http://www.cuvillier.de

Mineo, L. (May 8, 2020). *For Native Americans, COVID -19 is 'the worst of both worlds at the same time'.* The Harvard Gazatte. https://www.news.harvard.edu/gazette/story/2020/05

Mole, B. (May 1, 2020a). *NIH abruptly cuts coronavirus research funding, alarming scientists.* Ars Technica. https://arstechnica.com/science/2020/05

Mole, B. (June 24, 2020a). *White House ordered NIH to cancel coronavirus research funding, Fauci says.* Ars Technica. https://artechnica.com/science/2020/06/

Mooney, C. (2005). *The republican war on science.* Basic Books.

Murphy, B., & Stein, L. (July 9, 2020). *Trump attacks his own CDC scientists over how to reopen schools safely.* USA Today. https://www.msn.com/en-us/news/us/trump-attacks-his-own-cdc-scientists-over-how-to-reopen-schools-safely/ar-BB16ylNj?ocid=mailsignout

Netburn, D. (June 22, 2020). *Mike pence says America is 'winning the fight' against COVID-19. Is he right?* Los Angeles times. https://www.latimes.com/science/story.2020-06-22/

New York Times. (June 23, 2020). *Some republicans Aren't in denial about the virus. Trump still is.* Editorial https://www.nytimes.com

Norris, P. (April 10–13, 2000). The worldwide digital divide: Information poverty, the internet and development. In *Paper for the annual meeting of the political science studies association of the UK, London School of Economics*. Harvard University, Kennedy School of Government. http://hks.harvard.edu/fs/pnorris/Acrobat/psa2000dig.pdf

Newell, T., Reeher, G., & Ronayne, P. (2008). *The Trusted leader: Building the relationships that make government work*. CQ Press.

Norris, P. (2001). *Digital divide: Civic engagement, information poverty, and the internet*. Cambridge University Press.

Nuorgam, A. (2020). *COVID-19 and Indigenous peoples. Statement by chair of the United Nations Permanent Forum on Indigenous Issues (UNPFII)*. United Nations: Department of Economic and Social Affairs. https://www.un.org/developmnet/desa/indigenouspeoples/covid-19.html

Obeng, P. J. (2007). *Shaping membership, defining nation: The cultural politics of African Indians in South Asia*. eBook. Lexington Books.

O'Leary, R. (2020). *The ethics of dissent: Managing guerrilla government*. Sage.

Onoma, A. K. (2020). Epidemics, xenophobia and narratives of propitiousness. *Medical Anthropology, 39*(5), 382–397. https://doi.org/10.1080/01459740.2020.1753047

Oppenneer, M. (2009). *ICT4D: Seeking the spaces in between*. The ethnos project. Ethnos blog. http://www.ethnosproject.org/about/mark-oppenneer/

Ortiz, J. (2002). Tribal governance and public administration. *Administration & Society, 34*(5), 459–481.

Ostler, J. (March 2, 2015). *Genocide and American Indian history*. Oxford Research Encyclopedias, American History. https://doi.org/10.1093/acrefore/9780199329175.013.3

Otenyo, E. E. (2017). Being left behind amidst Africa's rising imagery: The Maasai in the world of information and communication technologies (ICTs). *Australasian Journal of Information Systems, 21*. https://doi.org/10.3127/ajis.v21i0.1526

Otenyo, E. E., & Lind, N. S. (2011). *E-Government: The use of information and communication technologies in administration*. Teneo Press.

Otieno, A. (2020). COVID-19 pandemic response: Using reflection to sustain critical engagement during a global public health crisis. *Practicing Anthropology, 42*(3), 27–32.

Packer, G. (June, 2020). *We are living in a failed state*. The Atlantic. https://www.theatlantic.com/magazine/archive/2020/06/underlying-conditions/610261/

Padovani, E., & Young, D. W. (2012). *Managing local governments: Designing management control systems that deliver value*. Routledge.

Pandey, A. (2020). *Digital Inequalities at the focus of Public Policy*. https://medium.com/@avinashmp73/digital-inequality-at-the-focus-of-public-policy-99ffbf76ce01

Parker, A., Dawsey, J., & Abutaleb, Y. (September 16, 2020). *Ten days: After an early coronavirus warning, Trump is distracted as he downplays threat*. Washington Post. https://www.washingtonpost.com/politics/trump-woodward-coronavirus-downplay-ten-days/2020/09/16/6529318c-f69e-11ea-a275-1a2c2d36e1f1_story.html

Pence, M. (June 16, 2020). *There isn't a coronavirus 'second wave'*. WSJ OPINION. https://www.wsj.com

Pence, M. (June 16, 2020). *There isn't a coronavirus 'second wave': With testing, treatments and vaccine trials ramping up, we are far better off than the media report*. Com/WSJ Opinion. https://www.wsj.com/articles/there-isnt-a-coronavirus-second-wave–11592327890

Péquignot, S. (2020). From 'Afro-Indians' to 'Afro-global' networking: contemporary identification and unification processes among Siddis. *South Asian History and Culture, 11*(4), 449–464. https://doi.org/10.1080/19472498.2020.1827599

Perley, B. (2014). Living traditions: A manifesto for critical indigeneity. In L. R. Graham, & G. H. Penney (Eds.), *Performing Indigeneity: Global Histories and Contemporary Experiences* (pp. 31–54). University of Nebraska Press.

Perper, R. (April 24, 2020). *China is injecting millions into WHO as the US cuts funds. Experts say Beijing to boost its influence over the agency and its "deeply compromised" chief.* Business Insider. http://www.businessinsider.com

Pink, B., & Allbon, P. (2008). *The health and welfare of Australia's Aboriginal and Torres Strait Islander people.* Catalogue no. 4704.0. Canberra: Australian Bureau of Statistics.

Power, T., Wilson, D., Best, O., Brockie, T., Bearskin, B. L., Millender, E., et al. (2020). COVID-19 and Indigenous peoples: An imperative for action. *Journal of Clinical Nursing.* https://doi.org/10.1111/jocn.15320

Quaintance, Z. (May 29, 2020). *Will COVID-19 cause long −term tech changes for courts?.* https:www.govtech.com

Rahhal, N. (August 7, 2020). *More than 100 Americans have died after taking hydroxychloroquine for COVID-19 despite Donald Trump insisting people had 'nothing to lose' by trying the unproven drug.* Daily Mail. https;//www.dailymail.co.uk

Ramdev, D. (May 13, 2015). *Racing to the finish: Siddi tribal woman in TCS World 10 K run. Nation.* Current Affairs. http://www.deccanchronicle.com/150513/nation-current-affairs/article/racing-finish-siddi-tribal-woman-tcs-world-10k-run

Renshaw, J. (2020). *Biden says Trump is 'surrendering' to coronavirus, putting lives at risk.* Reuters. https://www.reuters.com

Reuters. (April 6, 2020). *UPDATE 1-Trump says Americans are praying for British prime minister.* https://www.reuters.com/article/health-coronavirus-britain-trump/update-1-trump-says-americans-are-praying-for-british-prime-minister-idUSL1N2BU2PA

Risen, J. (April 9, 2020). *Under cover of Covid 19, Donald Trump ramps up his war on truth tellers.* The Intercept. https://theintercept.com/2020/04/09/coronavirus-trump-firings/

Rodrigo, C. M. (May 7, 2020). *Digital contact tracing is becoming available, but is it effective?* The Hill. https://thehill.com/policy/technology/496498

Rogers, E. M., & Agarwala, R. (1976). *Communication in Organizations.* Free Press.

Running Bear, C., Terrill, W. P. A., Frates, A., Peterson, P., & Ulrich, J. (2021). Challenges for rural native American students with disabilities during COVID-19. *Rural Special Education Quarterly, 40*(2). https://doi.org/10.1177/8756870520982294

Samaras, K. (2005). Indigenous Australians and the digital divide. *Libri, 55*(2−3), 84−95.

Samuels, B. (July 7, 2020). *Trump administration moves to formally withdraw US from WHO.* The Hill. https://thehill.com/homenews/administration/506214-trump-administration-withdraws-us-from-WHO-

Sayeed, V. A. (July 24, 2020). *Siddi community of African Descent gets its first representative in Karnataka legislature.* Frontline. https://frontline.thehindu.com/dispatches/siddi-community-of-african-lineage-gets-its-first-representative-in-karnataka-legislature/article32184152.ece

Schraer, R., & Lawrie, E. (April 15, 2020). *Coronavirus: Scientists brand 5G claims "complete rubbish".* BBC. http://www.bbc.com

Serhan, Y., & Mclaughlin, T. (March 12, 2020). *The other problematic outbreak. As the coronavirus spreads across the globe, so too does racism.* The Atlantic. https://www.theatlantic.com/international/archive/2020/03/coronavirus-covid19-xenophobia-racism/607816/

Serwer, A. (March 24, 2020). *Trump is inciting a coronavirus culture war to save himself.* The Atlantic. https://www.theatlantic.com/ideas/archive/2020/03/

Shah, A. M., Tamang, R., Moorjani, P., et al. (July 15, 2011). Indian Siddis: African descendants with Indian admixture. *The American Journal of Human Genetics, 89*(1), 154−161. https://doi.org/10.1016/j.ajhg.2011.05.030

Shastri, V. (May 16, 2015). *Bengaluru Marathon: Siddis can race but have no sponsors.* http://www.deccanchronicle.com

Shear, M. D., & Jacobs, A. (May 19, 2020). *W.H.O members reject Trump's demands, but agree to examination of global body's coronavirus response.* New York Times. http://www.nytimes.com

Sheth, K. (2014). *A certain grace: The Sidi, Indians of African descent*. New Delhi: Photoink. https://www.youtube.com/watch?v=s5ikcz125qE

Shodhan, A. (March, 2015). The Sidi Badshah of Western India: The politics of naming. *Journal of African Diaspora Archaeology and Heritage, 4*(1), 34–49. https://doi.org/10.1179/2161944114Z.00000000021

Shroff, B. (2008). Indians of African descent: History and contemporary experience. *Souls: A Critical Journal of Black Politics, Culture, and Society, 10*(4), 315–326. https://doi.org/10.1080/10999940802523836

Shroff, B. (March, 2017). *India's forgotten Africans*. Documentary. Down to Earth https://www.youtube.com/watch?v=2I4vUBTHc_4

Skelley, G. (May 22, 2020). *The gap between democrats and republicans is growing on the coronavirus crisis*. Pollapalooza https://fivethirtyeight.com/features/democrats-and-republicans-are-increasingly-split-on-the-coronavirus-crisis/

Slotnick, D. (June 26, 2020). *Europe will bar Americans traveling from the US as the coronavirus pandemic spirals out of control in the states*. Business Insider. https://www.businessinsider.com/

Smith, A. (April 14, 2020). *Trump backs down after Cuomo, other governors unite on coronavirus response*. NBC news. http://www.nbcnews.com

Snow, N. (2010). My fellow blogging Americans: Weblogs and the race for the white house. In J. Allen Hendricks, E. Robert, & Denton, Jr. (Eds.), *Communicator –in-Chief: How Barack Obama used new media technology to win the white house* (pp. 67–82). Lexington books.

Sohail, R. (August 8, 2018). *Pakistan's first lawmaker of African descent raises hopes for Sidi Community*. BBC. https://www.bbc.com/news/world-asia-45099970

Solanki, S. (2020). *The ethnomedical system of the Siddis of Gujarat, India. The dynamics of healing in an African diaspora community*. PhD. Thesis. University of Ken Academic Repository https://kar.kent.ac.uk/80351/

Spring, M. (May 26, 2020). *Coronavirus: The human cost of virus misinformation*. BBC. http://www.bbc.com

Stengel, R. (2019). *Information wars: How we lost the global battle against disinformation and what we can do about it*. Grove Atlantic.

Stuart, A. (April 22, 2018). *Inside a lost African tribe still living in India today*. Short film. National geographic documentary http://www.youtube.com/watch?v=B_a1WS5ncDk

Sullivan, M. (April 18, 2020). *The Trump administration is muzzling government scientists. It's essential to let them speak candidly to the press again*. The Washington Post. https://www.washingtonpost.com/

Tatz, C. (2001). *Aboriginal suicide is different: A portrait of life and self-destruction*. Aboriginal Studies Press.

The Lancet. (January 16, 2021). *New leadership at the US CDC*. https://doi.org/10.1016/S0140-6736(21)00082-9

Thomas, B. (March 20, 2016). *Bengaluru: Disappearance of Siddi tribal triggers panic*. Deccan Chronicle. http://www.deccanchronicle.com/nation/current-affairs/200316/bengaluru-disappearance-of-siddi-tribal-triggers-panic.html

UNESCO. (2019). *India, Country Information*. UNESCO Institute for Statistics. https://en.unesco.org/countries/india

United Nations. (February 21, 2017). *The Siddis: India's forgotten Africans*. Documentary https://www.youtube.com/watch?v=Q4NsgwXsys

Vartabedian, R. (May 5, 2020). *Scientists say a now-dominant strain of the coronavirus could be more contagious than original*. Los Angeles Times. https://www.latimes.com

Vick, K. (March 30, 2020). *The digital divide: A quarter of the nation is without broadband*. Time. http://www.time.com/4718032/the-digital-divide/

Vizenor, G. (1999). *Manifest manners: Narratives on postindian survivance*. University of Nebraska Press.

Walker, S. G, & Simon, S. W (1987). Role sets and foreign policy analysis in southeast Asia. In S. Walker (Ed.), *Role Theory and Foreign Policy Analysis* (pp. 141–159). Duke University Press.

Weixel, N. (June 15, 2020). *Trump on coronavirus: "If we stop testing right now, we'd have very few cases, if any."*. The Hill. https://thehil.com/policy/healthcare/

West, D. M. (2018). *Air wars televising advertising and social media in election campaigns, 1952–2016*. Sage.

Westrope, A. (April 3, 2020). *Virus shutdown: Time for digital government to prove itself*. Government Technology. https://www.govtech.com/

White House. (May 15, 2020). *New members of the white house coronavirus task force announced*. http://www.whitehouse.gov/briefings-statements

Whitehead, A. (November 27, 2000). *The lost Africans of India*. BBC. http://www.bbc.com

Wilkie, C. (June 30, 2020). *Joe biden rips Trump as coronavirus surges: 'The wartime president has surrendered'*. CNBC. https://www.cnbc.com/2020/06/30/

Willis, S., & Bruce, T. (2006). Beyond the "digital divide" internet diffusion and inequality in Australia. *Journal of Sociology, 42*(1), 43–59.

Wolfe, P. (2006). Settler colonialism and the elimination of the native. *Journal of Genocide Research, 8*(4), 387–409. https://doi.org/10.1080/14623520601056240

Woods, A. (June 19, 2020). *Florida gov blames coronavirus rise on 'overwhelmingly hispanic' workers*. New York Post. https://nypost.coom/2020/06/19/desantis-blames-covid-spike-on-overwhelmingly-hispanic-laborers/

Woodward, B. (2020). *Rage*. Simon & Schuster.

World Bank. (2016). *Development report. 2016*. Digital dividends. World Bank.

Yang, J., Lane, S., & Fritz, M. (April 28, 2020). *The dangerous global flood of misinformation surrounding COVID -19*. PBS News Hour. Arizona State University. https://www.pbs.org/newshour/show/the-dangerous-global-flood-ofmisinformation-surrounding-covid-19

Youth Ki Awaaz. (June 9, 2020). *The invisible tribal lives of India and their experiences with racism*. HT digital India. Digital Streams Ltd.HT Live.com.

Zakaria, F. (June 14, 2020). *GPS show. Is America still the shining city upon a Hill? Ju10 am. Global public square (GPS) show*. CNN. https://www.cnn.com/audio/podcasts/fareed-zakaria-gps

Zalkind, S. S., & Costello, T. W. (1962). Perception: Some recent research and implications for administration. *Administrative Science Quarterly, 7*, 218–233.

Zucker, J. R., et al. (March 12, 2020). Consequences of under-vaccination — measles outbreak, New York city, 2018–2019. *New England Journal of Medicine, 382*, 1009–1017. https://doi.org/10.1056/NEJMoa1912514

CHAPTER 3

COVID-19 policy responses and feedback in Navajo country

To understand the impact of COVID-19 in Navajo country one must begin by reading accounts from people who lived through those first months. Sunny Dooley, a traditional Diné storyteller, wrote a powerful piece for Scientific American in July called "Coronavirus Is Attacking the Navajo 'because We Have Built the Perfect Human for It to Invade'" (2020). Later, using similar imagery, Diné Historian Jennifer Nez Denetdale spoke about this experience and conceptual framework in her lecture, *Dikos Ntsaaígíí Building the Perfect Human to Invade: A Diné Feminist Analysis of the Pandemic and the Navajo Nation*. Our accounts here do not compare to the depth of knowledge and scholarship of Diné leaders like these, nor do we aim to tell that story. Instead, we provide a brief comparative context to frame the analysis of information and communication technologies (ICTs) and public health. We urge readers to seek out Indigenous scholarship and authors to gain a more holistic understanding of the on-the-ground impacts of the pandemic and how Native scholars expand theoretical frameworks through Native-led scholarship and storytelling.

Returning to the focus of this text we view the initial work on governance within the Navajo Nation as not fully realizing possibilities of e-governance systems which form the paradigm through which modern public health policies are managed. Building robust e-governance was also not a priority considering the immediate need to deal with the lack of respect and adherence to sovereignty by the federal government. Part of this obligation is the unfulfilled promise to adequately fund the Indian Health Service (IHS), which has been around since 1954. As a result, the agency's efforts to provide leadership in preventive medical care have been wanting. Still, the Navajo Nation (Diné), the largest tribal land in the United States (covering 27,000-square-mile land base), is a huge self-governing land mass following its own public service ethos. Important work on the evolution of governance structures and systems going back to 1923 when the Chapter House system and modern public bureaucracy was solidified exists (for example, Navajo Nation, 2011; Wilkins, 2002, pp. 91–129).

Further, the vast body of work, including Benally (2006), Jorgensen (2007), and Wilkins (1987), on self-governance is indicative of tribal responsibility to Diné people. One would assume such obligations shape and inform the organizational structures and bureaucracies

The Inequality of COVID-19. https://doi.org/10.1016/B978-0-323-99867-3.00003-X

that manage the affairs of Navajo people. Yet, rich scholarship also describes and analyzes the betrayal of Indigenous aspirations through violent state actions, unconstitutional federal maneuvers, and failure to adhere to the rights of sovereign nations (Brruyneel, 2007; Corntassel & Whitmer, 2008; Deloria, 1985).

Although the governance structure of the People (Dinètah) Navajo has changed from its traditional historical form, there are still strong relics of its norms. Diné communities live on and off tribal lands that cover multiple states and as with all people today there are Diné people living all over the world. There are a multitude of identities and experiences of people who are Navajo including different belief systems, religions, and ways of life. Many people continue to practice aspects of Diné cultural traditions to different degrees that emphasize the place of class and kinship even within the new political organization led by a president and council. Principles of leadership that have been passed from generation to generation guide the nation's approach to managing the COVID-19 crisis. The Navajo president remains the functional leader who has to enlist public confidence and act in ways that benefit the public interest. Likewise, the expectation is that all council members, by Navajo traditions, share governance principles of accountability to Navajo people. Native-led organizations also operate on tribal lands to meet the needs of local people, as is the case during the pandemic.

To deal with the pandemic the Navajo President Jonathan Nez and the Council met to chart the way forward. The meetings, according to some attendees, had all the symbols of the ancient traditional *Naachid* or gathering to discuss war. At the meetings, the president and his administration presented a pile of proposals and offered solutions to tackle the scourge. Council members also took turns to speak about the pending calamity. In much of the discussions at the leadership level, the call was to drive out the disease, which in Navajo language was described as the "big cough" or *Dikos Ntsaaigii*-19.

At this tribal government level people narrated COVID-19 in different ways. Most importantly, the picture was that of fighting or raiding the enemy. COVID-19 seemed to provide meaning for pandemic-specific centralized command. And thanks to the proliferation of new ICTs, in addition to the presidency and council, space for centralized messaging was available. Social media provided room for a virtual modern *Naachid* to take place. Now, more than ever, a mechanism for bringing people together for their common interest was available to a significant number of citizens. And the community found voice to debate and share stories of grief and victory as shaped through the COVID-19 experience.

It is important to note that as nonnative scholars we do not discuss details or even general descriptions of Dinè practices here out of respect for the privacy of the tribe and individuals. Instead, we focus on the policy aspects of pandemic management in this setting. And while we focus on the initial community responses, we acknowledge that there were indeed many untold stories of these gallant efforts in the entire community. The picture that emerges from these initiatives is that of sacrifice, courage, and resilience. It is a testimony that native organization and self-supporting mechanisms made a huge difference in COVID-19 mitigation strategies.

A large nonprofit organization rose to the top of social media communications at the inception of the pandemic led by Ethel Branch, former District Attorney of the tribe, who took a leave from her legal practice to run this large-scale direct-action organization for the protection and saving of Navajo people. In a formal personal communication with the author Ethel stated that their organization was working well within the tribe because the leaders were also Diné and they were working to save their elders. She said that the risks of COVID-19 created a target on the backs of elders which was a target on the backs of their culture and noted that many organizations rushed in to help at the beginning of the pandemic but, nearly a year later, their organization was one of the few that still remained. Their work is based on understanding the needs of elders and families and designing packages for isolation and quarantine based on the kinship patterns of households and their knowledge of what families would need including appropriate foods. These actions are important in a place where people live their lives traversing between tribal lands and neighboring border towns and experience and invoke traditional tribal customs and lifeways including medical pluralism of allopathic and medicine person-based healing, while also navigating life outside of tribal boundaries.

ICT: shrinking the digital divide

Arguably, ICTs are the avenue through which modern governance forms and government functions are enabled. It is impossible to picture the management and policy responses to COVID-19 outside of modern ICTs. When COVID-19 arrived on the Navajo Nation, the community was still grappling with issues of the digital divide. Like other tribal communities, many areas within the Navajo Nation are isolated, have poor housing conditions, and lack broadband internet (Theobald, 2020). Hundreds of people in Indian country live in crowded houses, partly for traditional reason and because the federal government restricted construction thereby causing shortages (McCarthy & Kestin, 2020). It was not surprising that President Jonathan Nez and his administration hoped to allocate the Coronavirus Aid, Relief, and Economic Security (CARES) Act of 2020 monies on addressing issues of the digital divide and housing issues.

The search for digital equity for the Navajo Nation has persisted. It was embedded in the nation's bid to modernize its managerial systems. For example, in 1991, reorganization of government streamlined the executive branch and introduced the Office of Telecommunication Regulatory Commission, Office of Management and Budget among other units involved in managing ICTs. The Department of Information Technology (DIT) emerged from a later reorganization in 2003. Although part of the motivation for systematic reorganizing was to counter the inroads by the federal government's unending interferences in tribal administration, a key goal was to find structural solutions to address the needs of the tribe in a globalizing world, especially leveraging benefits from the knowledge economy for Indigenous governance. But several hurdles remained a huge burden on the Navajo Nation.

One must add, the challenges of breaking down barriers to achievement of digital maturity and successful e-governance have been well documented in various reports and scholarly assessments (see, for example, Anderson, 1999; Bill & Melinda Gates Foundation, 2004; Bissell, 2004; Davis & Trebian, 2001; Dorr, 2001; Dorr & Akeroyd, 2001; Education Development Center, EDC, 2004; Gordon, Gordon & Dorr, 2003; Groski, 2002; Hayes, 2003; Johnston, 2001; Rural Assistance Center, RAC, 2008; Warren, 2000). A key limitation is that of weak managerial capacity, financial resources, and a host of enduring structural limitations brought about by the effects of colonialism and marginalization.

Yet, in spite of the limitations, the spirit of resilience and leveraging existing governance systems to provide development opportunities for Navajo people remain a strong resolve of Navajo leaders, who also know that new technologies are an important tool for addressing a myriad of pressing policy problems in public health, education, and other aspects of human services of which COVID-19 is just a new one. To shrink the gap in availability of modern communication technologies was, therefore, not just about modernization but a pragmatic response to the need for global connectedness and problem-solving.

After 1991, tribal leaders also moved onto the global stage to bring the attention of some of the problems that community members faced in their search for solutions to the problem of digital inequality. For example, in 2005, then President Joe Shirley, Jr., captured the spirit of embracing change through e-governance when he addressed the World Summit on Information Society held in Tunisia, Morocco. His delegation included chairs to the Navajo Nation Council Delegates, the Government Services Committee, and the Resources Committee who are key members of the tribe's government. At that level, the Navajo Nation was, in point of fact, an acknowledgment that Indigenous groups wished to be a part of the United Nations Millennium Project initiative to use ICTs in the eradication of extreme poverty. The act of "going global" also offers an opportunity for the tribe to share experiences with other Indigenous people in this area which tribal members have done successfully in many ways for years.

President Shirley, Jr., subsequently signed a memorandum of understanding to support infrastructures that enable information society to thrive (Hardeen, 2005). His other initiatives included signing a memorandum of understanding with the International Telecommunications Union (ITU) and UNESCO's the Observatory for Cultural and Audiovisual Communication to establish an OCCAM office in Window Rock. The office supports the global community's efforts at creating a global Indigenous portal. These and other initiatives provide a basis for engaging e-governance organizational systems.

Even as the political class moved ahead with plans to integrate people living on the Navajo Nation into the digital world, a few members of the political class resisted change. Evidently, influential tribal political leaders, on the "soft" (nontechnological) side of leadership, complained that the internet was going to cause disruptions in cultural values that may erode the nation's moral fabric. And there were also those who accused the Navajo Nation's political and administrative class of using information for political positioning and not for long-term societal transformation goals. The different political groups agreed, leveraging ICTs for public safety operations. For example, the political leadership promulgated resolution CJY 3492 that outlined the tribal ICT policy.

Although back in 2006, there was no expectation that ICTs would be a pivotal tool in solving economic and social problems brought about by a lethal pandemic, it turned out that e-government played a key role in the communication and management of COVID-19. Through its departmental portals, and in partnership with the federal and state governments, much of the reporting and awareness programming facilitated transmission of information relevant to the crisis. To build this robust e-government citizen service delivery system, much work had to receive support from public and private investments including from the Bill and Melinda Gates Foundation. The foundation funded an initiative, the Native American Access to Technology Program in an effort to narrow the digital divide among Southwestern Indigenous communities (Gordon et al., 2003).

The private initiatives built on work that the federal government has started in an effort to reduce the digital divide in the country and promote e-governance, in general. There were similar targeted initiatives within the different state governments as well. For example, in 2009, Governor Janice Brewer also endorsed the Navajo Nation's application for federal stimulus grants. These efforts had been in place since the enactment of the High-Performance Computing and Communication Act of 1991. The groundwork had been laid through the National Research and Education Network supported by the Bill Clinton—Al Gore administration. The administration also expanded use of ICTs through the Access Native American Reinvention laboratories which were an essential part of the Reinvention of Government movement.

Later, President George Bush's administration supported the efforts by calling for all classrooms and libraries to be connected to the internet. His administration, through the Bureau of Indian Affairs (BIA), provided funding which enabled more than 185 schools in Indigenous communities to be connected to the internet. Other initiatives were started within the framework of the Department of Interior's wide area networks, the Federal Communications Commission's (FCC) E-Rate, and several private sector programs. Since

1998, the E-Rate program has provided affordable access to telephone and internet services to most K-12 schools and public libraries in disadvantaged communities.

In 2009, then-governor Governor Jan Brewer also endorsed the Navajo Nation's application for federal stimulus grants. As a result, the U.S. Department of Commerce's National Telecommunications and Information Administration (NTIA) Broadband Technology Opportunities Program awarded the Navajo Tribal Utility Authority (NTUA) a federal grant of $32.2 million through the American Recovery and Reinvestment Act of 2009. The estimated number of beneficiaries stood at over 30,000 households and 1000 businesses. The grant, along with partial matching funds, would provide middle mile and last mile broadband infrastructure access to the Navajo Nation.

Also, in recognition of the need to move toward wireless solutions, new broadband players have emerged and changed the ICT landscape in the Navajo Nation. For example, the tribal government, working with Stutler Technologies of Kansas, completed a feasibility study to install Airspan's advanced technology WiMAX solutions to provide high-speed internet and VoIP to the Navajo. Areas that include Sacred Wind will benefit from this technology at an affordable cost. WiMAX will be easy to deploy, and access is no longer a function of the existence of telephone lines that are nonexistent in most of the area (Airspan, 2010). Although additional support was provided by the Obama administration, the question of whether leaders in the Navajo government embrace best industry practices remained pertinent to understanding the digital divide (Skow, 2004). Specifically, although Chief Information Officers (CIOs) are available and often contracted out, the overall supervision remained problematic as leaders took to contingency solutions and requested for investments only when needs were dire. As one leader put it, "The Navajo Nation is behind in making total use of IT due to lack of understanding of 'Best Practice', thus many are using 'Best Effort'" (Skow, 2004).

However, despite these efforts, the digital divide problem continued to have negative effects on quality of life on the Navajo territories. Without broadband capabilities some regions missed out on information resources enabled through ICTs. Such benefits include use of emails, social media, online banking, telemedicine, jobs, education, weather, and interaction with government officials, especially those concerned with managing COVID-19. Those connected to the internet benefited from other culturally relevant E-Government Solutions such as the Woven Information of Navajo Data (WIND), e-democracy projects involving local areas, which displays archival data such as maps, E-learning Solutions, and other resources. There are still key structural problems that have not been resolved. These include low capacity, organizational limitations, and funding of units such as the Navajo Nation DIT, among other stakeholders. DIT favored the expansion of wireless devices because of the cultural discomfort with laying out cables under the surface, which for traditionalists was an affront to the earth.

Logistical challenges

Among the enduring logistical and organizational problems is that the central government that houses the three branches (legislative, executive, and judicial) is in Window Rock, Arizona. Window Rock is geographically close to the State of New Mexico, about 20 miles away from the border and has stable internet connectivity there. This explains why the more remote areas are less well served. In terms of management of COVID-19, there were specific challenges derived from the fact that at the organizational level, the Navajo government is situated within three state jurisdictions, Arizona, Utah, and New Mexico, which have different internet platforms. During the design of e-governance systems, there were marked problems of inoperability. The enabling designs had ignored the human side of the projects. Law enforcement platforms include the DIT-T1 and Utah-based OnSat Network Communications Inc. There were claims that the Navajo Nation's Law Enforcement Division Director instituted OnSat unilaterally without the broad organization's input. Other units of government, including the Navajo Housing Authority, Dinè College, and IHSs, had their own centralized ICT systems. The tribal government outsourced services from different vendors including OnSat and through cable networks such as DSL through Frontier Network and AT&T.

Due to settler colonialism, the capacity of the nation to manage ICT services was deeply undermined. And for a long time, there has been lack of external support and resources to solve many preexisting problems associated with weak organizational and technical systems. In July 2008, for instance, internet suppliers cut services for nonpayment contractual failure, interrupting the tribal government's law enforcement and public safety network. Considering the size of the Navajo Nation territory, without ICT capabilities and stable functionality, citizens' emergency and public safety needs were in jeopardy. News reports revealed the withholding of 2.1 million dollars of reimbursement funds resulting in nonpayment to SES satellite communication service provider by OnSat Network Communication Inc., which bought bandwidth from SES Americom (Associated Press, 2008). The federal government working with the Universal Service Administration Company (USAC), which administered the funds under the aegis of the FCC, later resolved the issue. The public embarrassment showed contractual failures and an acknowledgment of management failures. The Utah-based OnSat Network Communications Inc.'s contract failure during summer 2008 was not the first in Navajo country. Earlier, in April, a majority of the Navajo tribe's local chapter houses (centers of local decision-making) lost internet connections after USAC withheld funds over concerns of the tribal government's audit of OnSat. USAC had not received a satisfactory audit report from the tribal government and claimed that OnSat had overbilled for the service and that the Navajo tribe officials failed to comply with procurement rules and had offered OnSat a contract without evidence of competition.

These situations should not be seen as a unique case of managerial failure. The lack of sufficient internal contract management capacities is a not a problem unique to the Navajo Nation but affects many communities. There is strong evidence that there is not enough capacity in governments across the world to manage contracts (Goldsmith & Eggers, 2004, pp. 48–49). The challenge is compounded by the political nature of the contracting out process (Savas, 2000, p. 144). Further, lack of full support and commitment by external governments to provide effective and efficient processes to maintain public functioning create additional barriers that become insurmountable when complicated systems begin to fail. Since the Navajo Nation, like other governments, had sought partnerships in its IT projects, it was necessary that during a crisis situation, there be suitable protocols to ensure smooth and flawless delivery of broadband services. To the extent that tribal leaders requested for additional broadband capabilities to help workers stationed at home, implied that some of the enduring gaps in access had not been resolved.

According to the Navajo Nation Division of Economic Development's census data collected since 2000, the number of housing units is estimated at around 70,000 of which over 30% lack of sufficient pluming and less than 20% have computers (Navajo Nation, 2004). In some areas of the reservation such as Sacred Winds, only 25% of the households had access to telephones or the internet (Airspan, 2010). The current satellite capabilities leave out vast areas of the Navajo Nation, thereby disadvantaging those who live in remote places.

Regardless of the limitations encountered, the Navajo Nation made great strides to work with other stakeholders to offer internet connectivity on the reservation. By the time the COVID-19 pandemic was reported on the Navajo Nation, not all managerial and technical capacity challenges had been overcome. However, the leadership was aware of the need to provide ICT infrastructures as a means to solve other pressing policy problems, especially in education, health, and human services. And that is what prompted President Jonathan Nez and other leaders to include CARES Act financial support for expansion of broadband capabilities as a strategy for fighting COVID-19.

The administration was also able to secure support from Northern Arizona University (NAU), which provided Wi-Fi hot spots for the Navajo and Hopi Nations. Due to the COVID-19 pandemic, NAU suspended in-person classes and reached out to students from Indigenous communities whose learning was affected by the coronavirus. NAU's Information and Technology Service set up Wi-Fi devices in Window Rock's Navajo Nation Museum facilities and at the Good Shepherd Mission center at Fort Defiance, among other centers (Table 3.1). Students drove to the respective parking lots to access the internet, using any computer-based appliances, including laptops, smartphones, and other handheld devices. Students from K12 to college programs were able to use the free service.

Table 3.1: NAU hot spots for Navajo and Hopi Nations.

Towns/Centers	Wi-fi hot spot parking lot locations
Chinle	NTUA Chinle District Office, Hwy 91, Navajo Technical University
Dilkon	NTUA District Office, Hwy 60
Tsaile, Crownpoint	Dinè College Request access
Holbrook Arizona	Holbrook High School, Holbrook Jr. High School, Hulet Elementary School, Park Elementary School, Indian Wells Elementary School, District Office
Window Rock	Library and Museum
Tuba City	Tuba City Chapter House
Shiprock South	Dinè College access
Fort Defiance	Good Shepherd Mission, Kit Carson Dr.
Hopi Nation	Peace Academic Center, Polacca Community Center, Village of Tewa

From field observations in Northern Arizona University (NAU). (April 27, 2020). NAU provides Navajo and Hopi Nations with Wi-fi hot spots, LJ the Lumberjack. *https://www.jackcentral.corg/news/, Flagstaff and Navajo Hopi Observer. (July 13, 2020).* Northern Arizona University creates Wi-Fi hotspots on Navajo and Hopi reservations to help students continue classes. *https://www.nhonews.com/news/2020/apr/14/northern-arizona-university-creates-wi-fi-hotspots/.*

Additional cooperation and support came from corporate sponsors including major service providers who gave a pledge to the Navajo Nation Telecommunications Regulatory Commission to keep the Navajo Nation connected during the duration of the pandemic (March through May). Among the service providers that offered support were T-Mobile, Verizon Wireless, AT & T/Cricket, Cellular One, Choice NTUA Wireless, Frontier, Naked Mobile, and Sacred Wind Communications.

COVID-19 stories and government action

The first confirmed positive cases of COVID-19 on the Navajo Nation were reported some 51 days after those reported in the state of Arizona. After the reports, the number of confirmed positive cases surged to a point where the infection and death on a per capita basis on the Nation were among the highest in the entire United States. In fact, the toll was more than 4.4 times higher than that in the states of Arizona, Utah, and New Mexico combined (Scher, 2020). On April 18, there were 1197 reported cases of COVID-19. 44 persons had lost their lives. By mid-May, the Navajo Nation had reported 2304.41 cases of COVID-19 per 100,000 people compared to New York State's rate of 1806 cases per 100, 000 (Silverman et al., 2020). These data were drawn from Johns Hopkins University which tracked COVID-19 cases globally (Johns Hopkins University, 2020).

As in other Indigenous communities, health professionals posited correlations between existing health inequalities and limited quality care and medical insurance issues to high risk though there are complicated reasons for this. Public health directives on the Nation were in many ways better and more aligned with national and international health knowledge than in neighboring states, yet there was a need for people to commute outside of the Nation for jobs and basic supplies (Kellogg, 2020).

In addition to the inability to adhere to directives due to economic and structural survival needs people deal with health risks regularly. These groups live with ongoing oppressive forces that make it difficult or impossible to access healthy foods, healthy homes, and other necessary instruments of health and wellness. They must also content with environmental racism as is true for Indigenous people globally who must live in areas with environmental contamination from irresponsible spills and the burden of serving as areas for environmental refuse. These health challenges are clearly associated with settler colonialism, which continues to extract wealth from these areas and not reinvesting profits accrued. Thus, for example, indoor air quality on tribal lands has caused devastating health challenges and as a result the communities report a much higher rate of asthma and related respiratory conditions. This situation is exacerbated by lack of electric grid, absence of local access to building and electrical supplies, difficulties in obtaining licensing in electrical and contracting skills, and reliance on coal burning promoted by local coal mining operations. Although these conditions persisted amid push to close down the mines in the wake changes in the energy sector, at the onset of the pandemic, the damage had already occurred.

Prior to the closure of the coal mines, Peabody Energy, a coal mining corporation, drained the Navajo aquifer of millions of gallons of drinking water, contributing to a long history of violent exploitation of tribal lands and the people who live there. It is estimated that by August 2019, the time Peabody Corp stopped coal mining, it had already depleted the Navajo Nation's aquifer of 45 years' worth of fresh drinking water (Scott, 2020). There were no efforts by the corporation to provide suitable water resources for the communities where it conducted its mining operations. The corporation had not even conducted the necessary cleanup and restoration required by the law. There have also been serious spills as the result of mining operations that have resulted in severe health and life challenges and the pollution of water supplies for decades. The negative outcomes of corporations exploiting resources from the Navajo Nation without supporting development of water infrastructures remain a huge data point in the weakness contributing to high rates of COVID-19 cases. And it does not help that decades of underfunding by the federal government made prevention efforts much more difficult to actualize. Water and water rights had always been one of the priority areas identified by Navajo elders, even before the pandemic hit the Indigenous community (Bitsoi, 2015). Navajo elders listed other high priorities to be housing, education, human services, governance, public safety, natural resources, and judicial needs (Bitsoi, 2015). Clearly, if housing and water were better funded, COVID-19 would have been less severe. President Nez said,

> *I hear a lot of people saying, close the borders, shut down, shut down ... Our folks are supposed to be helping get water for the livestock, water for the household. You shut all that down, how can our elders wash their hands with soap and water if there's no water available for them? (President Nez, quoted in Fonseca, 2020).*

An estimated 30 to 40% of residents of the Navajo Nation did not have running water, which posed a challenge in managing a virus such as COVID-19 (Lane, 2020; Scott, 2020). Another possible cause, according to Navajo President Jonathan Nez, was that the nation is a "food desert, which means that more people are at the few grocery stores, convenience stores and gas stations to purchase food" (quoted in Silverman et al., 2020).

Communications and decisions made

Before the first cases of COVID-19 were reported in early March 2020, the leadership took immediate steps to reduce the spread of the virus. Early on February 3, the Navajo Department of Health established its own internal working group on coronavirus. President Jonathan Nez, in conjunction with Navajo Department of Health, Executive Director, Dr. Jill Jim first held a public radio (KTNN) forum on February 20, 2020, to provide tips on prevention of the disease. On February 27, 2020, Jonathan Nez, the Navajo Nation president, announced the formation of the Navajo Nation COVID-19 Preparedness Team to address coronavirus precautionary efforts. The team was drawn from various government agencies. In his announcement, President Nez asserted:

> *Working together with the Navajo Department of Health, Bureau of Indian Education, Indian Health Service, State and local health departments, we are taking this proactive measure of establishing the Navajo Nation COVID-19 Preparedness Team to continue raising public awareness and plan and prepare for the potential spread of the coronavirus in our region. We're bringing these entities together to ensure that we are communicating and coordinating efforts to educate the public. There are no cases of the coronavirus on the Navajo Nation, but this is a very serious health concern, and it's vital that we continue to provide information with everyone, including students, elderly, and community members (President Nez, in Touchin & Curley, 2020).*

Within weeks health authorities and government agencies and nonprofit groups mobilized in efforts to prevent further spread of the disease. The task force was meant to take proactive or firefighting actions including communications and raising public awareness. At that time, February 27, there were already over 78,497 cases reported in China. And the CDC sent out signals that the United States had experienced limited spread within its borders. In fact, regionally, only a single case of a person who had not been to China's epicenter of Wuhan, Hubei Province, was known to have contracted the disease (Quintero, 2020a). Besides, communication from the Navajo Department of Health and IHSs sent out word that COVID-19 risks on the reservation were "low" (Quintero, 2020a).

After the initial relatively low incidence of the pandemic on the reservation, April turned out to be a different story and soon, the Navajo Nation was reporting among the highest numbers of per capita infections in the country (DeSantis, 2020; Rodriguez & O'Neal, 2020; Truong, 2020). By mid-April, all the major news outlets covered the crisis as it unfolded on tribal communities. There was real official fear that the spread of the virus could devastate the southwestern Indigenous communities all together (Kakol et al., 2021). There was also a disturbing trend in journalistic curiosity over the impact of the virus on Native Nations. Some journalists covered the stories responsibly, however. Amy Goodman and Dean Seneca, among others, highlighted the lack of water and other resources as worsening the pandemic in Indian country (Goodman, 2020; McCarthy & Kestin, 2020). In mid-April, 16% COVID-19 deaths in Arizona were those of Native Americans, yet they were only around 6% of the total population. At around the same time, in New Mexico, Indigenous people accounted for nearly 37% of the state's confirmed cases, yet they were only 11% of the population (Associated Press, 2020a). Dean Seneca told *Democracy Now!* "the Navajo Nation is definitely hit hardest." Asked if the Trump Administration reacted promptly to the spread of the virus in the Navajo Nation, Dean Seneca's response was,

> *Well, as you can tell, you know, right from the very beginning [sic], I mean, [Trump] didn't make this pandemic a priority. He did a lot of mixed messaging in the very, very beginning when he started to talk about this. And you see that he's trying now—in his recent reports, trying to justify that, "No, we were on top of this right from the beginning." And that's far from the case. You know, his mixed messaging is what was really critical. At times he would say, "Well, hey, this virus is just going to go away. And we'll wake up one day, and it won't be here." You know, people listen to this information, and that is the wrong thing to send (Seneca, in Goodman, 2020).*

The relative inaction at the federal level took another form. Through not providing necessary funding for water and other sanitation projects, the federal government contributed to the predicament. This claim was echoed, albeit, in different context by President Nez who openly expressed his frustration and complained the federal government ignored Indigenous people. "We feel that the United States government once again has ignored or even left out the first residents, the first people, the first citizens of this country: Indigenous people" In, Miller (2020). Similar remarks were widely shared in reference to the federal government's underfunding of health services in the Navajo Nation's health centers (Truong, 2020). There was also additional pressure from the Governor of New Mexico, Lujan Grisham (D), who informed the White House of the existential threat to communities within Navajo Nation, if support was delayed. The Navajo president's claims were, however, rejected by officials from the BIA who claimed they had been communicating with the tribal communities. The Bureau revealed that the $2 trillion COVID-19 stimulus fund would allocate some of the money to Indigenous communities. However, as we note, the funds arrived late and that became another subject of contention, this time within Navajo government's public bureaucratic forums.

Further actions and reactions

Response to the pandemic included political and on-the-ground efforts to save lives. The Navajo Nation government established an official fundraising and donation office whose work was coordinated from the command center through the Nation's Department of Justice and Office of the Controller. The fund not only accepted money but also in-kind donations from across the world. There were several acts by various public officials, most going beyond their call of duty to help alleviate the suffering from COVID-19. One example deserves mention. Lucinda Charleston, deputy commander of the Navajo Nation Incident Command Center, aggressively mobilized efforts to trace and isolate anyone who tested positive while also delivering aid (Goodluck, 2020). She did this during the peak of the pandemic and, even though a fearful undertaking, moved to Chilchinbeto Chapter House, church, and school. This area, with more than 400 houses, was one of the most devastated.

The Nazarene Church at Chilchinbeto had been identified as having hosted a "superspreader" event on March 7. Some analysts claim it was the first epicenter of the disease on the Navajo Nation (Kidd, 2020). The Navajo Department of Health announced more cases from the same area on March 19, 2020 (Allen, 2020). The zone rally ran concurrently with a church event at the Chilchinbeto Chapter House dubbed "Day of Prayer" (Allen, 2020; Fonseca, 2020). Among congregants were attended from other churches including Cameron Church. Charleston used her familiarity with the region, knocking doors and helping. Mother Jones reported the heroic actions of her team (Goodluck, 2020). Such brave acts were also reported in reference to other first responders, doctors, nurses, firemen and women, and police.

A street-level bureaucrat, like Michael Lee, with 29 years on the Navajo Police Department force, lost his life in the line of duty (Associated Press, 2020b). At the time of his passing, he was attached to the Chinle Police District of the Navajo Police Department. Officer Lee became the first police officer to lose his life from COVID-19 on duty for the Navajo Nation. President Jonathan Nez and Vice President Myron Lizer hailed Lee for his dedication to the Navajo community. President Nez and Vice President Lizer's joint press release said "It's a sad time for the entire Navajo Nation and, to Officer Lee's family, we offer our prayers. Our police department lost a truly honorable and distinguished person." The Navajo Nation also lowered flags in his honor (Office of the President and Vice President, Navajo Nation, 2020). Arizona Governor, Ducey too declared that flags be lowered to fly half-staff in his honor. Governor tweeted on June 19, "Arizona is grateful to all our frontline health and safety workers keeping us safe during this pandemic. In honor of Police Officer Lee's life and service, flags at all state buildings will be lowered to half-staff tomorrow from sunrise to sunset" (Ducey, 2020a).

Firemen too joined in the fight against COVID-19 on Navajo tribal lands. They were among the first street-level bureaucrats to arrive at "hot zones" to conduct "special ops." As part of the special ops, members of the Navajo Nation Fire Department delivered supplies, water, food, medicines and gathered medical information about people needing to fill prescriptions. Like other first responders, firemen too bore the brunt of the pandemic and many were infected (Quintero, 2020b). As many of the first responders battled against infections, it became necessary for the Navajo government to invite additional medical support personnel. Reinforcements came from across the country, including from Michigan (Murray, 2020). As more lives were claimed, the regional hospitals were transformed from acute care facilities into ambulatory care clinics. The nation, with only 170 hospital beds and 13 ICU beds, 52 isolation rooms, and 28 ventilators, was quickly overwhelmed, forcing a call for support by the IHS. In addition, more citizens were mobilized into action to reach out to elders and vulnerable members of the community and help provide water and food. And there were additional government mandates.

The Navajo government directed all Diné people to wear protective masks in public and also within weeks, curfews and shelter-in-place restrictions were imposed. Curfews were imposed on all weekends to prevent continued rise in COVID-19 numbers. The Navajo president issued a series of Executive Orders closing and extending closure of government offices. Executive Order 003-20 extended the declarations of the state of emergency and pushed curfew regulations to May 17 (State of Arizona, 2020). IHSs, while not always well funded, conducted regular contact tracing that far exceeded contact tracing efforts in neighboring border towns (Indian Health Service, 2020, 2021). These efforts were deployed quickly and effectively.

The government also announced some of the strictest stay-at-home regulations in the United States, which included a 57-h weekend curfew. Among the orders was the ban on residents from leaving their homes unless they had an emergency or were essential workers. Those who left homes for work were also required to provide documentation with verifiable contacts of their supervisors and managers. While restrictive, subsequent orders like the Public Health Emergency Order 2020–006 allowed Navajo residents to tend to livestock during Weekend Curfews (State of Arizona, 2020b). The orders confined operating hours for gas stations and numbers of persons allowed in any retail stores at a time. These business premises also were required to institute social distance requirements and monitor occupancy limits. For maximum outreach, these instructions were relayed on Facebook, Instagram, Twitter, and other social media outlets.

Massive campaigns to educate the public on the disease, its symptoms, and ways to prevent infection were instituted. The Navajo Nation also requested for support from the federal government. President Jonathan Nez held several social media town hall meetings in which he communicated various aspects of the COVID-19 pandemic. There were many other public service announcements broadcast on various ICT platforms, including Facetime, Mobile Phones Instant Messaging, and teleconferences, from various Navajo leaders. Many of the announcements were about curfews and lockdowns and other strategies for reducing the spread of the disease. However, there was criticism that some chapters did not receive detailed information about meeting needs for the elders. One chapter leader, David John, complained, “when the chapters were closed due to the state of emergency declaration, there was no plan for providing essential services or clear communication to chapter officials and staff, yet they feel an obligation to meet the needs of their community with little support” (Krisst, 2020a).

One of the challenges was to communicate about job security and continuation of Navajo lifestyles and traditions in the midst of the pandemic. Emotionally difficult messages from national media outlets and, especially, numbers tracking the incidence and deaths from COVID-19 became a major feature on formal and informal news broadcasts and media outlets. And the questions about schools reopening and new norms to govern economic activities on the Navajo Nation became a staple news item.

And since the Navajo Nation also falls within different states, actions from government officials in Arizona, New Mexico, and Utah impacted the disease trajectories. As Professor Lisa Hardy and her colleagues demonstrated, the early reopening of the state of Arizona contributed to an uptick in COVID-19 cases in the month of June 2020 (Hardy et al., 2020). Governor Doug Ducey (Republican) had announced, albeit prematurely, that the statewide stay-at-home order should expire on May 15, 2020. The Executive Order in the form of press release of May 12 from his office (which was translated into Spanish language) read, in part:

> *Since the start of this pandemic, Arizona has taken a calm [sic] and steady approach to protecting health and slowing the spread of COVID-19. Today, our hospitals have capacity to provide care to those who need it; our businesses are implementing and adapting to new physical distancing measures; and data shows Arizona is headed in the right direction. It is time to move forward with the next steps of Arizona's economic recovery—while continuing to make health and safety our number one priority. I'm grateful to all Arizonans for their partnership and cooperation during these trying times. By continuing to follow the data and recommendations of public health officials, we can continue to move forward safely and responsibly together (Ducey, 2020b).*

The net result of the increase in COVID-19-related deaths disproportionately impacted the state's Indigenous communities. About 18% of the fatalities in Arizona involved Indigenous community members. This is a high number considering that their proportion in state population is only 4.5% (Hardy et al., 2020). According to figures from the Navajo Nation Department of Health, as of May 17, 2020, the Navajo Nation had lost 140 individuals, and over 4002 community members had been infected (Goodluck, 2020).

Reactions: volunteer workers during COVID-19 lockdowns

One of the most remarkable responses to the COVID-19 pandemic was the outpouring of self-help and other community-based initiatives of members of Indigenous societies helping each other. Organizations like the Indigenous Environmental Network led efforts to distribute supplies, food, and money for community members as they stayed at home during the curfew and stay-in-place executive orders (Benallie, 2020). Organizations like Red Feather Development Group, led by Boards composed of American Indian leaders, quickly responded to needs by delivering resources. Red Feather along with other organizations in the area began delivering water stations and providing tools and knowledge to people in regions without sufficient water supplies on how to build handwashing stations. Red Feather, established in 1995, joined hands with other partners to find solutions to the COVID-19 pandemic. It built more than 100 handwashing stations and installed trash bins operated with foot pumps within those communities (Craft, 2020; Sevigny, 2020). Single mission organizations that had been previously associated with narrow goals expanded their missions to join in the effort to combat COVID-19. Individual groups of women also made homemade face masks and distributed these to community members. Such groups also offered prayers and songs to provide the much-needed support for grieving families.

Private individuals also stepped up and affixed billboards and messages at strategic places warning people not to visit each other of specific locations. Individuals also left signs on doors to remind the community of the need to shelter-in-place and avoid gatherings. Alicia Chee, a graduate of the University of Denver, returned home in 2017 to serve her community as an Assistant to a Navajo Council member once again went the extra mile to give back to those in distress (Kidd, 2020). Using her own money and resources drove from corner to corner to supply food, soap, water, and other necessities to families in remote parts of the reservations. Zoel Zohnnie, too, on his own accord, through a group Water Warriors United, purchased a water tank and hauled water to elders across the Navajo Nation (Robledo, 2020). There were many more volunteers like them who reached out to their neighbors, delivering groceries and water.

There were also transnational and intrastate efforts to raise funds for mutual aid. Among these were: The New Mexico Children Foundation Dinè COVID-19 Relief Fund raising on Facebook, the GoFundMe Navajo and Hopi Families COVID-19 Relief Fund, and the Food Flashlights in Rural Areas Harness program. One of the important features of these mutual aid programs was the spontaneity and later the sustained continuation of efforts. In addition, the virtual fundraising techniques were a reminder of the age of communication networks which these communities are part of. In the case of the Navajo-Hopi fund, fundraising was conducted through drive-thru and drop-offs, which were designed to reduce human contacts. These fundraising groups raised millions of dollars to supplement income security and well-being of members of the Indigenous communities.

One of the best publicized initiatives was the Adopt-A-Native-Elder program, which was popular within the Navajo Nation. Founded and led by Linda Myers, the program sent medication funds and food coupons to over 850 elders. The program was crucial in giving a lifeline to elders, many of whom lived in remote places where there were few or no groceries. The Navajo and Hopi Families COVID-19 Relief Fund raised at least $254,707 from 3517 donations by early March. At the end of July, the figure had risen to $5,676,444 of a goal set at $7 million. The charity organization airlifted and car-delivered food and supplies (sanitizers, and water included) to Chinle, Arizona, and other places in rural communities of Black Mesa, Blue Gap, Hard Rock, Low Mountain, Whippoorwill, Salina Springs, Cottonwood, Pinion, and Tsaile. Over 18,000 households benefited from the organization's support (Begay, 2020).

Support also came from philanthropic organizations including ISTAT Foundation which offered a grant (Begay, 2020). Their support helped alleviate the basic needs of members of the Indigenous communities in the Southwest. During the COVID-19 crisis, ISTAT Foundation implemented an air bridge operation out of Phoenix, Arizona, and transported critically needed humanitarian relief supplies to the Navajo Nation.

Church-based organizations like Cedar City's Knights of Columbus provided much support for the Navajo community living in Utah. The group working with the Church of Jesus Christ of Latter-day Saints gathered donations of canned goods, medicines, and other health supplies for use among Navajo settled in Blanding, Utah (Watson, 2020). Businesses also offered some relief. Bank of America directly gave support to American Indian communities through its Native American Community Development Financial Institutions (CDFIs) and several philanthropic grants to nonprofits that focused on meeting health and hunger needs in the Navajo Nation. It also donated Personal Protective Equipment (PPE) and hand sanitizers to the community (Bank of America, 2021).

Likewise, Communications for Air Serv helped airlift food to remote areas in conjunction with the Navajo Air Bridge Program teams (Begay, 2020). Bashas, an Arizona-based food store, created a safer shopping environment for many of the Navajo elders. Through its Operation First of the Month endeavor, Bashas set aside shopping hours from 6 to 10a.m. at the beginning of the month for exclusive shopping for elders. There were additional hours for the senior citizens above 65 on Wednesdays 6 to 7a.m (Benallie, 2020).

Private donors also joined in the effort to fight COVID-19 on the Navajo Nation. Forager, a Portland, Maine-based digital organization, collaborated with Iskashitaa, a Tucson-based nonprofit organization to supply fresh foods to Navajo residents (Yurth, 2020). Forager's mission is to link communities of farmers, producers, grocers, restaurants, and consumers who collaborate to grow the local food economy. Forager's concept was based on the idea of transferring surplus foods to areas where there were food deserts. And there was more, inspired Navajo and Hopi volunteers helped their neighbors and community. Some worked 14 h a day, "inspired by the power of k'e" (interrelationship of all things). Through this spirit, food reached to vulnerable elderly grandmothers and others in isolated homes, some lacking running water (Dailykos.com., 2020).

A remarkable example of business efforts to support the Navajo community to combat COVID-19-related problems came through an offer of 1-month pro bono cleanup services. A company provided expertise and technical support to decontaminate areas of the Navajo Nation adversely affected by the virus. Working in collaboration with the Navajo Nation's COVID-19 Health Command Operations Center and Southern Solutions Environmental, a Tennessee-based company, provided industrial products called Decon7 or "D7" to disinfect spaces contaminated with COVID-19. The product had been used in Italy and China to kill the coronavirus by destroying its RNA and is approved by the Environmental Protection Agency guidelines (Krisst, 2020b).

Compassion from San Francisco

Among the volunteer groups that arrived in the Navajo Nation to help the community combat COVID-19 were nurses and physicians from the University of California San Francisco (UCSF) who jetted in to work alongside local medical workers in April and May 2020. These teams were mainly specialists in critical care, intensive care, and emergency medicine care that were critically needed to support the Nation's health-care fraternity. The volunteer medical workers responded to a call from President Jonathan Nez for health workers' reinforcements. The volunteers were a part of the teams that responded to the call from Navajo leaders and traveled to IHS hospitals in Chinle, Arizona, and in Gallup, New Mexico, to care for the sick. Another group of UCSF volunteer also served the community at local hospitals in Ft. Defiance and Tuba City, Arizona.

Navajo politics and budgeting priorities

Initially President Jonathan Nez and his administration appeared to rally community to join hands and fight the COVID-19. President Nez, initially, received national coverage by all major print and electronic news organizations as he highlighted the plight of the Navajo and other Indigenous American nations (see, for example, Silverman et al., 2020). He was eloquent in pressing the federal government for funds to help mitigate the effects of the pandemic. President Nez invoked the spirits of his predecessors who demanded federal obligation to the tribes and the need to fulfill the promises made. The federal government had been slow at providing funds under the Coronavirus Aid, Relief, and Economic Security Act (CARES Act). Tribes waited for 6 months before receiving the first tranche of 60% of the CARES Act funds (Goodluck, 2020). It took a court action by US District Court Judge Amit Mehta to force the US Treasury to release all the funds without delay. A separate account outside the Navajo General fund framework was established for these funds (Office of the Speaker, Navajo Nation, 2020).

Initially, there was wide support for the May 15, 2020, Resolution CMY-44-20, which outlined the contours of the Navajo Nation CARES Fund and its enabling CARES Act. A December 31 sunset date was written into the law, thus mandating faster implementation of programs to respond to COVID-19. However, the politics of the Navajo's COVID-19 management turned rocky after the Council and the Navajo resident disagreed on modalities and priorities for spending $714,189,631.47 funds allocated for the COVID-19 mitigation and management (Office of Speaker, Navajo Council, 2020). The funds to the tribal governments had been approved by President Trump on April 27 but arrived at the Navajo Nation Office of the Controller late June 2020. The US Treasury and the Department of the Interior had determined the allocation levels to the different American Indian and Alaska Native communities.

Colloquially, COVID-19 morphed into political football. The *Navajo Times* described the political disagreements as a battle between the president's office and the Navajo Nation Council. The main issue of contention was the authority to spend funds beyond the emergency needs of the Navajo community. The President and the executive had proposed allocation money for water, agriculture, solar power, and infrastructure projects. These included proposals to use the funds for telecommunication and broadband services. In addition, President Nez listed scholarships and Navajo businesses, housing, PPE, and compensation for frontline warriors as requiring immediate attention. The chapters disagreed with the executive's comprehensive plan insisting on local control of COVID-19 mitigation funds. Brinksmanship was therefore one of the reactions to COVID-19 pandemic. Disputes heightened after the Nez administration line-vetoed $73 million coronavirus response spending which the Council had approved in June in legislative bill Nos. 132—20 and 116—20, respectively. The president vetoed monies that were direct relief to the local 110-chapter houses. Through numerous online Council hearings in the Health, Education and Human Services and Naabik'iyati committees, evidence of dissent was recorded. Some council delegates had demanded decentralization of mitigation efforts.

President Jonathan Nez's administration eliminated relief amendments that would have provided money to other line items such as the Hataalii Association, which supports work done by traditional practitioners, medicine men, and provides cultural and educational outreach during the COVID-19 pandemic. There was strong feeling that the Dinè Hataalii Advisory Council was established under the aegis of the Navajo Nation Historic

Preservation Department and recognized by the Council merited funding. The nonprofit association was established in 1970 with a core mission to "protect, preserve and promote the Dinè cultural wisdom, spiritual practices, and ceremonial knowledge for present and future generations of Dinè" (Navajo Times, 2020). However, the opposition view was that the association was in character, similar to faith-based groups, which was disputed by Lorenzo Max, a medicine man and in-residence elder at NAU's Applied Indigenous Studies program, who regarded the Navajo belief system to be a nonreligion but way of life.

One million dollars had been earmarked for the association. This amount included proposals for staff emoluments and salaries, consultancy, outreach, radio broadcasts, and other publicity and outreach communication interventions. The Association also proposed to use funds for ceremonies, which it planned to host using CDC guidelines. Another line item that was vetoed was for the Judicial Branch ($10 million) and a further $7 million earmarked for burnout assistance for families without shelter (Krisst, 2020c). The vetoed items were included in the Amendments Bill No. 116–20. Debates over line items resulted in much lost time and fed into the narrative that governance in the Indigenous community was ill-prepared to manage the pandemic. Analysts regarded the political bargaining process as rooted in mistrust between the executive and legislative branches of government. Still, there was the view that the executive was frustrating the efforts to localize the funds. Here is how a chapter leader, Damien Augustine expressed the frustration:

> *Day in and day out, many of us have been busy collecting essential items for our communities ... Oftentimes, we are travelling off the reservation and putting our lives at risk. It's surprising that our president, being a former chapter official himself, would indicate that our needs are not a priority (Krisst, 2020c).*

The differences between the Navajo president and members of Council and civil society were also evident in President Nez's rejection of providing funding to the Council backed Navajo Managed Care Organization. The Navajo Managed Care Organization had been touted as the first -of-its kind Indigenous-run effort to serve the community. It was designed to replace the traditional Medicaid. The organization was to be run jointly by the Molina Healthcare and Naat'aanii Development Corporation (NDC), a business arm of the Navajo Nation.

The search for an Indigenous-run health-care system was enabled initially after the Barack Obama administration pushed forward the American Reinvestment and Recovery Act of 2009. This legal framework set contours for tribes to control their managed care systems. Additional work on this possibility came after the New Mexico Human Services Department requested support from the Centers for Medicare and Medicaid Services to operate a managed care unit on tribal lands. This request, actualized in 2017, applied to Navajos living in New Mexico, many of whom had already been enrolled in Medicaid and the Children's Health Insurance Program (CHIP) and used a fee-for-service approach.

NDC's request for funds was rejected after Molina Healthcare had sent the president a letter stating that it was ready to supply 5000 COVID-19 testing kits and N95 masks and other supplies to IHSs. President Nez's administration rejected to allocate funding for the outfit terming it as a "scam" (Becenti, 2020a). He remarked, "We've been cautioning our people about companies attempting to scam the public during the COVID-19 pandemic, and the actions of the Naat'aani Development Corporation and Molina Healthcare fit that category" (President Nez, in Becenti, 2020a). President Nez argued that the health service providers ought not have started managing services by "ransoming medical supplies." And he understood NDC's role as that of fostering and providing support to the development of diverse businesses and not to provide health-care services (Becenti, 2020a). This infuriated many of Council members who had thought the COVID-19 pandemic was the right time to move forward with creation of an Indigenous mechanism to fund health care on the reservation. The NDC board chairman, Dr. Manley Begay felt that President Nez's administration was showing poor leadership. In their estimate, approximately 75,000 Medicaid-eligible residents from the tribe would be beneficiaries (Hamway, 2020). By this estimate, every Navajo in the state of New Mexico would be a part of the new organization.

As December edged by, there was a modicum of spending priorities. Three areas which had not been subject to President Nez's line-item veto of resolution No. CJY-67-20 were agreed upon. These three areas were broadband, water lines, and electric power infrastructure (Becenti, 2020b). The Council and President came to an agreement on allocation of funds for the broadband and telecommunication needs of the society. Council Speaker Seth Damon voiced the consensus that broadband capabilities would enhance the Nation's ability to enjoy better telehealth services, distance learning, and offer Navajo people opportunities for remote working (Becenti, 2020b). Broadband was seen as a critical element in public health solutions during the pandemic.

Saving lives and healing

Community actions that we have shared, albeit, in anecdotes were exemplifications of first responders embracing principles of *Hozho doo K'è*. These are deeply rooted principles of care and connection. The level of support and togetherness reflected embracing solutions that focus on wholeness of life. These principles offer strength and knowledge for understanding current events. While there is a vast diversity in beliefs in the region an analysis of messages delivered during the pandemic demonstrates these principles in some ways. Many Navajo elders have frequently called for a return to relevant aspects of Navajo culture and society prior to the pandemic. After the nation was notified that it would receive funds to help it cope with the vagaries of COVID-19, members of the Diné Hataalii rekindled the call for a return to ways that would help the community heal from the physical and emotional trauma caused by the pandemic. There was lobbying for the association to receive $1 million of the CARES Fund to support such activities. The Hataalii made presentations to support the teachings of Diné cultural healing practices and use of Indigenous knowledge to reduce the mental and physical health impacts of the disease. Among the tenets of the wellness strategies is the use of ceremonial activities. The organization promotes the concept of sharing the Diné philosophy of life for promotion of self-care and healing. Its historical role has always been pronounced during times of anxiety, fear, and societal turbulence. We do not include further detail here out of respect for Diné ways of life and privacy.

COVID-19 and tele-mourning

In the wake of the COVID-19 pandemic, Navajo people like everyone around the world faced recalibrating grieving traditions. Communities are customarily required to gather for 4 days of mourning before burial, where this was possible sacred stories were communicated and tribal elders gave lessons on coping. All this was done in rushed graveside services and some families opted for cremation of loved ones (Fonseca, 2020). However, mourning shifted online. Carlson and Frazer (2015) observe that the "sorry business" which includes practices such as notifications of deaths and funerals; condolence offering, grieving, and healing; and messages of support was a prominent feature in Indigenous online presence (Carlson & Frazer, 2020). In the case of Navajo people in mourning, notwithstanding digital divide challenges, social media platforms like Facebook became the venue where photographs and lives of those who were taken away by COVID-19 were shared. Additional mourning was also done over text messages, three-way phone calls, and videoconferences (Fonseca, 2020).

Messaging challenges

A number of the residents of the Chilchinbeto Community were disappointed that messaging concerning the area being a "hot spot" was not properly communicated. More significantly, many Navajo people in the area wondered aloud that the government at Window Rock did not notify them about the shelter-in-place and other restrictions which affected school children. Some learned of the rules from Facebook and not even chapter officials were kept abreast of the outbreak. There were also issues concerning framing of the pandemic's effect on local schools. Some parents felt the Chilchinbeto school was negatively portrayed by others (Allen, 2020). Community members worried about government's ability to handle the panic that crippled the church communities in the area.

Another important impediment to communication during COVID-19 crisis was the translation of message from English to Navajo language. Government communication required clarity, which in most cases was ensured but there were also reported cases where elderly non–English-speaking Navajo had to find suitable ways to communicate the vast amount of information that was generated about the novel coronavirus. Because national leaders at federal and Arizona state levels gave imperfect communication cues regarding the spread of coronavirus, there remained inevitable ambiguity in what was really going on. Furthermore, the Dinè language has sacred origins and manifestations, enduring notions that some of the thoughts and behaviors can only be conveyed by an inner wind, "*Nilch'ih,*" which is holy. Some of its words cannot, therefore, be translated with precision into English. It was apparent that the lack of specificity in information about disease progression and cure implied additional burdens on Native speakers of Dinè.

COVID-19 patients write and speak

Personal stories are an excellent tool for showcasing reactions to challenging policy problems. To solve existential problems, the role of communicating inspirational messages adds to the policy tool kit. Motivation from influential community leaders inspires others and central to policy implementation because words have power. And personal narratives matter. In one of the most moving stories of hope and resilience Alastair Lee Bitsoi of To'ahani clan, former *Navajo Times* reporter, penned a long personal note titled My COVID journey. Having survived the disease, her essay captures the strength of Navajo people who have death with the pandemic. As an effective health reporter and communicator, she felt the need to give hope to others. Hope, in her own words, is cited in many "different ways and forms through singing, prayers and rituals. It's beautiful, or hózhó" (Bitsoi, 2020). To quote her:

> *We have overcome many plights. During primordial creation, we emerged from underworlds to the Hero Twins fighting the ravaging monsters in an effort to restore order from chaos. From the Long Walk, through droughts, the Spanish Flu, Hantavirus, the 1979 Church Rock uranium spill, overgrazing and livestock reduction, the influence of colonial cultures and systems imposed upon us, the Gold King Mine Spill, the rollback of Bears Ears National Monument, and now the World Health Organization's declaration of the entire world being in a COVID-19 pandemic, we have endured (Bitsoi, 2020).*

In her long expose, Bitsoi credits her recovery from COVID-19 to Navajo cultural identities and courageous inspirations from various relevant stories of hope. She notes that having Indigenous knowledge and practicing it helped her. In her view Diné, knowledge supplements scientific measures such as taking prescribed medicines, doing breathing exercises, washing hands, and sheltering-in-place. She also did what many other COVID-19 survivors did, that is, used cultural herbs and respiratory medicines including sage, cedar, and juniper, among others. Historians also record that there were previous pandemics that devastated tribal communities, including the cases of smallpox in the 1930s. And just how did the Navajo beat such scourges? The answer is in Navajo traditions which focus on treating the whole person, mind, and body. In addition, praying and curing ceremonies involving family and community have been at the heart of the coping strategies. Her essay was, in a nutshell, a celebration of Navajo culture and the powerful healing spirits embedded in traditional knowledge.

Signs of hope and reopening plans

First signs of hope were evident in mid-August after the tribe experienced nearly 2 months of decline in COVID-19 cases. By August 17, for 52 days, there were less than 100 new daily cases being reported (Curtis, 2020). And the numbers of COVID-19 in New Mexico were much lower than the national average (KOB Web Staff, 2020). In Arizona, the opposite was true but within the Navajo community, figures from the Navajo Department of Health noted that in mid-August, there were 6948 recoveries of the 9412 cases and the IHSs together with the Navajo Epidemiology Center were seeing fewer community-acquired cases. The nation administered over 97,644 COVID-19 tests through August (Becenti, 2020c). And a total of 478 COVID-19 deaths were confirmed (KOB Web Staff, 2020). After months of agony and pain, the Navajo community begun to experience some success in reduction of coronavirus cases. The ABC News reported:

> *The Navajo community suffered one of the highest COVID-19 infection rates per capita in the country just after three months ago, but with the implementation of strict public health measures such as curfews and implementation of strict public health measure such as curfews and mandatory mask-wearing, Navajo Nation has curbed the spread of coronavirus (Brantley-Jones, 2020).*

President Jonathan Nez remarked that the Navajo people together had worked hard to achieve a remarkable fete. The Navajo Department of Health and the Navajo Area IHS announced no new cases were reported in early September, signifying that the Navajo nation had turned around the COVID-19 situation (Horn, 2020). This result, according to President Nez, showed that Navajo people utilized tribal teachings and culture and public health knowledge to fight the common enemy. The community's triumph was also attributed to the cooperation of the population. There were no "resistance" rallies against wearing of masks or calls to prevent government from imposing curfews and lockdowns. The President continued to hold Facebook Townhalls, bringing with him the administration's health professionals who fielded questions from the community. The Navajo example, by far, was remarkable, given the intensity by which the community had been impacted. The debate over use of federal CARES grant was resolved, at least as far as there was agreement that some of the funding would go toward increasing broadband capabilities to support e-learning.

The leadership announced that it would open up the economy based on advice from health professionals. The reopening of the economy, including schools, was done in phases, subject to adherence of the public health protocols. The command center at the Department of Health was diligent at monitoring implementation of the reopening plans. The Community Development officials also maintained communications with the 110 Chapter and advised on the reopening plans.

Conclusion

In sum, in all their efforts, people of the Navajo Nation worked to ensure safety and health in the face of a crisis. The messages for reopening retained the emphasis on core principles and CDC guidelines of hand washing, wearing masks, social distancing, and saying no to visitors. These measures continued through to end of year 2020 when a new wave of the coronavirus pandemic emerged. In many ways, the structures and governance systems that firmly institutionalized the COVID-19 response on tribal lands held strong employing both modern and traditional communications mitigation approaches.

References

Airspan. (February 13, 2010). *Navajo reservation using WiMAX to bridge rural digital divide.* http://www.airspan.com/case-study/2010/navajo-reservation-using-wimax-to-bridge-rural-digital-divide/.

Allen, K. (March 25, 2020). *Virus strikes at rally: Cilchinbeto church gathering may be source of outbreak.* Navajo Times. https://navajotimes.com/coronavirus-updates/two-deaths-in-western-may-have-been-covid-virus-spread-at-church-rally/.

Anderson, R. (1999). *Native Americans and the digital divide.* The Digital Beat. http://www.benton.org/DigitalBeat/db101499.html.

Associated Press. (July 22, 2008). *Navajo internet shutdown postponed.* http://www.msnbc.msn.com/id/25798376/ns/technology_and_science-internet/.

Associated Press. (June 19, 2020b). *Navajo Nation police force loses first officer to COVID-19.* Associated Press. https://apnews.com/0ade591e30ca5142963f17107daaa72c.

Associated Press (AP). (April 20, 2020a). *Coronavirus: If you jump the gun…you're gonna set yourself back.* Indian -Country Today. https://indiancountrytoday.com/coronavirus/coronavirus-if-you-jump-the-gun-you-re-gonna-set-yourself-back-FVMxPcQgDU6TgsZN5K9I4Q.

Bank of America. (January 14, 2021). *Bank of America directed more than $ 13 million to Native American communities hardest hit by coronavirus in 2020.* https://newsroom.bankofamerica.com/content/newsroom/press-releases/2021/01/bank-of-america-directed-more-than–13-million-to-native-america.html.

Becenti, A. (March 28, 2020a). *Nez accuses NDC of trying to profit from epidemic.* Navajo Times. https://navajotimes.com/coronavirus-updates/.

Becenti, A. (August 20, 2020b). *Nez, council agree on some projects.* Navajo Times. https://navajotimes.com/rezpolitics/nez-council-agree-on-some-projects/?fbclid=IwAR3feuZd0MAdTIqTAj5R6VUYQQla1ampzMCaxyBiuuBs0NxznWutplhPJBI.

Becenti, A. (September 17, 2020c). *No new cases.* Bavajo Times. http://navajotimes.com.

Begay, C. (March 15, 2020). *Navajo and Hopi families COVID-19 relief fund.* https://www.gofundme.com/f/NHFC19Relief.

Benallie, K. (July 4, 2020). *Relief efforts for Indigenous people 'don't have months'.* Indian Country Today. https://www.indiancountry.com.

Benally, A. (2006). *Dine binahat'a' Navajo government.* Arizona State University.

Bill and Melinda Gates Foundation. (2004). *Toward equality of access: The role of public libraries in addressing the digital divide.* Seattle: The Bill & Melinda Gates Foundation. http://www.Gatesfoundation.org.

Bissell, T. (2004). The digital divide dilemma: Preserving native American culture while increasing access to information technology on reservations. *Journal of Law, Technology & Policy*, 129–150. http://www.jltp.uiuc.edu/archives/bissell.pdf.

Bitsoi, A. L. (July 2, 2015). *Branch chiefs identify 9 priorities*. Navajo Times. https://www.navajotimes.com.

Bitsoi, A. L. (April 30, 2020). *My COVID journey*. Guest essay. Navajo Times https://www.navajotimes.com.

Brantley-Jones, K. (August 13, 2020). *Navajo Nation turns the corner on coronavirus*. ABC News. http://www.abcnews.go.com/US/navajo-nation-turns-corner-coronavrus/story?id=72336050.

Bruyneel, K. (2007). *The third space of sovereignty: The postcolonial politics of U.S.-Indigenous relations*. University of Minnesota Press.

Carlson, B., & Frazer, R. (2015). "It's like going to a cemetery and lighting a candle:" Aboriginal Australians, sorry business and social media. *Alternative: An International Journal of Indigenous Peoples, 11*(3). https://doi.org/10.1177/117718011501100301

Carlson, B., & Frazer, R. (2020). "They got filters": Indigenous social media, the Settle Gaze, and a Politics of Hope. *Social Media + Society*. https://doi.org/10.1177/2056305120925261

Corntassel, J., & Whitmer, R. C. (2008). *Forced federalism: Contemporary challenges to indigenous nationhood*. University of Oklahoma Press.

Craft, M. (August 11, 2020). *Red feather development group distributes sanitary washing stations to native communities*. Navajo-Hopi. https://www.nhonews.com/news/2020/aug/11/red-feather-development-group-distributes-sanitary/.

Curtis, C. (August 17, 2020). *Navajo Nation ends stay -at-home order but remains closed to tourists*. Arizona Republic. https://www.azcentral.com/story/news/local/arizona/2020/08/17/n.

Dailykos.com. (April 9, 2020). *Inspired Navajo, Hopi volunteers working to save isolated grandmothers during COVID-19 crisis*. EricAZ. https://www.dailykos.com/stories/2020/4/9/1936001/-Inspired-Navajo-Hopi-Volunteers-Working-to-Save-Isolated-Grandmothers-During-Covid-Crisis.

Davis, T., & Trebian, M. (May 23, 2001). *Shaping the destiny of native American people by ending the digital divide*. Educause Review. http://net.educause.edu/ir/library/pdf/ERM0112.pdf.

Deloria, V. (1985). *Behind the trail of broken treaties*. University of Texas Press.

DeSantis, R. (April 20, 2020). *Navajo Nation has lost more lives to COVID-19 than 13 States combines*. People.com. https://people.com/human-interest/navajo-nation-lost-more-lives-coronavirus-than-13-states/.

Dorr, J., & Akeroyd, R. (2001). New Mexico tribal libraries: Bridging the digital divide. *Computers in Libraries, 21*(9), 36–42.

Ducey, D. (June 19, 2020a). *Governor Ducey orders flags at half-staff to honor Navajo police department officer Michael Lee*. News Release. https://azgovernor.gov/governor/news/2020/06/governor-ducey-orders-flags-half-staff-honor-navajo-police-department-officer.

Ducey, D. (May 12, 2020b). *Governor Ducey announces next phase of Arizona recovery*. News Release. https://azgovernor.gov/governor/news.

Education Development Center, The, (EDC). (March 23, 2004). *E-government for all: Ensuring equitable access to online government services*. Newton, MA: The EDC Center for Media & Community & The NYS Forum. http://www.nysforum.org/documents/docs/reports/egov4all/eGovforAll.doc.

Fonseca, F. (May 11, 2020). *'The grif is so unbearable': Coronavirus takes toll on Navajo*. Fox 10. https://www.fox10phoenix.com/news/the-grief-is-so-unbearable-coronavirus-takes-toll-on-navajo.

Goldsmith, S., & Eggers, W. D. (2004). *Governing by network: The new shape of the public sector, innovations in American government*. Brookings Institution Press.

Goodluck, K. (May 21, 2020). *Every corner of the Navajo Nation has been hit by COVID-19*. Mother Jones. https://www.motherjones.com/politics/2020/05/every-corner-of-the-navajo-nation-has-been-hit-by-covid-19/.

Goodman, A. (April 15, 2020). *COVID-19 & Indian country: Pandemic exposes Navajo Nation's water access crisis & health disparities*. Democracy now. http://www.Democracynow.org.

Gordon, A. C., Gordon, M., & Dorr, J. (2003). Native American technology access: The gates foundation in four corners. *The Electronic Library, 21*(5), 428–434.

Groski, P. C. (October 9, 2002). Privilege and repression in the digital era: Rethinking the socio-politics of the digital divide. *Race, Gender & Class, 10*(4).

Hamway, S. (February 19, 2020). Navajo -backed medicaid program progresses. *Albuquerque Journal*. https://abqjournal.com/1422106/navbacked-medicaid-program-progresses-ex-tribal-managed-care-entity-would-be-the-firts-of-its-kind–in-the-nation-html.

Hardeen, G. (November 14, 2005). *Press release, Navajo Nation President Joe Shirley, Jr., to address world Summit on information society in Tunisia, Africa, communications director office of the President and Vice President, the Navajo Nation*. http://www.navajo.org/images/pdf releases/George Hardeen/nov gh/.

Hardy, L. J., Saul, G., & Smith, S. M. (June 30, 2020). *As Arizona coronavirus cases surge from reopening, indigenous nations suffer not only more COVID-19 but also the blame*. The Conversation. https://theconversation.com/.

Hayes, H. (2003). *Digital divide in Indian country: Obstacles and opportunities for native American communities*. http://students.washington.edu/aliss/silverfish/archive/april2003/hayes.pdf.

Horn, M. (September 8, 2020). *Zero new COVID-19 cases on the Navajo Nation; the first time since March*. AZ family.com. https://www.azfamily.com.

Indian Health Service. (June 5, 2020). *IHS*. Update for Navajo Area https://www.ihs.gov/sites/navajo/themes/responsive2017/display_objects/documents/NavajoArea_COVID-19Report.pdf. Indian Health Service. (February 2, 2021).

Indian Health Service. (February 2, 2021). *IHS receives an additional $ 1 billion for the continued COVID-19 response*. https://www.ihs.gov/newsroom/pressreleases/2021-press-releases/ihs-receives-an-additional-1-billion-for-the-continued-covid-19-response/.

Navajo Nation. (2004). *An overview of the Navajo Nation -Demographics. Fast Facts*. Division of Economic Development. http://www.navajobusiness.com/fastFacts/demographics.htm.

Johns Hopkins, University. (2020). *COVID-19 dashboard by the Center for Systems Science and Engineering (CSSE) at*. Johns Hopkins University (JHU). Coronavirus Resource Center. https://coronavirus.jhu.edu/map.html.

Johnston, R. (2001). *Connecting Indian country: Tribally driven telecommunications policy*. Washington, D.C.: National Congress of American. http://www.ncai.org/main/pages/newsreleases/NCAI%20Telecom%20Policy%20Report%20Fina.

Jorgensen, M. (Ed.). (2007). *Rebuilding native nations: Strategies for governance and development*. University of Arizona Press.

Kakol, M., Upson, D., & Sood, A. (2021). Susceptibility of southwestern American Indian tribes to coronavirus disease 2019 (COVID-19). *The Journal of Rural Health: Official Journal of the American Rural Health Association and the National Rural Health Care Association, 37*(1), 197–199. https://doi.org/10.1111/jrh.12451

Kellogg, J. (May 7, 2020). *Gallup, near Navajo Nation, locked down for a 2nd weekend*. Farmington Daily Times. https://www.azcentral.com/story/news/local/arizona-breaking/2020/05/07/gallup-near-navajo-nation-locked-down-2nd-weekend/3092217001/.

Kidd, D. (July 16, 2020). *Dispatches from Navajo Nation and its battle with COVID-19*. Washington, DC: Governing Magazine. https://www.governing.com/.

KOB Web Staff. (August 15, 2020). *Navajo Nation reports 50 consecutive days of COVID cases below 100*. Albuquerque, NM: KOB 4. www.kob.com/new-mexico-news/navajo-nation-reports-50-consecutive-days-of covid-cases-below-100/5827974/.

Krisst, R. (August 21, 2020a). *Mexican water chapter waits on water, COVID-19 relief*. Navajo Times. http://wwwnavajotimes.com.

Krisst, R. (September 10, 2020b). *Cleanup crew fights COVID-19 on Navajo*. Navajo Times. https://www.navajotimes.com.

Krisst, R. (July 10, 2020c). *Line-item vetoes 'send wrong message': Nez nixes CARES act funds for chapters, Hataalii group*. Navajo Times. https://ww.navajotimes.com.

Lane, R. (2020). *The impact of COVID-19 on Indigenous peoples*. Policy brief no. 70. United Nations.

McCarthy, J., & Kestin, O. (May 22, 2020). *The Navajo Nation's lack of water has fueled a COVID-19 crisis.* Global Citizen. https://www.globalcitizen.org/en/.

Miller, H. (April 1, 2020). *Navajo Nation President: U.S government ignoring native Americans amid coronavirus.* Huffington Post. https://www.yahoo.com/huffpost/navajo-nation-coronavirus-135958037.html.

Murray, C. (July 7, 2020). *Quincy firefighter deploys to Navajo Nation for COVID -19 mission.* Hillsdale Daily News. https://www.hillsdale.net/news/20200707/quincy-firefighter-deploys-to-navajo-nation-for-covid-19-mission.

Navajo Hopi Observer. (July 13, 2020). *Northern Arizona University creates Wi-Fi hotspots on Navajo and Hopi reservations to help students continue classes.* https://www.nhonews.com/news/2020/apr/14/northern-arizona-university-creates-wi-fi-hotspots/.

Navajo Nation. (January 3, 2011). *Official site of the Navajo Nation government, history.* https://www.navajo-nsn.gov/index.htm.

Navajo Times. (June 25, 2020). *'We must return to the teachings of Hozho': Hataalii Assoc. poised to receive $ 1M for coronavirus response.* Navajo Times. https://www.navajotimes.com.

Northern Arizona University (NAU). (April 27, 2020). *NAU provides Navajo and Hopi Nations with Wi-Fi hot spots, LJ the Lumberjack.* https://www.jackcentral.corg/news/.

Office of the President and Vice President, Navajo Nation. (June 19, 2020). *Nez-lizer offer condolences to the family and colleagues of Navajo police officer Michael lee.* Communications Director: Press Release. https://www.opvp.navajo-nsn.gov/.

Office of Speaker, Navajo Nation. (2020). *Navajo Nation receives total $714,189,631.47 in federal CARES Act funding.* The 24th Navajo Nation Council. Press release. http://www.navajonationcouncil.org/pressReleases/2020/JUN/Navajo_Nation_receives_total_714189631_dollars_in_federal_CARES_Act_funding_PR.pdf.

Quintero, D. (February 29, 2020a). *Navajo Nation creates COVID-19 team.* Navajo Times. https://navajotimes.com/reznews/navajo-nation-creates-covid-19-team/.

Quintero, D. (March 22, 2020b). *Farmington firefighter tests positive; first responders take precautions.* Navajo Times. https://navajotimes.com/reznews/farmington-firefighter-tests-positive/.

Robledo, J. (August 10, 2020). *Navajo "water warrior" drives miles during COVID to deliver water to those in need.* Daily Sun. http://www.azdailysun.com.

Rodriguez, C., & O'Neal, N. (May 11, 2020). *Cases rise to 3,122 on Navajo Nation, 100 total reported cases.* https://www.kob.com/albuquerque-news/cases-rise-to-3122-on-navajo-nation-100-total-reported-deaths/5726308/.

Rural Assistance Center, (RAC). (2008, October 12). *Expanding telecommunications access in Indian country, (FCC sponsored).* http://www.raconline.org/info_guides/technology/.

Savas, E. S. (2000). *Privatization and public private partnerships.* Chatham House Publishers Seven Bridges Press.

Scher, I. (April 20, 2020). *The Navajo Nation's coronavirus infection and death rates are 10 times higher than the neighboring state of Arizona.* Business Insider. https://www.businessinsider.com.

Scott, C. (June 19, 2020). *Report: Largest coal producing mine in U.S. drained aquifer o coronavirus –ravaged Navajo Nation. 12 News. Arizona.* https://www.12news.com.

Sevigny, M. (July 16, 2020). *Nonprofit group brings handwashing stations to Hopi, Navajo.* National public radio (NPR) KNAU. Morning edition. https://www.npr.org

Silverman, H., Toropin, K., & Sidner, S. (May 19, 2020). *Navajo Nation surpasses New York state for the highest Covid-19 infection rate in the US.* CNN. http://www.cnn.com.

Skow, H. (June 30, 2004). *Navajo Nation information technology plan, "digital Navajo Nation".* The Department of Information Technology, Window Rock, AZ. https://www.dit.navajonsn.gov/forms/NAVAJO%20NATION%20Strategic%20Information%20Technology%20Plan%201.3.pdf.

Navajo Nation initiates additional weekend curfews, outlines provisions for essential businesses AZEIN. (April 16, 2020). State of Arizona. https://ein.az.gov/emergency-information/emergency-bulletin/navajo-nation-initiates-additional-weekend-curfews-outlines.

State of Arizona. (2020). Arizona emergency information network (AZEIN). In *Navajo Nation extends closure of government until May 17 to reduce the spread of COVID-19*. https://ein.az.gov/emergency-information/emergency-bulletin/navajo-nation-extends-closure-government-until-may-17.

Theobald, B. (June 4, 2020). *Electoral Inequality detailed in expansive study*. The Fulcrum. https://thefulcrum.us/voting/native-american-voting-rights-2646153121.

Touchin, J., & Curley, S. C. (February 28, 2020). Office of the President and Vice President. The Navajo Nation. In *Navajo Nation COVID -19 preparedness team established to address coronavirus precautionary efforts*. Press Release. https://www.opvp.navajo-nsn.gov/Portals/0/FILES/PRESS%20RELEASES/2020/Feb/FOR%20IMMEDIATE%20RELEASE%20-%20Navajo%20Nation%20COVID-19%20Preparedness%20Team%20established%20to%20address%20coronavirus%20precautionary%20efforts.pdf.

Truong, K. (April 24, 2020). *The Navajo Nation's COVID-19 death toll is higher than that of 13 states combined*. https://www.yahoo.com/lifestyle/navajo-nations-covid-19-death-173000233.html?bcmt=1.

Warren, G. S. (2000). *Closing the gaps: The digital divide and native Americans*. Victoria University of Wellington: Sherson Media. http://www.shersonmedia.com/grant/cv/items/papers/Closing_the_Gaps.pdf.

Wilkins, D. E. (1987). *Dine bibeehaz'dhanii: A handbook of Navajo government*. Navajo Community College Press.

Wilkins, D. E. (2002). Governance within the Navajo Nation: Have democratic traditions taken hold? *Wicazo Sa Review*, *17*, 91–129.

Yurth, C. (August 13, 2020). *Nonprofits work to get surplus produce to quarantining Dinè*. Navajo Times. http://www.navajotimes.com.

CHAPTER 4

Internal and external Maasai communications and management of COVID-19

The first case of COVID-19 in Africa was reported on February 14, 2020. In Kenya, the first case was reported on March 13. Patient zero had arrived from the United States via the United Kingdom. And it did not involve members of Indigenous communities. Early reporting on the pandemic's potential unimaginable devastation of African communities was the story of the western media. Fueled by a video clip of millionaire William (Bill) Gates commenting on Africa's ill-equipped health-care systems, pundits predicted hundreds of thousands would be dead by April. The pundits believed that Africa had been associated with several episodes of viral attacks, including HIV/AIDS, Marburg virus, and Ebola, and, therefore, was vulnerable and ill-prepared. Social anthropologist Alex de Waal and University College London emeritus Professor of Anthropology Paul Richards pointed out that Africa's burden of diseases was different to other continents and that millions of people with tuberculosis or with HIV were at risk of being affected by COVID-19 (de Waal & Richards, 2020).

It did not help that Africa had not successfully tackled problems associated with treatable diseases such as Cholera, Malaria, Meningitis, Tuberculosis, Lassa fever, and Typhoid, among others. If there is any lesson from Ebola, it was that Africa's health systems were demonstrably weak. It took the efforts of the World Health Organization (WHO), USAID, French Red Cross, and other international actors to deliver supplies necessary to combat the recent epidemics. However, much work remains to be done as flare-ups remain a possibility.

This doomsday scenario was countered by another narrative, which is, Africa has a relatively youthful population and, ironically, has learned from management of other pandemics like HIV/AIDS and Ebola, which would be valuable for contact tracing and community-based surveillance projects (Farge, 2020). Thus, for Africa, fighting COVID-19 was to use the words of Professor Christian Happi, Infectious Disease Institute at Nigeria's Redeemer's University, "business as usual" (Happi in AFP, 2020). Stemming from the above narrative, there was a sense that millions of African peasants who do not leave their homes and practice subsistence forms of living would probably wither the storm.

The Inequality of COVID-19. https://doi.org/10.1016/B978-0-323-99867-3.00006-5

Projections of mass deaths in urban settings were probably not widely off the mark, if we consider that slum urban settings such as Kibra, Mathare, Mukuru, Kawangware, KwaNjenga, or Korogocho in Nairobi lack basic sanitary infrastructures and are overcrowded. The next sections show in a few short months, how the national government has managed the rising coronavirus pandemic. This discussion reflects on the government's actions and policy bargain with implications to the Indigenous Maasai.

Part I: Situation in Kenya's Maasai universe

Evidence from urban settings also suggests that Maasai, who according to Otenyo et al. (2020) are frequently hired as watchmen, are included in populations who are increasingly experiencing high rates of noncommunicable diseases that include diabetes and heart diseases. These opportunistic diseases are associated with changes in food diets. Although rural Maasai people eat lots of meat and drink blood and milk, they do not have high rates of cholesterol as urbanites do. Maasai people living in rural areas maintain subsistent lifestyles as pastoralist whose houses or Kraals have no sanitation and deal with issues of child mortality, undernutrition, sanitation, and infectious diseases. Poor nutrition is linked to their high consumption of milk, at 66% of their 2500 kcal per day. Some authors proclaim the existence of a Maasai Paradox and wondered, as have researchers, why that is the case (Arimi, 2011). Among the popular theories posited in numerous studies conducted between 1960 and 1980 is the preponderance of physical activity among the Maasai. Maasai people move a lot for various reasons, including searching for firewood, grazing, and in ceremonies. There is a cultural belief that inactive Westerners and African urban dwellers have to walk 19 km more per day to burn calories at the rate of rural Maasai (Arimi, 2011).

A second theory, which has also been enhanced in Maasai response to COVID-19, is the use of traditional medicines and herbs. Maasai people, even those living in the United States, reportedly ordered familiar herbs to use overseas. These herbs are frequently added to diets. A study by scientists from McGill University, Canada, found 19 plant food additives in Maasai diets (Arimi, 2011). The compounds had properties that reduced cholesterol. Among the chemicals found in the diets were saponins, flavonoids, antioxidants, and other water-soluble dietary fibers with potential role in lowering cholesterol, the health benefits of which occurred for Maasai people from regions in Kenya and Tanzania. The most common herbs added to Maasai foods are from the bark of the magic gwarra (olkinyei), jacket Plum (oldimigomi), the East African green heart (Olsokonoi), and thorn mimosa (Olkiloriti).

Maasai territory covers at least 160,000 square miles of mainly regions of the Rift Valley of Tanzania and Kenya. A higher percentage live on Kenyan territory but census tracking is unreliable. With specific reference to Maasailand, which in Kenya is mostly concentrated in Kajiado, Narok, Laikipia, and Nakuru regions, reports about mothers dying while giving birth at the hands of traditional birth attendees were a common news item. In villages of Kajiado county, some of the deaths occurred because telephone and ambulance eservices were rare. Mobile phones hardly have enough network in Pakase village near Lake Natron, which is also along the Kenya–Tanzania border. For one to make a phone, they have to climb Mt. Pakase to find intermittent signals (Okoth, 2015). This was in spite of one in every three persons owning a mobile phone, which is also used for mobile money transfers (Rutten & Mwangi, 2012). Without reliable telephone coverage, outreach efforts by available ambulance services from organizations such as Amref are difficult to access. There are also well-documented cases of cattle raiding–related security and climate change challenges that complicate medical care in parts of Maasailand.

Root causes of poverty: Maasai colonial and neocolonial context

Maasai people are the original settlers of the North Tanzania and South Kenya savannahs, and their displacement mirrors that of other Indigenous groups. Maasai communities have been displaced both by colonial and new independent leaders of Africa, which established national parks and hunting reserves for tourist dollars or conservation (Wisner, 2013). Ruling elites from larger tribes have not ceased land grabbing in areas which are known to be ancestral homes of the Maasai community.

Using legal jargon of constitutionalism, land agents linked to high-ranking government officials acquired Maasailands. And even when Maasai leaders demand compensation from State Corporations, like the Kenya Electricity Power Transmission Company for using tribal lands; the compensation rates are dismal or nonexistent (Githaiga, 2015). To complicate these matters, power lines are not meant to serve the Indigenous communities.

Maasai people have also suffered disproportionately from effects of climate change, which has taken the toll on livestock as well. One perennial challenge that came to the fore during the COVID-19-related curfews and lockdowns was an increase in clashes between wildlife and Maasai livestock. Since the theft of Maasai rights to flora and fauna in regions where they reside, compensation from Kenya Wildlife Service (KWS) authorities was not commensurate with loses meted to Indigenous peoples. At times, there was no compensation for livestock killed by wildlife. Maasai tribesmen appeared on various media platforms discussing difficulties of livestock loss to wild animals. Some of the Morans (youthful warriors) threatened to kill lions because the KWS was not listening to them. One group operating in Kajiado, Samburu, and Transmara Narok told Sauti Ya Mwananchi program on CITIZEN TV that they would take matters in their own hands.

The loss of Indigenous rights and weak political bargaining is responsible for the environmental degradation and lack of investments in regions where Maasai live. As such, availability of water, which is crucial for fighting the spread of COVID-19, is a huge problem. Electricity, which supports information and communication technologies (ICTs), is unavailable in remote villages but scanty in settled areas such as Kajiado Town and in Mara Narok are lodges visited by tourists.

Kenya policies and communications approaches

Like other developing countries, the use of ICTs was a major tool in the management of COVID-19. Without a robust ICT infrastructure, it is inconceivable that any community would be effective at provision of information about the novel coronavirus. Prior to the pandemic, many investments had been made in terms of expanding power into rural areas. Kenya also supported Green Solar Power initiatives, which reached many county headquarters including Narok and Kajiado. From a policy standpoint, much progress has been made in leveraging new ICTs to drive development agendas. Kenya made calculated decisions to establish a Ministry of ICT and sought to transform the country's digital services offerings. It unveiled an ICT Policy document that laid out various goals. Among the legal frameworks governing the ICT policy is the Computer Misuse and Cybercrime Act (2018) which prohibits spreading falsehoods on social media. The fine for publication of false information carries a maximum fine of KSh. 5 million or maximum imprisonment of 10 years in prison or both. This law has turned out to be difficult to enforce but it served to deter bloggers and activists on social media who turned the pandemic into an opportunity to spread misinformation and other forms of "infodemics" on COVID-19.

The country moved to digitize its National Health Insurance Funds (NHIF) and identification processes. The government also supported the health sector in other ways. For example, new applications appeared, thanks to the infrastructures and road map provided by the government. Besides the telemedicine and telehealth applications, there are chatbots to answer health-related questions. One example applied in Kenya is Sohie Bot, for matters of reproductive health.

In addition, a poorly managed launch of one-laptop-per child (Digital Literacy Program, DLP) policy was launched to enhance e-learning activities (Otenyo, 2017, pp. 122–23). It was a failed political project as laptops did not reach all corners of the country rendering moot any dreams for e-learning. In urban settings, e-learning gained much presence during the curfew and lockdowns associated with government efforts to lower the spread of COVID-19. But many rural teachers did not succeed with the government's plan and the school year was basically canceled.

Kenya's approach to narrowing the digital divide included establishment of the Communications Authority of Kenya, which acknowledged the role of ICTs in the new economy and sought to use these new technologies to realize national goals of information equity. The authority ought to narrow the digital divide through a Universal Service Fund and also the Education Broadband Connectivity Project, whose goal was to include 8500 public schools to high-speed networks. During the 2017/18 financial year, the authority funded connectivity to 896 schools (Wangusi, 2020).

From government pronouncements, digital services were also a direct method of reducing infections, mainly through use of digital money (M-PESA) and online e-government services. Safaricom, Kenya's M-PESA applications, has been hailed as an innovative model for low-income banking and remittances. By 2009, studies reported that 40% of adults relied on M-PESA for in-country remittances and one out of six used the application to store money while traveling (Aron, 2009). Government, therefore, took steps to require transactions be made using cashless money (Koskei & Kipruto, 2015) Kenya already had a robust cashless media, in which over KSh. 164 billion is transacted per month (Daily Nation, 2015). In addition, the Maasai use M-PESA apps for purchasing and selling cattle. Besides, mobile devices have become popular for herders who want to find out where to bring their cattle for water. The availability of mobile telephones is considered an essential tool for information sharing since a Maasai herder can inform others where to find water instead of the old ways in which those in remote areas would roam around in search for pasture and water for the cattle. In 2020, Kenya's path to digital maturity was evolving but still problematic. ICTs, which would enable services such as e-learning and telemedicine, were crippled with various challenges including inadequate hardware and broadband.

Government policies of impoverishment denied Maasai people of opportunities to experience the marvels of telemedicine which were instituted in the affluent area of Kenya. Kenya telemedicine infrastructure was enabled by Telecom Kenya and Valentis Health, a private venture that gave patients access to several licensed doctors through its App "Dial Daktari." Patients called a care manager through a toll-free number who then assigned a session doctor. Telemedicine was one of the initial responses to COVID-19. The WHO encouraged telemedicine as a way of reducing risks of spreading the coronavirus in hospitals.

A surprise announcement from the political leadership appeared to raise hope that broadband capabilities would be enhanced to help manage the COVID-19 communications platforms. In March 2020 President Kenyatta announced the coming of Google Internet solar powered balloons which, initially, Kenyans on the internet criticized for not being a priority. Government insisted that the Google solution would increase broadband capabilities to enable Kenyans to work from home and also increase capacity for telemedicine. Alphabet Inc. (Google's parent company) launched the first commercial high-speed internet balloons that float around the stratosphere, beaming 4G internet signals. The 4G capability would allow download speeds of 18.9 Mbps (megabits per second) with upload speeds of 4.7 Mbps, making it possible for thousands of rural Kenyans to stream videos and surf the internet (Emecheta, 2020).

Although the government invested funds to provide for internet connections, the charges for connection were unaffordable for most of the Indigenous communities and the low-income groups. More specifically, mobile-based internet connections appeared cheap in terms of one shilling per Megabyte (MB) of data. However, for most uses like telemedicine or e-learning during the lockdowns, much more data were required, often to the tune of 5 GB and above of data. The average cost of 5 GB of data is roughly KSh. 5000 (US$ 50), which is out of reach for most Kenyans including middle-level civil servants. Many Maasai who work as watchmen in urban centers earn as little as KSh. 10,000 per month, which would mean they would have to spend half their incomes on bundles of data for internet services! Thus, provision of online services to most parts of Maasailand was frustrated by both weak signals and unaffordable pricing.

The national government's installation of Google Internet balloons was not the only gesture instituted to address problems associated with the digital divide. There were additional efforts to provide free bundles of data and airtime to frontline workers, including nurses and doctors. Sponsored by Safaricom, the country's largest internet service provider, the resources included 195 terabytes of data, 10 million voice call minutes and 33 million SMS resources to health workers. There were more data resources offered to the Ministry of Health (MOH) and other related COVID-19 command and control centers (Wafula, 2020).

The incomplete e-governance project meant a heavy reliance on bureaucratic norms. The growth of government departments has been a feature in African governance systems. Kenya, in particular, suffers from a bloated bureaucracy, which has caused a huge wage bill. This growth is entirely rooted in politics of identity and representativeness. The same mentality upholds an ideology of a spoils and patronage system where senior civil service jobs are political rewards. This mentality has come with the unintended consequence of decisions being made almost entirely on political reasoning. A problem of this mentality is that it is unhelpful in managing national crisis such as the COVID-19 pandemic.

Communication from leaders and bureaucracy

This draws our attention to the language and impact of communication which originated from four strategic agencies that took lead in combatting the spread of the coronavirus: the MOH, Office of the Presidency, National Police units, and the heads of the devolved county governments, including health officials and the governors. The national task forces established to lead policy coordination efforts were hardly atypical. They, too, were staffed with individuals molded in civil service cultures and showed an inclination to be defensive when called upon to be accountable.

Once borne, the bureaucratic units assigned with tasks of combatting the spread of the coronavirus repeated lines that were familiar and politically correct. Politics remained the wellspring of the process in messaging for control of the pandemic and its effects on the economy. There was no communication strategy and template for reducing the spread of the disease. And, because there is little drive to employ officially binding paperless memos, the files directing policy were shared both manually through electronic messaging systems. Kenya lacks a secure intra-agency intranet system for communications on critical and operational tasks. There are also unresolved issues concerning applications of the colonial and archaic Official Secrecy Act in the era of ICTs. Within the different ministries and agencies, the paper trails were long. Table 4.1 is the gist of the over a dozen such memos. For Kenya, official paperwork reflected the enduring nature of a highly hierarchical system. Letters sampled show some evidence of gobbledygook-jargon and circumlocution.

Kenyatta's messages were designed to provide hope and showcase that Kenyans were "responsible." In parts of his speech, the president was more candid and accepting that there was no need for celebrating. In his own words, "While we mourn the 14 that have succumbed to the disease, we also celebrate the 98 recoveries. I am encouraged by the progress we have made but we are not yet VICTORIOUS. That is the reason why I want to announce the following additional measures." At this point, Kenya announced cessation of movement in and out of the Nairobi Metropolitan area and also in the high tourist regions of Kwale, Mombasa, and Kilifi counties. A dusk to dawn curfew was also extended for 21 days. In addition, Kenyatta emphasized the need to stick to scientific protocols in the nation's policy responses to COVID-19.

Regardless of the efforts from Kenya's leadership, there were still several areas of concern. The months of April—May proved to be a challenge in the nation's management of the

Table 4.1: Samples of widely circulated official COVID-19 executive memos.

Parts of Presidential Speeches (Uhuru Kenyatta).
The fifth presidential address on the corona virus pandemic at state house, Nairobi, Saturday, April 25, 2020, at 14.30 p.m. This speech was given on the 44th day since the first confirmed case of COVID-19 in Kenya. Kenyatta noted that: "over that time, we have demonstrated the resilience, courage, and ingenuity required to defeat this invisible, lethal enemy. This corona virus has affected the whole world, some nations far more severely than others. We have been spared the more terrible human cost so far because the Government followed the advice of our medical experts and acted quickly to stop this disease spreading and killing large numbers of Kenyans. At the beginning of this pandemic, our Medical experts had projected 10,000 people would be infected by end of April. However, due to DETERMINED mitigation measures—we are now at 343 confirmed cases. I thank the millions of Kenyans for showing DISCIPLINE and RESPONSIBILITY. Your attitude is saving thousands of families from grieving for those who would otherwise be lost to the coronavirus."

Source: Press Release, Office of the President, State House, Kenya.

coronavirus pandemic. Amid use of country's National Intelligence Service (NIS), the country's use of force to enforce rules and militarize the operations to prevent the spread of coronavirus was criticized (Kwamboka & Omondi, 2020). The Standard editorialized the corruption in the police force as unacceptable (Standard, 2020). A study conducted between March and June found that during the curfews, the police violated human rights of thousands of Kenyans. Reports indicated numerous cases of sexual assault, shootings, inhumane treatment, and even robbery perpetrated by policepersons (Apollo, 2020). The Independent Policing Oversight Authority (IPOA) reported that the scores of citizens had filed complaints against the police, which they described as the "weakest link in the war against the virus" (Apollo, 2020).

The government's demand that those quarantined pay for their own upkeep at designated hotels and other facilities was also considered a harsh and unrealistic measure (Wesangula & Korir, 2020). The country ranked as one of the most corrupt nations witnessed (on social media video clips) reports of misuse of public funds initially allocated for the COVID-19 containment effort. In one bizarre of corruption, the government spent thousands of shillings on buying tea for officials (Mwaura, 2020). Journalist George Kegoro (2020) claimed the government was losing control and leadership of the country's responses to the COVID-19 pandemic.

It is useful to recognize that since the COVID-19 management process was fluid, all government ministries, in their own unique ways, joined efforts to communicate to stakeholders and offer some degree of direction on mitigation efforts. Undeniably some of the communication lacked clarity and was essentially routine, thereby leaving much discretion to unit leaders to organize for exigencies emerging from the increasing COVID-19 cases. Government snail mail continued to be the backbone of spontaneous reactions initiated to manage the crisis. Some of the correspondence is provided to illustrate the policy debates as they unfolded in various departments (see Table 4.2 on reopening Universities). Table 4.3 was a circular on Public Service Operations and Working From Home Leave Rules. Tables 4.4 and 4.5 provided guidelines for industries, businesses, and small enterprises activities during COVID-19. These circulars provided the overriding principles upon which all subsequent county-level policy decisions affecting the Maasai were actualized.

Pressure to reopen the economy and society

Workers at the globally famous Maasai Mara National Reserve (MMNR) would be tested for COVID-19 in preparation for the gradual reopening of tourism in the region (Sayagie, 2020a). The Maasai Mara is the biggest revenue source for the county government of Narok and is a key source of employment for local residents. In the words of Maasai public service official, Francis Ole Nkako, "The Mara is an awesome natural wonder, a

Table 4.2: Re-Opening od Universities After COVID-19 Pandemic.

Vice Chancellors of Universities Principals of Constituent University Colleges RE: Re-Opening of Universities After COVID-19 Pandemic The above subject matter refers. Following the report of the COVID -19 in the country on March 15th, 2020, His Excellency the President directed all learning institutions to be closed immediately. The choice was part of the measures put in place to control the spread of the disease. The Ministry of Education in collaboration with other Government Agencies is working on the modalities of reopening all institutions of learning. Modeling projections by the Ministry of Health (MoH) on when the peak of infections is anticipated/projected to be in August/September, and therefore, it is expected that fewer cases will be reported after September 2020. In this regard, the Vice Chancellors and Principals of Universities are hereby directed to come up with reopening plans for their institutions. in the plan, ensure that provision of the minimum requirements for prevention of COVID-19 transmissions is provided for which include: a) Clean running water and soap for hand washing b) Sanitizers c) Frequent cleaning and sanitizing of often-touched objects such as doorknobs, light switches, and stair railings with disinfectant. d) Adequate space for social distancing. Universities may put signs to require staff, students, and visitors to keep social distance as well as hand washing and personal hygiene. e) Mandatory use of masks or face shields for learners, lecturers, and non-teaching staff. Universities will ensure that all learners and staff have masks. The universities may provide these at a subsidized rare to students and staff. and f) Thermo guns for monitoring body temperatures of all persons entering the institution and specific public spaces on campus. Institutions whose facilities are still currently being used for isolation, quarantine or as treatment centers, are supposed to liaise with Ministry of Health (MoH) to cease any further admissions and to transfer patients to MoH facilities. Once the patients are transferred, you must ensure that the facilities used during treatment are properly fumigated for safe use as per the guidelines from the Ministry of Health. Universities are to ensure that contingency plans are provided for those students who might need assistance to get back to their institution s as well as those who might not be able to get back. Universities should therefore make sure that students and staff who might not be able to get back are facilitated to continue with teaching and learning remotely. Principle Secretary

SOURCE: Excerpt from Ministry of Education, Republic of Kenya, State Department for University Education and Research, June 17, 2020.

place where Maasai warriors share the plains with hunting lions, a place of mighty herds and timeless cycles of life, death and regeneration" (quoted in Wahinya, 2009). Tourists missed the wildebeest migration spectacle, dubbed the Great Wildebeest Migration which occurs from July to October. It is an event in which the wildebeest cross the River Mara from the Serengeti in a region that is also ecologically vital for national water catchments and forests (Ochami, 2009, p. 8). The river is also home to thousands of crocodiles, nearly 2 million wildebeest, zebra, and other wildlife (Wahinya, 2009, p. 14). The pressure to reopen the Maasai tourist areas increased as cases of illegal logging, charcoal making, and

Table 4.3: Guidelines on working from home and utilization of staff annual leave for FY 2019/2020 during the COVID-19 pandemic.

Excerpt of circular Ref. No. OP/CAB.26/1/3A Guidelines on working from home and utilization of staff annual leave for FY 2019/2020 during the COVID-19 pandemic Reference is made to this office circulars ref. Nos. OP.CAB 39/1A and OP/CAB.26/1/3a dated March 17, 2020 and 26th March 2020, respectively on government business continuity measures in response to the COVID-19 Pandemic. It has however, been noted that majority of staff are facing challenges with the conditions of working from home. It has therefore, been found necessary to review the guidelines and direct that: All employees who have been working from home should apply in full for their unutilized leave days for the FY 2019/2020 with immediate effect. There shall be no leave days for the FY 2019/2020 with immediate effect. In State Agencies where the concept of working from home has been successfully implemented, such arrangements should continue with clear targets and deliverables as earlier guided by the previous circulars. Authorized Officers will ensure these guidelines are implemented in a manner that will enable services in their respective MDAs to continue. Please bring the contents of this circular to the attention of all officers under your purview and ensure compliance with the same.

Source: The Presidency, Executive Office of the President/Head of Public Service Nairobi, Kenya.

bushmeat poaching increased. Staff that managed the tourist resources were laid off, perhaps fueling insecurity and an increase in poaching and bushmeat hunting (Brown, 2020). The lockdowns took toll on poor communities which eked out food resources from the forests (Brown, 2020). While the Office of the President had in March sent a letter (referenced OP.PA1/94 A dated March 30, 2020) to all Regional Commissioners directing them to develop plans for food distribution for possible 21 days lockdown, the exercise of supplying food rations to subdistricts failed its mission. The State Department for Devolution lacked the will and technical capacity to go beyond symbolic gestures. There was little evidence that the exercise was based on data-driven analysis. The state was also skeptical of individuals providing assistance to the poor because of the tendency to view such activities as political gestures calculated to promote individual politicians.

Church as supporter of government efforts

According to various Pew Research polls, Kenyans are religious, and the church plays a particularly strong role in national affairs. In addition, weekly attendance in worship centers is very high in Kenya (Pew Research Center, 2018). The government had shut down churches and places of worship as Part of the policy to contain the spread of COVID-19. Initially, religious leaders gave support to the government, but as COVID-19 fatigue creeped in, much anxiety was evident from comments posted on social media on the need for churches to be opened. Church leaders, including those in areas of Kajiado, Laikipia, and Narok, urged national leaders to consider the shutdown procedures. After the first series of presidential addresses urging citizens to comply with MOH directives on

Table 4.4: Guidelines for business operations during COVID-19 (excerpt).

Republic of Kenya. Ministry of Industrialization, Trade, and Enterprise Development. Guidelines for business operations during COVID-19. June 2, 2020. Cabinet Secretary
Forward by the Cabinet Secretary (pp. 5–6)
The outbreak of Novel Coronavirus leading to the COVID-19 disease, which begun in China towards end of 2019 has since become not only a global public health pandemic, but also an economic crisis of unquantifiable proportions. There is no global economy that has been left untouched by this crisis, Kenya included. Interruption of global distribution networks has affected sources of intermediate inputs in Kenya's manufacturing sector and their ability to supply both the domestic and regional markets; the horticulture sector has suffered from lack of critical inputs, imports of finished consumer products affected, stoppage of international travel, closure of hospitality facilities, among other effects. To contain the spread of the virus, Kenya, just like other global actors, responded by putting together in place containment measures including stay-at-home measures, prompting closure of some business while others scaled down operations. Most business activities slowed down while some others totally closed down. In responding to these and to ensure that distribution networks for essential goods and services did not grind to a halt, the government, through the Ministry of industrialization, trade and enterprise development, came up (with)market guidelines as well as a list of essential services. These have served the economy fairly well over the first three months of implementation of containment measures. It is however noted that no one is certain about the end of the pandemic and for this reason there is need to put in place a mechanism for gradual re-opening of the economy. The "Guidelines for living with COVID-19" are aimed at allowing business to reopen while at same time strictly observing the social distancing and other containment measures. The guidelines, though not exhaustive provide a framework within which individual operations can be safely run. The guidelines have been developed in consultation with various sector associations and business member associations. Individual businesses, companies, office premises, etc., are expected to derive Standard Operating Procedures from the guidelines for the smooth operation of their enterprises. Each actor is expected to strictly observe the implementation of the SOPs for which they will take responsibility. The guidelines and SOPs will be subject to regular updates to be issued from time to time by the Ministry of Health.
Cabinet Secretary (p. 5)
KAM, KNCCI, KEPSA, PERAK, Federation of Jua Kali Association of Kenya, Kenya National Hawkers association (Nairobi County) among others collaborated.

Source: Republic of Kenya. Ministry of Industrialization, Trade, and Enterprise Development. Guidelines for Business, Nairobi, Kenya.

combatting COVID-19, church leaders offered support to the government plans. The support came through different forms, including pulpit declarations and prayers. In addition, there were pastoral letters.

Pastoral letters

The Most Reverend Dr. Jackson ole Sapit, Archbishop of the Anglican Church of Kenya, who hailed from the Maasai community, issued pastoral letters concerning COVID-19. The letter dated May 17, 2020, released during the shutdown activities was addressed to the nation and community leaders in all of Maasai districts where the Anglican Church was strong. The letter's Preamble stated in part:

Table 4.5: Guidelines for business operations during COVID-19 for different business spaces.

Republic of Kenya. June 2, 2020. Guidelines for business operations during COVID-19 for different business spaces Introduction (pp. 7–8) Introduction COVID-19 being a very new disease still not very well understood. Major modes of transmission of COVID-19 have been found to be through respiratory droplets and contact from an infected person. Droplets from an infected person when inhaled by a healthy individual have high chances of developing COVID-19 disease. Close contact with an infected person or touching surfaces and items contaminated with the coronavirus then touching may also spread the disease. The Ministry of Industrialization, Trade and Enterprise Development (MoITED) Situation room has produced these guidelines to offer direction to different players on how to minimize the risk of infection with Corona Virus leading to COVID-19 disease. These guidelines are meant to help businesses operate safely during COVID-19 pandemic and reduce the rate of COVID-19 spread across different workspaces. Measures proposed herein will entail ensuring: (i) Designation of staff or teams responsible for in-facility enforcement of compliance with the guidelines (committee or single staff depending on the establishment) (ii) Social distancing amongst members of staff and as well as customers/clients; maintain a physical distance of at least 2 m (6 ft). (iii) Regular hand washing using running water and soap. This is a sure way of minimizing many other communicable diseases besides COVID-19. In the absence of handwashing facilities, use alcohol-based hand sanitizers. Use of gloves does not replace hand washing. (iv) Disinfection of surfaces such as tables, chairs, counters, tools and other equipment regularly. (v) Donning of masks whenever in public places including workplaces. (vi) No handshaking/hugging as a means of exchanging greetings (vii) Proper management of waste arising Personal Protective Equipment (PPE) such as used masks and gloves. These guidelines provide the framework against which business entities/will develop SOPs to suit their individual business premises in order to ensure compliance with these measures to cut the transmission and prevent the spread of coronavirus. The sections included in these guidelines are: Jua Kali, itinerant traders, waste collectors, and aggregators, grooming sector, restaurants, wholesale and retail transport and logistics; air, rail, and ground terminals; e-commerce, cyber cafes, warehouses, extractives, construction sector, agriculture, bars and gyms. The guidelines may need review and update from time to time as new evidence emerges about COVID-19. Responsibility Business owners/managers are expected to adhere to these guidelines and put measure sin place to ensure that each and every worker and client comply. The SOPs will be implemented together with, and in addition to the various Coronavirus containment measure pronounced by the Ministry of Health from time to time. The responsibility of implementing and ensuring compliance rests with business owners(s)/managers/supervisor/contractor or foreman. The name and contact of the responsible person shall be deposited with the local Public Health/MoH representative and the nearest local security agency for prompt action. In case businesses have any questions or feedback regarding SOPs related to their spaces they should contact the moited situation room through our 24-h hotline and email. Principal Secretary Trade and Enterprise Development Principal Secretary, Industrialization

Source: The Ministry of Industrialization, Trade and Enterprise Development (MoITED) Situation Room, Nairobi, Kenya.

Preamble

> *Fellow Kenyans, we continue sending you greetings and good will messages as we seek ways and means for coping with the COVID-19 and other National challenges, including the unprecedented floods and locust invasion. The COVID-19 pandemic has greatly disrupted the way we work and live. We empathize and condole with families that have lost their dear ones through these two challenges. We also pray for those whom the floods have displaced; whose property has been destroyed. We pray for resilience and restoration. We take this opportunity to encourage you to take heart. The Apostle Paul reminds us that indeed, the whole creation has been groaning as in the pains of childbirth right up to the present time, but we can still have hope in the God who cares. God is with us in this journey and will not leave us alone. Take courage (Ole Sapit, 2020).*

In principle, the church supported the MOH in its focus on preventing the spread of COVID-19. The church supported the government's withdrawal of charges against those subjected to isolation and/or quarantined and also supported the COVID-19 prevention and treatment programs and the COVID-19 emergency fund committee for raising money to buy personal protection equipment. The church also called for accountability and transparency in use of public funds.

The church outlined its role in managing the COVID-19 pandemic. It emphasized its strategic role in praying for health workers, and direct support of faith-based health facilities. These, the church added would be important in helping the nation's efforts of taking care of the afflicted. The church also urged the government to resolve labor disputes in county health-care facilities. In addition, the church articulated its role in helping members heal from collective stigma of COVID-19 stigma of those who had been exposed, quarantined, or treated from the virus. A very important reminder in the pastoral letter was the request that people who had the disease be addressed appropriately with compassion and not as "cases" or victims. The church demanded an end to stigmatization of those who suffered from COVID-19 at their workplaces and in society.

President Uhuru reacted by announcing the formation of an Interfaith Council to study the matter and offer solutions to safely opening up church and other places of worship (Table 4.6). The Interfaith Council released its recommendations, which were immediately approved by government officials. A key feature of the council's recommendation was that a phased reopening would commence on July 14, 2020 (Table 4.7).

The Government of Kenya endorsed most of the recommendations and released an official document, Interfaith Council on the National Response to the Coronavirus Pandemic: Guidelines of in-person Congregational Worship signed by the Cabinet Secretary Ministry of Interior and Coordination of National Government, Dr. Fred Matiang'i, Cabinet Secretary, Ministry of Health, Sen. Mutahi Kagwe, and Archbishop Anthony Muheria, Interfaith Council.

Table 4.6: Appointment of Interfaith council on the National response to Corona pandemic.

Public annoucement in Kenya Gazette
Special Issue The Kenya Gazette Republic of Kenya, Nairobi, June 12, 2020. Vol. CXXII-No. 113 Gazette Notice No. 3976 Interfaith council on the National response to the Corona Virus pandemic appointment Recitals Acknowledging the supremacy of the Almighty god of all creation: Cognisant of the solemn place of religion within our god-Fearing Nation: Affirmation the contribution of Places of Worship to the common good of our nation—as places of solace, comfort, stability and dignity; and noting the greater need for them as Kenya combats the uncertainties of the COVID-19 Pandemic. Conscious that our Nation and the whole world shall have to contend with COVID-19 until such as time that a vaccine is developed. Honouring the sacrifices made by kenyans of all faiths; and noting their enormous patience and forbearance in being unable to exercise their Religious Rights in the normal way; and Recognizing the need to develop protocols and that will guide the re-opening of places of Worship for in-person congregational gatherings in a safe, secure and sustainable manner. Now therefore, and pursuant to the Presidential Directive issued during the 8th State Address on the Coronavirus Pandemic delivered on June 6, 2020; it is notified for the general information of the public that the Cabinet Secretary for Interior and Co-ordination of National Government and the Cabinet Secretary for Health do jointly appoint The Interfaith council, consisting of: Chairperson Most (Rev.) Archbishop Anthony MuheriaMembers: Rt. (Rev.) Bishop John Obala Rt. (Rev.) Bishop Joseph Obayi Rt. (Rev.) Moderator Julius Mwamba Rev. (Can.) Chris Kamau Kinyanjui Rev. (Can.) Rosemary Mbogo Pastor (Dr.) Samuel Makori Al Hajj Hassan Ole Naado Sheikh Sukyan Hassan Omar Sheikh Abdulatif Abdulkarim Rev. (Fr.) Joseph Mutie Bishop (Dr.) David Oginde Rev. Connie Kivuti Sujata Kotamraju Rt. Rev. (Dr.) Samuel Thiongo Wmangi Sheikh Ali Saidi Samojah Joint Secretaries Kepha Ombacho (Dr.). Francis Kuria Kagema (Dr.) Paul Famba The Terms of Reference of the Interfaith Council are to— Develop stringent protocols for the phased re-opening of the Places of Worship Develop guidelines towards the gradual and progressive resumption of in-person congregational worship that adheres to physical and social distancing guidelines. Develop protocols for the celebration of weddings and other religious ceremonies in the places of worship in ways compliant with physical and social distancing guidelines. Develop protocols for religious funerals and gravesides/crematoria last rites and send off; Develop a program for Public sensitization and capacity building of all religious leaders and clergy on the protocols Develop a prioritized implementation matrix, clearly stating the dates for re-opening and the numbers—capping applicable and Conduct any other matter ancillary to or in furtherance of any of the foregoing Terms of Reference. (1) Maasai In Kenya & II Maasai in Tanzania. (2) The Council shall be answerable to His Excellency the President of the Republic of Kenya, through the National and County—Governments Co-ordinating summit. (3) The Council shall regulate its own procedure. (4) In performance of its functions, the Council may (a) solicit, receive and consider any written memoranda/information from the public and (b) conduct any other actions or activities required for the effective discharge of the mandate. (5) The Council may co-opt any other persons as may be required to assist in the discharge of its functions. (6) The mandate of the Council shall lapse on December 31, 2020; but may be extended at the discretion of His Excellency the President. 7. The Secretariat of the council shall be located at the Ministry of Interior and Co-ordination of National Government. Dated the June 11, 2020. Fred O. Matiang'i Cabinet Secretary for Interior and Coordination of National Government Mutahi Kagwe Cabinet Secretary for Health

Source: Republic of Kenya, The Kenya Gazette, Nairobi, June 12, 2020.

Table 4.7: Recommendations of interfaith council—highlights.

Recommendations of interfaith council: Press statement
Press statement interfaith council on the National response to Corona pandemic Dear brothers and sisters, dear religious leaders of all faith traditions, dear Kenyans. When the Corona pandemic hit our country Kenya, we came together as religious leaders from all faith traditions invited by our President Uhuru Kenyatta, and raised our hearts to god in prayer, a king for his protection and wisdom so as to know how to face up this threat to our people, and to our Nation. You will recall that the entire Nation came to a standstill as we received the blessings to courageously face this new enemy, that had already devastated other Nations. In order to contain its spread, containment measures were instituted. included the closing of all gatherings, including public worship. This was a big blow to all of us, but our kenyans have faithfully adhered to this requirement, in spite of our great desire to come together in our places of worship. Yesterday the July 6, 2020, President Uhuru Kenyatta, released the norms going forward, in which many lockdowns restrictions were lifted and the responsibility given to institutions and individuals, in order to live up to the Indications and Measures given laid down by the Ministry of Health, to contain the Covid-19 and reduce the spread of the infection. We thank the President who appointed this interfaith Council, in order to prepare guidelines and assist all faiths, open the places of Public worship without endangering the congregants or favoring the spread of the virus. We have worked hard in order to prepare suitable guidelines that can apply across the board to all faith traditions. We did so, first by consulting widely with a very wide base of Religious leaders in eight (8) meetings we held on the June 24, 2020, and collecting suggestions submitted earlier in other regional meetings. Besides listening to the views of how we can reassure safety in our congregational gatherings, we also sought the partnership and the leadership of all Religious leaders, in encouraging and pleading with our Congregants, and indeed all Kenyans, to fully comply with the Ministry of Health directives, as then can we keep our dear sisters and brothers, and especially the elderly safe. Yesterday July 6, 2020, we presented to His Excellency the President, the guidelines for a phased reopening of Public Worship, together with the interim report of the interfaith Council. The President in his statement referred in part to these Guidelines. Further, he has asked this interfaith Council to take up the task of sensitizing and publicizing these guidelines, in order to allow resumption of Public Worship not later than three weeks from yesterday. We therefore wish to share with the Kenyans, the plans that we propose in order to resume public worship, within this framework. We are advising a Pre-opening state, where we will open the places of Public Worship in the next few days, while monitoring for the full phased reopening. This is in order to prepare our congregants for the Public Worship in the later phases. General Re-opening of Places of Public Worship The resumption of public worship will be as per the following guidelines: There must be sufficient stations of handwashing with soap or sanitization with Alcohol based sanitizer, so that all congregants sanitize their hands before and after the gathering. That all congregants must wear proper face masks over their mouth and nose throughout the congregational worship. The social distance of 1.5 m between all congregants must be strictly observed at all times. That all the maximum number of congregants should not exceed 100 persons. That we emphasize care and protection of the most vulnerable among the congregants, due to age or other ailments. Consequently, children under the age of 13 years and persons over the age of 58 are encouraged to keep at home in the first phases of re-opening. The time of duration of the congregational worship should not exceed 60 min. Other general guidelines are also given on mitigating within the worship any other avenues of possible infection. The regular cleaning and disinfection of the places of worship care in avoiding all unnecessary contact within the worship care in sanitizing the shared items within the worship special sanitization measures for all those presiding in the public worship Duration of the congregational gathering to be minimized at the initial phases Where possible, thermal guns be used to detect those who may have indicators of illness of

Continued

Table 4.7: Recommendations of interfaith council—highlights.—cont'd

Recommendations of interfaith council: Press statement
COVID-19 collections and other activities that involve passing around items be rethought to avoid contact. The re-opening of places of Worship will follow four Phases, which will depend on the situation of the Infection trend and spread, that is generally indicated by the daily positive cases, and the levels of compliance in our congregational gatherings. After due consideration, the interfaith Council now advices, that the places of public worship may be—reopened in the next 7 days from Tuesday July 14, 2020. We advise that all the congregations should prepare themselves adequately for this Phase. Planned sensitization of the interfaith Council we have planned together with a number of other religious leaders to visit the counties to share and get feedback of these guidelines in order to accommodate new ideas, and reassure all religious leaders are well versed with them We in the interfaith council believe that we as Religious leaders, now have to more effectively fight the COVID-19 monster, by taking the frontline in the advocacy and spiritual persuasion of our congregants, to take up the weapons we have at hand to fight it. These are the wearing of the face masks at all times while in public, the constant washing our hands with soap and water, the strict adherence to social distancing, and keeping away from any interaction if one is sick or vulnerable, or having any symptoms of COVID-19. Self-regulation and monitoring of compliance in our guidelines we are, advising the formation of interfaith Committee at the sub count, and county levels to help oversee the compliance with the guidelines in all places of worship. We have encouraged self-regulation, and we expect responsibility of the faith leaders in ensuring this is followed for the safety of our people and the effective fighting of the pandemic. Each congregation is advised to form a COVID-19 response committee to help implement and oversee these guidelines. Conclusion: call for more prayers: Letter signed by Archbishop Anthony Muheria Chairman interfaith council July 7, 2020.

Source: Interfaith Council of the National Response to Corona Pandemic, Press Statement, Nairobi, Kenya.

Church and the argument against singular solutions

Although the religious communities from different backgrounds perceived their role as not only providing a moral campus for the society, by condemning acts like domestic violence which were reported on the rise during the lockdowns, they also understood their role as a collaborative body to help solve a national crisis. The council saw itself as fulfilling an essential service at a time of calamity and crisis. The religious members were in agreement about their role as leaders but after the release of the recommendations old schisms emerged. There was vigorous opposition from a number of church communities, which rejected the recommendations and even accused the government of using the COVID-19 crisis to impose a singular religious order in the society. There were numerous complaints raised as well. Advise that those over 58 years and above were not to attend church was particularly denounced as was preventing young children from attending Sunday School. Some pastors, including Bishop Jonah Kariuki of the Pentecostal Alliance of Kenya, stated that limiting worship time to 1 hour, "was an insult to God and should not be entertained in a country where the Constitution allows the right to worship"

(Daily Nation Team, 2020). The Interfaith plan was also contested for assuming that worship services only take place on Sundays. Some church leaders wanted churches open at times convenient to each church group or every day.

Leaders of Islamic faith too raised concerns over the Mosque reopening plans claiming that reopening would lead to divisions because it limited the numbers of people who would be allowed to attend mosques. For example, leaders of the largest Mosque in Nairobi, Jamia, declined to reopen worship services because of the logistics of managing crowds of over 5,000, which is the sitting capacity of the mosque. The regulations required only 100 persons to be allowed in the pews but making arrangements for different sitting shifts was considered impractical. If evidence of the first reopening day on Friday, July 17, 2020, is a signal of community reactions to government policy, then the directions did not work in predominately Islamic regions like Mandera where social distancing in mosques a challenge was (Daily Nation Team, 2020).

On Sunday, July 19, 2020, as the country reported over 12,750 COVID-19 infections and 225 associated deaths, worshippers streamed in various churches in Kajiado and other parts of Kenya. Evidence suggests that although attendance was low, not all congregations followed the strict guidelines ordered by the authorities. There were reports that, for the most part, police did not interfere to enforce the guidelines. Most significant, wearing masks and social distancing were not fully implemented as some churches preferred to "trust in God."

One important point to note was that in the generally affluent suburbs of Nairobi and the larger cities of Nakuru, Kisumu, Eldoret, and Mombasa, where internet services were available, church viewership in online formats remained popular. The latter echoed events happening in the United States where the churches had adopted similar reopening strategies (Bickerton, 2020). Thus, overall, the emergence of Zoom and online worship services became a norm during the global COVID-19 pandemic.

The so-called third wave of COVID-19 in Kenya which commenced in early 2021 saw a more frustrated response from the church community. Many church bishops and their congregations had now been directly affected by COVID-19. And unlike the first cautious response, the Interfaith Council while pledging to continue to comply with public health mitigation practices took to urge members of their faiths to accept vaccines. The leader of the Interfaith Group issued a press release to call support of government curfews and restrictions of worship in Kajiado and other areas where COVID-19 infections were high (Muheria, 2021).

Communication problems and reactions

Kenya took a highly centralized approach to messaging on COVID-19 pandemic. The Minister for Health, Mutahi Kagwe was on the news stations on daily basis, mainly providing updates on numbers of cases reported in Kenya. His briefings became too monotonous and sometimes devoid of analysis. After all, he was not a medical expert and only told the public what was in the interest of the political establishment. Critics argued that Cabinet Secretary should have given more time for technical staff, including the Health Director General Dr. Patrick Amoth to be the face of the coronavirus response (Tororey, 2020). Many argued that communication coming from trained medical physicians would be more inspiring and probably have better sense of compassion for patients and their relatives and friends.

There were many unanswered questions, especially with reference to accountability of funds available for managing the COVID-19 crisis (Tororey, 2020). The MOH gave no proper response to how it had used funds allocated by the Parliamentary Health Committee for combatting COVID-19. Nor did it offer an inclusive plan for spending donor funds provided through donor Rapid Credit Facility to help fight the economic effects of COVID-19 (IMF, 2020). COVID-19 provided an opportunity for the government to provide support for marginalized communities like the Maasai that had been underserved.

Although the list of public grievances was long, one that was relevant to public health was the reported theft of equipment donated by billionaire Jack Ma and the Alibaba Foundation. They had donated 1,000,000 face masks and 20,000 testing kits. These had been in short supply and there were indications that officials from the MOH colluded with unscrupulous local business interests to steal the consignment from China (Wasuna & Oketch, 2020).

There was also uncertainty with regard to numbers of the ruling elite who may have contracted the disease and issues of secret airlifting of individuals in and out of the country for all manner of reasons. Such secrecy is not uncommon in Kenya's long history of weak accountability governance structures. Much of the reported failure in effective communication was attributed to misinformation and the problems associated with overbearing social media posts. These posts, which came from both international and local content providers, were widely shared. Initially, many on WhatsApp, Tik Tok, Instagram, YouTube, Twitter, and Facebook were messages about the COVID-19 outbreak in Italy, China, Spain, and Great Britain. Later, as the disease moved into Africa, clips became more home-based and expressive of the unfolding crisis. Government officials and private citizens also used social media to share stories about COVID-19, its causes, manifestations, and dangers and mitigation. There were also government and religious messages targeted toward different audiences.

In a few county governments, including Nakuru where a sizable number of Maasai reside, there were efforts to increase communication about COVID-19 through nonverbal approaches. Although the Disability Act of 2003 required field medical staff to be available at local hospitals and interpret medical terms to the deaf, there had been hardly much push in terms of recruitment of those trained as interpreters. The County Assembly was among the first ones to pass a motion to compel the county to recruit sign language interpreters to be assigned duties at all county hospitals. This effort was meant to cater for deaf and other disadvantaged individuals who were not benefitting from the numerous messaging on the COVID-19 pandemic. The biggest challenge in the early months were conspiracy theories especially those linked to the eugenics movements and alleged plans to depopulate Africa. There were also numerous reposts about vaccines and solutions to the problem. Catholic apologist Gary Michael Voris, and his The Vortex, with many other Church organizations brought to public attention the controversial vaccine mania, supported by Bill and Melinda Gates Foundation. They shared video clips of Sara Cunial, a controversial Italian member of parliament calling for the arrest of Bill Gates on charges of crime against humanity for promoting the use of vaccines as a means of reducing population growth in poor countries and neighborhoods (Voris, 2020).

Cunial, a skeptic, referred to persons funding vaccines as a "vaccine criminal(s)" (Voris, 2020). In India, his foundation funded polio vaccines that led to at least 50 children being paralyzed. His trial Human Papiloma Virus vaccines was conducted on mostly tribal girls in Gujarat and Andhra Pradesh. The *Indian Journal of Medical Ethics* among many periodicals questioned the ethical merit of these trials. These trials resulted in deaths of girls from the Scheduled tribes, including Siddi who had not provided sufficient consent to allow the trials (Ray, 2010; Sharma, 2010). And in Busia County, Kenya, over 20 children were paralyzed for taking antimalaria vaccine injections (Wakhisi & Inyanji, 2020).

There were social media campaigns against testing vaccines on African populations. The Nigerian authorities turned down money to allow COVID-19 vaccines to be tried on their youth. Money from the Bill and Melinda Gates, which also supports WHO vaccination activities, was also subjected to much criticism on African medial platforms. Stories from the media especially covered WHO guidelines on managing COVID-19 and promoted the continuation of vaccinations in African countries. Country-specific clip on managing the pandemic was also prominently featured some of which showed images of Maasai leaders demonstrating how to use face masks and washing hands with soap.

National radio stations also joined in the campaigns by airing programs in Maasai language with messages on the ways to combat the coronavirus. Public Health Modeling behaviors, by leaders of counties and including the Governors in Kajiado who wore a mask at public functions, encouraged the public education campaigns. Public officials from the MOH, who led the efforts to combat COVID-19, prepared bulletins circulated by local chiefs to inform Maasai to wear masks as mandated by government orders. Kenya's Minister for Health made it mandatory for individuals across the nation to wear face masks while in public spaces. Not wearing a mask was declared an offense and those arrested would be fined KSh. 20,000 (about US$ 200) or a jail term of 6 months. The government in Kenya ramped up efforts to force citizens to wear masks. The media and the Civil Society Reference Group reported cases of human rights abuses and use of excessive police force even among Indigenous people (Gathara, 2020; Olingo, 2020). As in other countries, the mask quickly became the fashion item of the COVID-19 era. And on Maasai lands, the face masks frequently matched with their traditional red-colored shukas (Latoti, 2020). The government of Kenya also set up isolation units and quarantine centers for those who had traveled into the counties. In Kajiado and Samburu, where Maasai communities live in large numbers, there was serious concern about cases being brought from outside. Kajiado, being near Nairobi city, which was a hot spot, was of particular concern to county officials. The MOH issued guidelines that the county of Kajiado relayed to members of the Maasai community.

Kajiado was also a problem area because it is a border county. Kajiado shares border with the Republic of Tanzania, where Maasai also reside. The main concern with COVID-19 management strategies in the region is that they were not harmonized. This is not unexpected because over the years, the two countries frequently deprioritize Maasai people and instead act in their own national interest. Magufuli's administration has on occasion confiscated thousands of cattle from Kenyan Maasai that had crossed from Kenya into Tanzania for grazing pasture, disregarding the cultural unity of the Indigenous communities.

As we shall describe later, Tanzania and Kenya had different approaches to managing COVID-19. These differences affected cross-border movements of local Maasai. The border town of Namanga witnessed increased security operations as authorities from either side of the boundary disagreed on testing and other mitigation approaches. Normally the border was eventually temporarily closed as the two countries worked out ways of implementing agreeable mitigation plans. Much of the disputes were over truck drivers, many of whom raised issues with COVID-19 testing protocols. As Kenya's health officials resorted to mass testing, their Tanzania counterparts protested, often citing faulty testing kits and long delays in transmitting results. Health authorities threatened to block entry into Kenya, if one failed to produce a clearance certificate and/or tested positive for COVID-19 (Kabale & Ngotho, 2020).

The governor of Kajiado, Joseph Ole Lenku, who is a member of the Maasai tribe, supported the decision (People Daily, 2020). Going by Kenya's epidemiological history, truck drivers crossing in and out of neighboring countries have had among the highest rates of infectious diseases. The tension prompted some media outlets to editorialize the quagmire in economically amicable ways. The Star called for issuance of COVID-19 passports to be issued to truck drivers (Star, 2020).

There is need to observe that the differences in approaches taken by the two countries are deeply rooted in national political cultures, which have recalibrated to a small degree the world views of the Maasai subcultures. But the core traditional values of the Maasai have remained intact regardless of what side of the border they are in. Therefore, at the family levels, norms about managing diseases remained remarkably unchanged.

Community reacts on social media

Furthermore, church leaders in Kajiado encouraged wearing of protective face cover. These were also distributed to poor members of the Maasai community. We tend to think of musicians as entertainers but, during COVID-19 pandemic, it was common practice for singers in choirs and musicians to take the role of messengers of ideas to prevent the spread of the disease (Otenyo, 2020). Among Maasai artists, similar messages on the need for sanitizing (Enkishomi osiligi), wearing masks, and social distancing were conveyed in musical art forms and broadcast in various forms of digital media. Content analysis of the many messages reflect patterns elsewhere.

On a video tape by NOREKS Inspiration TV, Events available on Facebook, and Other outlets, commentators urge Maasai people to play a part to end the spread of the coronavirus. The messages were in Maa language. The words Maibooi Corona (in Kiswahili—Komesha Corona, English—stop Corona) were echoed several times. Another script tape ends with the words Ekiyooki Aamirisho (We Shall Overcome).

Role of Pastors and Religious Leaders in Education on Prevention of COVID-19 was also evident. For example, Pastor Eric Ole Muntet used Maasai proverbs to urge his community to be careful about Coronavirus. Many of his sermons to congregations at Enkishomi Osiligi Maasai Mara Church were distributed on WhatsApp and other online platforms.

Fear and anxiety

The concerns over the impact of COVID-19 on societal well-being were amplified through reported cases of misunderstanding of the disease, especially concerning those who recovered. In rural areas, including in Maasai areas, there was much stigma associated with COVID-19. Those who were in contact with persons known to have been exposed were at times harassed and shunned. The government's efforts to quarantine those arriving in the county and stories about the dangers of the disease raised much anxiety in communities, often making it hard to fight the pandemic (Cherono, 2020). Leaders in the Maasai community expressed fear that since Nairobi was the epicenter of the COVID-19 crisis, it was incumbent for the national government to support the Maasai in the neighboring Kajiado district. Leaders like Joseph ole Lenku, governor, and Senator Mary Seneta requested for supplies of masks from the national government (Marindany, 2020). Other leaders also brought about the real concern that the Maasai would be starved if they continued to be under lockdown. The fear of hunger was one of the most noted unintended consequences of the COVID-19 pandemic (BBC, 2020; Marindany, 2020).

Community reactions: constitutionality of government actions

As expected, there were challenges to the government's restriction on movement of people within the country. Kenya's famous citizen legal defender, Okiya Omtatah challenged the COVID-19 restrictions and curfews, terming them unconstitutional. The lawsuits were prompted by many reported cases of police abuse of authority and Kenya's experience in constitutional practices. The government reacted by notifications concerning dangers to public health. Health Cabinet Secretary Mutahi Kagwe warned that without restrictions, there would be more cases of COVID-19 (Ogemba, 2020). As *The Washington Post* article had mentioned, the government in Kenya treated the pandemic a law-and-order issue. With this practice of turning a public health problem into one of criminal justice, the population turned to finding ways of dodging police roadblocks and other measures put in pace to combat the spread of the virus. So many stories exist of persons who moved into rural areas, including to Kajiado, Laikipia, and Narok, even though an official ban of such movements had been instituted. There were numerous documented cases where police received bribes to let go of those who had flouted some of the stay-at-home (shelter-in-place) orders.

The country's administrative system failed, in part because it drafted poor policies and was also incapable of controlling corruption. For example, Gathara (2020) reported that the government implemented ill-thought put rule such as banning the sale of alcohol unless served as part of a meal. Bars and other beer drinking spots had been closed but restaurants were reopened from 5 a.m. to 4 p.m. "It turned out that patrons would order sausages with beer as a token meal. This tactic was used to get around the country's poorly crafted laws" (Gathara, 2020).

Solar panels initiatives

Maasai women swung into action and accelerated search for solar solutions to power needs. Groups like the Women and Entrepreneurship in Renewable Energy Project (WEREP) and the Osopuko-Edonyinap Group in Kajiado County helped improve on lighting in homestead. These initiatives were meant to increate ability of local villagers to power their mobile phones for improved communication during the COVID-19 crisis. But the lack of internet connection made it near impossible to have regular access for schooling purposes during the nationwide lockdowns.

Community water projects

In spite of policy on use of Equalization Funds to support development in marginalized areas, the areas where most Maasai live have been left behind. The Equalization Fund is a constitutional facility established in 2010 to create equity in resource allocations. The Fund is allocated for improving services such as water, health, roads, and electricity in marginalized areas. One noted reaction to the COVID-19 crisis was the increase in attention to water shortages in the arid and semiarid (ASAL) regions. Maasai, Samburu, Elgeyo-Marakwet, Narok, Turkana, and Pokot regions of Kenya had the lowest investments in water and sanitation resources. Cumulatively, prior to COVID-19 pandemic, Narok, where a vast number of Maasai live, received the third largest allocation after Turkana and West Pokot Counties, respectively (Koskei, 2018).

In Kajiado, community efforts to drill bore holes were announced. The president's first address on COVID-19 was poignant. The speech directed the Nairobi Metropolitan Authority and the Ministry of Water, Sanitation, and Irrigation, to provide free water to the most vulnerable urban poor, several bore holes were dug in the slum informal areas. Athi Water Works Development Agency and other community independent water supply providers also joined in to supply water tanks. The bore holes lowered the price of water from the KSh. 20 per liter jerrycan.

The scenario in the ASAL regions was not as elaborate, partly because of the costs of installing pipes across long distances. The vulnerable urban seemed to receive more government attention than the rural dwellers. In the arid regions, there were additional costs associated with higher rates of evaporation. Arid rural areas in Maasai lands also depend on natural rain for replenishment. This means that any successful solution to water shortages was bound to be multisector approaches. The problem was that reforestation faced grave challenges due to lack of political will (Jamah, 2009).

Fear and fake products in the community

Kenya has a particularly dark history of bureaucratic corruption and public mistrust of the ruling elite. Rumors about cures of COVID-19 were treated with a grain of salt. This is because, in the past, Dr. Arthur Obel, a Kenyan researcher working with members of the ruling elite, claimed to have discovered two drugs that cured HIV/AIDS. These drugs, which predate the antiviral drugs, were given the commercial names, Kemron and Pearl Omega but were not effective. The government provided facilities for the manufacture of these drugs in a scandal that involved officials with political connections to the country's ruling class.

The issue of false promise to cure COVID-19 gained prominence in the Kenyan media after the Kenya Medical Research Institute (KEMRI); the research body that had previously worked on HIV/AIDS announced that it was testing herbal medicines to treat COVID-19. KEMRI researchers had discussed the possibility of using herbs dubbed Zedupex, which treat herpes simplex, a common viral disease. Kenyans argued that herbs have been known to work in advanced countries and there was no reason for Africans to doubt their own remedies. Moreover, the oldest malaria treatment drug, quinine, which is chemically close to the Chloroquine or Hydroxychloroquine, the drug that President Trump touted, had for many years been extracted from bark of trees grown in Kenya.

Elders among Maasai people and other communities also invited government officials to test their herbal concoctions.

Fake news, in the form of cures to COVID-19, were peddled around on social media platforms. In some cases, there were interventions by government authorities to post messages that denied the authenticity of the erroneous messages. The WHO and the Africa Center for Disease Control also sent rejoinders urging Africans not to use untested remedies for coronavirus. In a typical way, members of the Maasai and other Kenyan communities were also forced to react to the proliferation of fake hand sanitizers. There were reported cases of unscrupulous traders using COVID-19 pandemic to sell merchandize that was substandard and unhelpful in combatting the disease. Such products were impounded in many towns in Kenya, including in Meru and Kajiado (Muthetia, 2020). Cases of selling fake virus removal cards from China had also been reported in the neighboring country of South Sudan. These cards were meant to be worn for protection.

Creativity in the community

COVID-19 ushered in an era of creativity among Kenyans with ICT competence. In particular, those residing in Ngong and Kajiado, which are closely Nairobi city, exposed Maasai living on the outskirts to applications that linked them to those who would provide food during lockdown and curfew periods (Ooro, 2020). The App, Adopt-a-Family, provided financial help and food vouchers to many of the urban poor including people who are Maasai.

Community reactions: diets and testing challenges

While the vast majority of the reported incidence of COVID-19 in Kajiado were not associated with the Maasai people, but with foreigners living in the region, there was also indication that a sizable number of Maasai men rejected the notion that they should be tested or even eat foods rich in vitamins. A popular refrain was that consuming fruits and vegetables was less manly and more associated with diets for children and women. The MOH worked hard to change this narrative and launched campaigns to educate the masses that healthy eating was an important tool in managing COVID-19 (Oketch, 2020). The government also distributed documents from the Ministry of Agriculture, Livestock, Fisheries and Cooperation, and the United Nations Food and Agriculture Organization (FAO) which gave advice on food charts and the need for healthy snacking to prevent weight-related disorders.

Maasai men have been known, on the contrary, to prefer eating meat and not sufficient vegetables and fruits. Thus, messaging, mostly in brochures, which was distributed in Ministry of Health Guidelines on Foods for COVID-19 was remarkably unpopular among the moran age sets. Reliable testing was not available in most of rural Kenya. And when available, the results were not dependable, especially if the tests were conducted in questionable laboratories (Joseph, 2020). Perhaps, lack of testing more than any other reasons accounted for lower rates of infection within Narok and in other Maasai communities. A story relayed by Maasai elders hints at another possibility of authorities inflating numbers as a means to attract more COVID-19 mitigation resources. In Maasai neighborhoods where numbers of infection were low, a few cases were reported in Narok, where deaths of persons who had been known to have been ill for a long time were ruled to be COVID-19 contrary to expert autopsy evaluations. These stories suggest that government coronavirus rapid response teams rushed to pronounce deaths as coronavirus related just to inflate numbers. The Cabinet Secretary for Health had confirmed that the government lab at KEMRI offered more reliable facilities than the Lancet Laboratories in Nairobi (Joseph, 2020). However, two things also undermined KEMRI's work. First, it too faced financial challenges and had not solved the management problems of Kenya's state-run institutions. KEMRI was reportedly in poor financial standing and was unable to meet most of its research obligations (Oketch, 2020).

Secondly, regional branch offices were poorly staffed and not adequately resourced to handle COVID-19 testing. Third, KEMRI's labs have also, in the past, been involved in other priority areas, such as malaria vaccination trials done by foreign entities including the US Walter Reed Army Research Institute and pharmaceuticals like GlaxoSmithKline (GSK) (Otieno & Olwenya, 2009).

Although there was an example of pool testing in Rwanda, the approach was not incorporated in county hospitals. Pool testing applied an algorithm-driven process whereby samples from a group were combined in a single tube for quick testing. The assumption was that if virus detection was successful, then the entire group would be placed in quarantine. And if negative results were obtained from the pool, there would be no further action. Without a systematic plan for testing, the country response was bound to raise more verification questions. The importance of testing was also undermined by messages from some doctors who observed that individuals testing without symptoms overstretched the limited testing kits and facilities. Instead, several members of the medical profession recommended adherence to safety guidelines like social distancing, washing hands, and wearing masks in public spaces.

But pool testing was used sparingly within Kenyan settings. In Kajiado and Narok where many residents are Maasai, there were other complications noteworthy of discussion. Many Maasai people regarded the COVID-19 as another form of *homa* (flu) and would not voluntarily test. It has been said that many villagers knew there was a difference between COVID-19 and other *homa* thus they took extra care by practicing their traditional distancing measures. The Maasai have traditionally reacted to flu epidemics by adopting their own ways of social distancing. In the past, they did so by not interacting with persons from areas where flu related to bovine infections had been identified. Maasai people have developed health strategies to self-quarantine in order to protect their communities and this principle played at in Narok and parts of Kajiado, Laikipia, and Samburu as well.

Admission of ineffective measures

In his ninth State Address on the pandemic, President Kenyatta brought forth the dilemma he and his administration grappled with, that was to choose between opening the economy or continuation of lockdown policies. He reminded the country that COVID-19 was both a health crisis and an economic crisis. What was worrying was that by June 7, cases of those infected were on the rise and projections were not good. In his remarks, President Uhuru remarked that, "After much reflection, my administration opted for the health argument over the economic argument. More so because we can always revive an ailing economy, but we cannot bring back to life those who die from this pandemic" (Kenyatta, 2020a).

Kenyatta went on to announce new measures to restrict movements in selected regions of the country. He, however, allowed phased reopening of certain sectors of the economy. President Kenyatta also urged Kenyans to be each other's keepers and also issued a national call to civic responsibility. He also accepted the recommendations of the Interfaith Council and reiterated that up to 100 participants were allowed at places of worship. The government allowed air transport with the country effective July 15, 2020. International air travel was scheduled for reopening on August 1, 2020. Restrictions on social functions such as weddings and funerals were continued as in previous memos.

As the situation deteriorated, the Chairman of the Council of Governors and other regional leaders prepared for worse-case scenarios. Some governors proposed to establish mass graves and ordered for body bags. State House released another statement on July 20 (Table 4.8).

On July 20, 2020, the Executive Office of the President issued another circular noting that many civil servants had not complied with provisions of other circulars concerning mitigation against the spread of COVID-19. In particular, the Office of the President referenced circulars that included those of March 17, March 26, April 23, April 29, among others, that offered guidelines on managing the crisis. In the July 20 memo, head of public Service Mr. Joseph Kinyua pointed out "the lack of strict observance of the issued guidance and measures meant to enhance the efficacy of the management of the disease and protect all Kenyans from its effects" (Kinyua, 2020).

He issued another set of rejoinders reminding public officials to promote social and physical distancing and take measures to reorganize their workplaces. In addition, the memo requires officials with preexisting medical conditions aged above 58 to work from home. The memo also reiterated the need for public officials to practice workplace

Table 4.8: Presidential action notification, July 2020 (excerpt).

Press Release Notification of Presidential Action (1) – 20/07/2020.
Extraordinary Summit to Review the Surge of COVID-19 Infections. His Excellency The President has on this twentieth day of July 2020 convened the Fifth Extraordinary Session of the National and County Governments Coordinating Summit, which shall be held on Friday, July 24, 2020. The session has been convened following a surge of COVID-19 infections across our Nation; with the disease having now manifested in 44 of our 47 counties. The session shall be a status conference that shall: (i) Assess the Counties state preparedness within the wider national response to the novel coronavirus pandemic and evaluate Kenya's overall national response and capacity, (ii) Review the efficacy of the containment measures in place and review the impact of the phased easing of the restrictions that were in place, and (iii) Consider the evolution of the disease and the medical models on how COVID-19 may propagate within our country over the months of August and September 2020. His excellency calls on all Kenyans to continue applying the simple but yet most effective individual and collective measures that are our Nation's best weapon against COVID-19. Kenyans should continue correctly wearing facemasks while in public spaces, apply correct face and hand hygiene at all times, and cooperate with the physical and social distancing guidelines and protocols that have been put in place to safeguard our individual and collective health and safety. Kanza Dena Mararo State House Spokesperson State House, Nairobi. Monday 20, 2020.

Source: Office of President, Press Statement, State House Nairobi.

hygiene and safety and ensure that their temperatures were checked at places of work. The memo also called for fumigation of office premises and government vehicles. Concerning working from home, the letter called for a "weekly duty rota covering minimum personnel numbers to ensure business continuity indicating the officers who will be in charge of operations" (Kinyua, 2020). An important element of the memo was a call to end stigma and discrimination of those afflicted by the disease.

a) Public Officers shall have access to adequate mental health and psychosocial support being offered by the Government;
b) Officers returning to work from COVID-19 quarantine and/or isolation shall not be discriminated or stigmatized; and
c) It is an offense to discriminate any person on grounds of actual, perceived, or suspected COVID-19 status in the workplace.

Most government ministries also issued numerous circulars giving orders on best practices for managing COVID-19. There were also circulars on scaling down of operations and services in various units. For example, the Ministry of ICT, Innovation and Youth Affairs, State Department of Broadcasting and Telecommunication circulated a letter dated July 14, 2020, in which services were reduced following reported fatality due to COVID-19 (Republic of Kenya, 2020). A government official had lost their lives prompting mass testing and fumigation and sterilization of the Ministry's Headquarter offices at Teleposta,

Postbank, and Uchumi House and other field offices (Republic of Kenya, 2020). Similar measures were repeated whenever COVID-19 cases were reported in government offices, including those at county levels (Matara, 2020).

As the death tolls from COVID-19 continued to rise, President Uhuru Kenyatta's administration released numerous other policy statements and executive addresses. The Tenth (10th) Presidential address, published on July 27, 2020, marked the president's 21st day since his previous communication (Kenyatta, 2020b). President Kenyatta's level of engagement with his country was by all measures remarkable and honest. He admitted that some hospitals had been overwhelmed and 280 Kenyans had lost their lives. As in his first address, he reiterated that Kenya was still at war. In his own words, "In war, no mercy is shown by either side. In war, survival is key. Self-preservation is the priority for all in a theatre of war" (Kenyatta, 2020b).

His address admitted that there was a surge of infections which required public attention. As of July 27, over 17,603 persons had tested positive for COVID-19 in Kenya. 44 of 47 counties had been affected and the average case Fatality Rate in the country was 1.6%, which was lower than the global average. The low deaths was a puzzle to researchers from Oxford University, Kenya Medical Research Institute (KEMRI), University of Warwick, and other institutions who hypothesized that Kenya had high rates of positive test detections, but surprisingly lower deaths attributed to COVID-19. The highest concentration was in Nairobi, Mombasa, and areas around those cities. Rates were high in Kajiado because of its proximity to Nairobi but remarkably low in Narok. The researchers speculated that Africa was probably spared massive death due to its youthful population and "some kind of immunity against severe COVID-19" (Gathura, 2020).

In the 10th address, the president ordered a nationwide curfew for another 30 days. He also banned the sale of alcohol in eateries and restaurants and ordered that all restaurants and eateries close by 7 p.m. All bars were also closed, and the police was empowered to withdraw licenses for bars breaching the order. The 11th address was published in August 2020 and contained similar administrative restriction on movement of people and reiterated the observance of public health protocols. The measures were not fully enforced and by the end of 2020, it looked like new virus strains were present and the country's infection rates were rising. Uhuru Kenyatta's administration lost ground by allowing political rallies to take place in both rural and urban centers. These rallies were a part of the process of a Building Bridges Initiative (BBI), which was meant to lower tensions as the country moved closer to the 2022 elections. It was also a program for constitutional change to expand the executive arm of government. Many of these rallies are blamed for the steady increase in COVID-19 in the urban areas. The more rural Narok area of the Maasai community faired relatively better and by early 2021, some of the county's ICU beds were empty. In addition, the Ololulunga Level 4 Hospital in Narok South reported few deaths (21 from 524 cases since June 13, 2020) from COVID-19 (Sayagie, 2021). Kajiado was one of the worst affected regions, perhaps due to its proximity to Nairobi, the epicenter of COVID-19 in the county. The government issued many circulars to try and control the spread.

By March 12, 2021, Kenya's leadership raised concerns over rising COVID-19 cases. President Kenyatta released the 14th Presidential address on the coronavirus pandemic where he observed that the virus had wounded the country and that the generic mitigation policies had established a new normal (Kenyatta, 2021). The 14th address reported that 1879 Kenyans had died from COVID-19 and acknowledged the economic devastation caused by lockdowns. The 14th address not only locked down Kajiado county and Nairobi city but extended the nationwide curfew to another 60 days. These measures were encoded in the Public Health (COVID-19) restriction of movement of persons rules, 2021, a part of the Public Health Act (Cap 242) of the laws of Kenya.

In conclusion, arguably, the government's leadership directly involved itself in defining the problem. The policies adopted were well scripted and generally effectively communicated to the Indigenous communities. By seeing images of what coronavirus did in other countries, a general mood of anxiety ensued. Maasai people and other Indigenous communities did not experience high incidence of COVID-19 cases. The highest rates were in urban centers where the "burden" of social distancing and wearing masks could not offset the poor health infrastructures or overcome the iniquities of poverty, manifest in housing and sanitation challenges. In part II, we take a look at the case of Tanzania.

Part II: The other story: Tanzania's COVID-19 policies and reactions

Colonial experience and narrative

Like their counterparts in Kenya, the Maasai in Northern Tanzania lost their vast territories to National Parks, conservation, and ecotourism enterprises (Dahir, 2018). They too were forced out of their fertile homelands and impoverished. Due to environmental degradation and forced movement into dry, less fertile savannah, much of their connection to cultural ways of life have been strained. Thousands of acres of traditional lands have been privatized and lost to commercial hunting groups and other corporate interests. One of the consequences of marginalization and forced removal is homelessness and settlement in squalor conditions in urban areas. This postcolonial experience is similar to that in Kenya.

COVID-19 enters and Tanzania's denial

In Tanzania, the first reported case of COVID-19 was reported in early March 2020. The patient arrived in Tanzania from Europe. Tanzania, like Kenya, has a weak health-care infrastructure. Soon, the virus spread throughout the country. There were reports that three members of Parliament had died from COVID-19 complications. Magufuli visited churches and asked Tanzanians not to shun away from attending church services. He compared the virus to Shetani (the Devil) suggesting that it could not survive in the body of Jesus Christ (Awami, 2020a). This declaration, however, was tantamount to denial. Although the government denied it, COVID-19 cases were informally reported by "citizen journalists" using modern technologies. In April 2016, an opposition party member of Tanzania's parliament representing Mbeya Urban constituency, Joseph Mbilinyi's comments went viral on his Twitter feed. Speaking in Kiswahili, he noted that the situation was so bad, and the government was presenting false statistics (Mbilinyi, 2020).

As reports reached his desk confirming a looming national crisis, the president resorted to his deep sense of Catholicism. And in mid-April Magufuli called for 3 days of national prayer to seek God's protection from the scourge. In early May, the Minister for Health, Ummy Mwalimu had told a gathering that there were few patients, ignoring that in the previous press briefings, over 480 cases had been announced for end of April, which was the last month that the Government of Tanzania released COVID-19 data to the WHO (Houreld & Lewis, 2020). There were claims that the Minister for Health, Ummy Mwalimu declined to publish COVID-19 case figures on May 8, 2020, and limited her contacts with international media.

In addition, the government also reneged on its pledge to join the WHO in surveillance missions around the country (Houreld and Lewis, 2020). Even after the donor community promised financial aid to support efforts to stop the spread of COVID-19, there was little cooperation from the Magufuli administration.

Reports had indicated that cases were on the rise and that deaths had occurred in Tanzania, including on the island of Zanzibar. On June 9, 2020, perhaps fearing national outrage and panic, he claimed victory to his East African nation. He urged pubs to throw parties to celebrate the end of COVID-19 (Said, 2020). In his remarks, he invoked the powers of God had caused the disease to leave the country. Health Minister, Ummy Mwalimu also invoked God's mercy and shut down most of the COVID-19 isolation health centers. The BBC quoted her saying, "we thank God that coronavirus is heading towards an end in Tanzania" (Awami, 2020a). There were 85 COVID-19 health centers that had been set aside to isolate and offer treatment for those who had been infected. By early July, there were only 11 such centers in operation (Awami, 2020b).

In mid-June, the country's Prime Minister Kassim Majaliwa told parliament that there were 66 active coronavirus cases, but provided no further details (Awami, 2020b). The president refused to release data on death because doing so would cause unnecessary panic. He continued to have rallies and attended several public functions in which he continued to encourage Tanzanians to continue living life, following what health officials advised but not blindly following rules from foreigners.

The government subsequently stopped releasing numbers of those affected by COVID-19. The exact numbers were unknown, and the government chose not to do its diligence in tracking the cases that had been reported. Because of commerce and travel, Maasai people were vulnerable, just like any other community in the region. By some accounts, Maasai cattle traders, whose merchandize was sold in urban centers, exposed their communities to coronavirus. Reported cases of black markets of all manner of commodities were also feared to be entry points for COVID-19.

Tanzania's approach to managing the coronavirus outbreak was marked with controversial decisions coming from the centralized political Chama Cha Mapinduzi (CCM) Party establishment under President Magufuli. Although there were certainly voices against Magufuli's handling of the crisis, he did not experience any serious pushback from the opposition or civil society organizations (Said, 2020). Unlike his regional neighbors, in Kenya, Rwanda, and Uganda, Magufuli refused to order nationwide lockdown and shelter-in-place policies. He ridiculed the actions taken by Kenyans and joked that it was time that Tanzanians overcharged Kenyans for food supplies. In addition, President Magufuli declined to cooperate with regional heads of states in development of a common strategy to combat COVID-19. He rejected the idea of coordination of efforts to prevent the spread of coronavirus and skipped two virtual meetings held for that purpose (Said, 2020). The May 22 meeting on coordination those efforts included Presidents Yoweri Museveni (Uganda), Uhuru Kenyatta (Kenya), Paul Kagame (Rwanda), and Salva Kiir (South Sudan) (Agutu, 2020). The late President Pierre Nkurunziza of Burundi was also absent from the meeting. His wife had been airlifted to Nairobi and treated of COVID-19. He also missed a meeting of the Southern African Development Community meeting in which President Cyril Ramaphosa of South Africa had personally invited him to attend. President Magufuli's absence from regional meetings was not new. He also turned down official invitations to the United Nation's General Assembly and to Western nations.

Prior to the emergence of the COVID-19 crisis, one of the characteristics of his style of leadership was his visibility in the media. Magufuli loved "going public" by holding well-advertised rallies in which he frequently fired or praised public officials for their public actions. The concept of "going public" was developed in Samuel Kernell's (1986) seminal work, *Going Public: New Strategies of Presidential Leadership*. This work offers some insights into the behavior of presidents who rather than bargaining with other political actors make public proclamations which undermine the ability of compromise to be reached. It is the sort of idea in which presidents act in dramatic manner due to popular pressure. In the case of Magufuli, his actions were meant to show him as fighting both imperialism and the coronavirus (Awami, 2020a).

Magufuli, on several occasions, handed money and other benefits to individual citizens who had sent him requests for assistance. Magufuli was also a regular attendee in many village-level church services and was unafraid of using the "bully pulpit" to articulate unilateral public policies. Earlier on, Magufuli had self-isolated in his hometown of Chato, Geita region, and attended church where he addressed fellow congregants on divine intervention (Said, 2020). Again, the idea of bully pulpit is borrowed from American scholars, namely Richard Neustadt (1980). Magufuli cherished publicity and was energized by public admiration and praises. He continued with this style during the entire coronavirus outbreak.

In his public COVID-19 pronouncements, Magufuli became a subject of international attention. And here is why. Tanzania's handling of the COVID-19 crisis was much different than that in its neighboring countries. Tanzania and North Korea were among the few countries that did not report cases of COVID-19. Amnesty International deemed these countries as hostile to freedom of the press. This was in spite of appeals from the Africa Centers for Disease Control, which is a body that had been established in 2017 at the behest of the African Union (AU). Tanzania's argument was that the disease was not present in their environs. It seemed odd that 188 countries reported infections and cases and not one of East Africa's most stable countries.

President Magufuli officially declared the pandemic over and urged citizens to resume work, even without wearing masks. The President's Office issued a press statement in which the government banned the wearing of face masks (See Table 4.9). On several occasions, including in a speech to Parliament on June 29, Magufuli urged Tanzanians not to wear face masks, frequently hinting those in advocacy groups including the so-called

Table 4.9: End of Coronavirus in Tanzania Memo.

End of Coronavirus in Tanzania
Jamhuri Ya Muungano Wa Tanzania Kurugenzi Ya Mawasiliano Ya Rais simu: 255—262961500/1 Nukushi: 255-262961502 Baruapepe: press@ikulu.go.tz Tovuti: Ofisi Ya Rais Ikulu, 1 Barabara Ya Julius Nyerere, Chamwino, S.L.P. 1102, 40400 Dodoma Taarifa Kwa Vyombo Vya Habari Rais wa Jamhuri ya Muungano wa Tanzania, Mhe. Dkt. John Pombe Magufuli Leo tarehe hii 24/05/2020 amepiga marufuku suala la uvaaji Barakoa kwawatanzania maana kitendo hicho kinaleta taharuki na watu kuendelea kuwa na hofu na kufunga biahara zao. Pia Mh: Dkt. John Pombe magufuli ameruhusu magari na mabasi yote kuanza kujaza watu kama zamani pamoja na mashule yote kufunguliwa. Tunamshukuru mungu tumeishinda corona tuendelee tu kunawa mikono maana ni usafi. Kwaheri Corona. Gerson Msigwa Mkurugenzi wa Mawasiliano ya Rais, IKULU Chato 24 Mei, 2020.

Source: Office of the United Republic of Tanzania, Dodoma, Press Release.

Magufuli Forum were misleading people. These groups had been distributing masks to Tanzanians. Magufuli suggested the masks were not safe and would hurt the people (Magufuli, 2020).

Journalist Ruud Elmendorp, who visited Dar es Salaam and filed his COVID-19 Diaries, argued verification of government claims was not easy or possible under the circumstance. The government refused to disclose the exact nature and spread of the pandemic in the country (Elmendorp, 2020). There was widespread fear that the government was hiding the true scale of the pandemic (Houreld & Lewis, 2020; Mwai & Giles, 2020). The U.S. Embassy, fearing that the Tanzania government's actions would endanger the international community, reacted by issuing travel alerts and claimed in various reports that hospitals in the country had already been overwhelmed. This caused the government of Tanzania, in early June, to summon the American ambassador for diplomatic scolding.

Tanzanian President Magufuli also endorsed the inhaling of steam therapy to prevent coronavirus was as confusing as the case of Donald J. Trump's "ingest bleach" remedy. Magufuli had suggested that viruses would melt when steam was inhaled. He claimed his own daughter had recovered from COVID-19 after using steam and other local remedies (Said, 2020). The alleged steam treatment was not supported by health specialists in Tanzania (Kombe, 2020). Tanzanian authorities also gave limited guidance about the virus and in some cases restricted people from going to seek medical attention, at times urging people with symptoms to treat the health challenge as simple flu. President Magufuli insisted on numerous occasions that virus had been defeated through prayers. This is even after 50 truck drivers from Tanzania tested positive by Kenyan authorities at the northern border. As mentioned earlier, this prompted authorities in Kenya to temporarily shut down the border in May 2020. Tanzania countered with its own accusations. There was a flurry of correspondence between the two countries, as exemplified in the press release below (Table 4.10).

The president refused to shut churches, mosques, and other gatherings at public spaces including at beaches and restaurants. His administration also questioned the accuracy of tests done by government laboratories. President Magufuli claimed the swabs themselves could have been tainted with the virus (Associated Press, 2020). He in dramatic terms shocked the world by saying on national TV that his administration had found out that the test kits were inaccurate and produced misleading results (Biryabarema, 2020). In his remarks, he pointed out that the tests returned positive results on random samples taken from a goat, and pawpaw fruit (Ng, 2020). These samples, including from quail, had been assigned human names and ages to trick the lab technicians. However, fact checkers ruled Magufuli's assertions as "unproven" and, perhaps, made in the context of his disdain for the country's opposition blaming him for mismanaging the coronavirus crisis (Lee, 2020). President Magufuli also on other occasions charged lab professionals for conspiring with

Table 4.10: Corona virus across Tanzania—Kenya border.

COVID-19: Tanzania accuses Kenya
The United Republic of Tanzania President's Office Regional Administration and Local Government Arusha Region. Telephone Contacts "REGCOM" Phone No. 254608/25–2289/2545870 Email address Website Upon responding, Please refer to: Office of Regional Commissioner PO Box 3050, Arusha May 20th, 2020 Press Release Following high infection rates of COVID-19 across the borders, Arusha region has decided to seriously put in place a system of collecting samples for testing from all inbound truck drivers crossing Namanga border from the neighboring country of the Republic of Kenya. All samples being collected by our health experts, are sent to our main National laboratory in Dar es Salaam for testing and subsequent release of test results. We have deliberately taken this decision in order to protect our citizens against any COVID-19 infections from neighboring country as well as making sure our general populations stand in good health as we prepare to receive tourists when high tourism seasons re-opens come this June 2020. The system for collecting Covid-19 test samples from truck drivers coming in from Kenya started on May 14, 2020, thus samples were collected on 14th, 16th, and May 18, 2020 and laboratory results are elaborated here-under, 1. Samples taken from 4 truck drivers from Kenya on May 14th, 2020, and results were released on May 16, 2020 whereby 14 drivers (11 from Kenya, 1 from Uganda and 2 from another country whose name shall be kept private in this document) all coming in from Kenyan side were positive. Thirty (30) drivers were negative. 2. On the 16th May, a total of 23 drivers from Kenya had their samples taken for testing and results were released on May 19, 2020. Results profiles were as follows: ten (10) drivers were positive for COVID-19 (All were from Kenya) and 13 were Negative. 3. On the May 18, 2020, more samples were taken from Eleven (11) drivers, all arriving from Kenya, and results will be given out immediately after release from our main National Laboratory in Dar es Salaam. In efforts to confirm reliability of Covid-19 test results. We took samples from 19 drivers from Tanzania who had been tested and declared positive by Kenya authorities border on Kenyan side. After these samples were submitted to our Main National Laboratory in Dar es Salaam, the results came back Negative for all these drivers from Tanzania. Arusha region is confident that this is a deliberate sabotage strategy designed by Kenya against our tourism industry in Arusha and Tanzania at large. Apparently, all drivers determined to be "Positive" are not allowed to either cross the border into Tanzania or enter Kenya. Arusha region will continue to take all necessary precautions to prevent any potential wave of COVID-19 infections that might cross into our country from Namanga border along neighboring Kenya. Once again, wish to express our sincere gratitude for the cooperation we are getting from our Kenyan neighbors living along Namanga border. Regional Commissioner, Arusha

Source: Republic of Tanzania, Office of Regional Commissioner, Arusha, Press Release.

imperialists to portray Tanzania in negative light (Said, 2020). However, the muddled messaging made it difficult for the country to manage the COVID-19 outbreak, which affected many areas where the Maasai live (Jerving, 2020). These, as we know, also happen to be areas of close proximity to high-end tourist zones.

Policy, limited government outreach, and public health education

Initially, the government in Tanzania took to announcing messages of mitigation against the spread of the virus. The strong party structure and influence of radio and other traditional means of communication echoed the now well-rehashed WHO template, wash hands, social distance, wear masks, and eat balanced diets. The messaging was amplified in regional district government offices including those along the border with Kenya-Arusha, Mara, Kilimanjaro, and Tanga, areas where a majority of Maasai resided. Many Maasai people learned about COVID-19 through radio, smart phones, and other community word of mouth communications. Maasai people in Tanzania have mobile telephone capabilities. And those in areas where tourism is heavy have internet surfing capabilities as well (Santos, 2010). Some of the messages came through radio stations owned by Churches and also the government. The messages that reached Maasai land included not greeting each other by hand and hand washing, but social distancing was and staying at home were not well articulated in the messaging. Blame for cultural practices showed up through concerns about contagion related to the Maasai practice of touching children's heads, as a form of greeting. Awareness of the disease in Tanzania followed similar patterns seen in Kenya: First, through local radios, mobile and smart phones, and word of mouth. The messages delivered to villages were about preventive measures to protect them from COVID-19. These included washing hands, not touching each other, and social distancing. There were also campaigns about putting a halt to various cultural activities. Some Maasai people also considered COVID-19 along with religious beliefs as was the case across the globe. Some heard of the disease but left it to God to protect them. Others saw the diseases as a punishment from God to human being who have failed to protect nature (Naingisa, 2020).

Barriers to COVID-19 risk reduction measures

Tanzania ranks poorly on the global Networked Readiness Index. It is at number 126 compared to Kenya at rank 86 (World Economic Forum, 2016). The ability of Tanzanians to enjoy broadband internet speed and leverage information on reducing developmental impacts COVID-19 was therefore constrained by poor ICT capacity. In addition, Tanzania's Communications Regulatory Authority restricted publication of pandemic-related materials that "ridiculed" the nation. The country ranked at 118 on the World Press Freedom Index making it one of the worst performers in allowing free flow of information critical of the government's handling of the COVID-19 crisis (Reporters Without Borders (RSF), 2020).

As an example, President Magufuli's government suspended a reporter for the Tanzania Daima newspaper for referring to a coronavirus patient. And TV operators from the Star Media, Multichoice Tanzania, and Azam Digital Broadcast were fined and ordered to issue apologies for reports that criticized the president refusal to order national lockdowns following a rise in number of COVID-19 cases in the country. The government ordered that content on coronavirus be published only through its approval mechanisms. As a result, information of the devastation caused by the coronavirus was largely censored. Being one of the nation's where control of the internet and media served to prevent the sharing of the challenges posed by the spread of the virus.

An additional barrier to communications and messaging was the lack of broadcasts in Maasai language. This was particularly noted in remote areas of Tanzania. Local stations tend to use Kiswahili and did not target non-Kiswahili–Maasai speakers (Naingisa, 2020).

Like everywhere, people living in Tanzania have social customs that make isolation difficult. In Tanzania, many share a strong sense of social interactions and reciprocity especially when community members share food. Likewise, ceremonies based on age-set systems brought many members into direct contact with one another. Social interactions among community members are strong and turning away visitors is frowned upon. Women who receive visitors in homes were most likely to be exposed to the pandemic. They were also among the most disadvantaged social group in both Tanzania and Kenya (Mbula, 2020). Rates of illiteracy are highest among Maasai women, who also have to face oppressive cultural practices including forced early marriages and female genital mutilation/cut (FMG/C). During the pandemic shutdowns, high rates of teen pregnancies were reported in the Narok area. 5500 teens were confirmed to have become pregnant after schools closed down in March because of COVID-19 (Sayagie, 2020b). It will be important in the future for people to understand more fully what this means and how it impacts the lives of these teens.

Communication barriers

One of the biggest challenges to effective communication was—there were no or limited public health and literacy broadcasts in Maa language. Only Kiswahili broadcasts were available in much of Maasai land (Naingisa, 2020). As was noted, without media that targets Maasai communities, access to information and solutions to the COVID-19 crisis were bound to be challenged or ineffective. Only Maasai who have gone to school spoke the national language of Kiswahili.

Misinformation and fear of loss of life and indigeneity

Just like elsewhere, COVID-19 came into Tanzania with lots of misinformation and propaganda, and fake news (Associated Press, 2020). At the official level, Tanzania's MOH worked collaboratively with WHO officials in developing protocols to limit the spread of the disease throughout the vast country.

Community reactions

Francis Shomet Naingisa, Maasai elder in Tanzania, observed that the Maasai community in Tanzania was equally endangered by the COVID-19 pandemic. The Maasai community in Tanzania was affected much like their counterparts in Kenya. They too suffered from loss of meat markets. Many Maasai sold livestock across the border as well. Monies from selling livestock product are used in purchasing manufactured commodities including bread, soap, processed oils, sugar, and teas. With restrictions in travel, there were remarkable disruptions to the intra-Maasai trade. However, some black markets did emerge, and limited commerce continued.

Loss and temporary suspension of cultural traditions

Like in other Indigenous communities, the Maasai of Tanzania too are communal and implementing some of the recommendations from the WHO and MOH were difficult to execute. For example, Maasai tend to be very welcoming of visitors, and many found this difficult to halt. Naingisa (2020) reported that this placed a huge burden on Maasai women who tend to house chores and serve visitors with food or drinks.

Herbal medicines and miti shamba: why not?

WHO data show that a vast majority of African governments actively promote the use of herbal medicines (Miti shamba), in addition to western medical pharmaceuticals (WHO, 2020). Tanzania was one of the first countries to publicly support the use of herbs-COVID-Organics (from Artemisia plant) developed in Madagascar's Malagasy Institute of Applied Research to treat COVID-19. President Magufuli wrote a letter requesting for the herbs from Madagascar. Aljazeera (2020) reported that a consignment of herbal concoctions from Madagascar arrived in Tanzania in May. This was against the wishes of WHO officials, who doubted the efficacy of the herbal medication. The WHO rejection was countered by activists who claimed the WHO and other western institutions created structures that reinforced the belief that African solutions were inferior to those prescribed by practitioners from the west. In a sense, a reinforcement of racism and undermining of institutions run by communities of color.

There were also other variants of herbal medicines touted by authorities in Tanzania. A variant dubbed "Bay Caro" was launched in Kibaha District by government officials. Government officials claimed it could cure COVID-19. The reliance on herbal medicines also was a key reaction within Africa's ruling elite classes. And this was extended to rural communities that include the Maasai villagers in Northern Tanzania. They, too, utilized herbs for both treatment and prevention of COVID-19 in their districts.

One may summarize the political management of the COVID-19 crisis in Tanzania as exemplification of expediency. 2020 being an election year in Tanzania, not surprisingly produced a different official response to COVID-19 pandemic. President John Pombe Magufuli took the fight against the pandemic to a different level than his counterpart in the East African regions. He, for the most part, was a COVID-19 denier. Fueled by his society's conformism and apathy, the outcomes were mostly unknown.

Conclusion

Although the Maasai in Tanzania and Kenya were both affected by the rise of COVID-19 cases along their common border, the actual cases within rural Indigenous communities were comparatively low. This is largely attributed to their use of herbal medicines and adherence to common sense public health mitigation measures. There was also the repeated claim that Africa, as whole, has a small population of persons 65 and older, who were at most risk from dying from the disease. The lower mortality rates in rural Maasai communities were also a function of the ability of community leaders to enforce their own control policies. In contrast, poor urban-based Indigenous persons faced the same risk factors as other slum dwellers and faced enormous economic stress, leaving them to severe health vulnerabilities. Urban health-care systems were definitely overstretched and incapable of handling increased patient numbers.

References

AFP. (May 19, 2020). *Virus fight just one front against epidemics in Africa.* https://medicalxpress.com/news/2020-05-virus-front-epidemics-africa.html.

Agutu, N. (May 13, 2020). *Magufuli skips EAC meet even as TZ virus cases rise* (p. 2). The Star.

Aljazeera. (May 8, 2020). *Tanzania gets Madagascar's anti-coronavirus drink disputed by WHO.* ALJAZEERA. https://www.aljazeera.com.

Apollo, S. (August 24, 2020). *Report paints grim picture of police service in fight against COVID-19.* Daily Nation. https://www.nation.co.ke.

Arimi, J. (February 22, 2011). *The Maasai paradox.* UCD, Dublin, Ireland: The Institute of Food and Health. www.arimifoods.com.

Aron, M. (July 16, 2009). *Study puts M-Pesa ahead in local money transfers* (p. 30). The Standard.

Associated Press. (May 22, 2020). *Tanzania says virus defeated through prayer, but fears grow.* http://www.voanews.com.

Awami, S. (June 18, 2020a). *Tanzania's John Magufuli-the man vowing to defeat coronavirus and imperialism.* BBC News. https://www.bbc.com/news/world-africa-52983563.

Awami, S. (July 6, 2020b). *Tanzania shuts centers handling COVID-19 patients.* BBC. https://www.bbc.com/news/topics/cjnwl8q4qdrt/tanzania.

Bickerton, T. J. (2020). *New York annual conference COVID 19 guidelines for church reopening.* The United Methodist Church, (Spring). https://www.nyac.com/files/pdfs/nyac-covid-19-guidelines-for-church-reopening.pdf.

Biryabarema, E. (May 3, 2020). *President queries Tanzania coronavirus kits after goat test.* Reuters. https://www.reuters.com/article/us-health-coronavirus-tanzania/president-queries-tanzania-coronavirus-kits-after-goat-test-idUSKBN22F0KF.

British Broadcasting Corporation (BBC). (May 6, 2020). *Coronavirus: Most Africans 'will go hungry in 14-day lockdown'.* https://www.bbc.com/news/world-africa-52557464.

Brown, W. (August 22, 2020). *Covid hunger: Kenyans forced to hunt giraffe for food.* The Telegraph. https://www.telegraph.co.uk/global-health/climate-and-people/giraffe-covid-threat-kenya-hunger/.

Cherono, S. (May 13, 2020). *COVID -19 linked stigma hinders pandemic fight* (p. 8). Daily Nation.

Dahir, A. L. (May 15, 2018). *Ecotourism is being used to displace one of East Africa's long-standing Indigenous people.* Quartz Africa. https://qz.com/africa/1278167/maasai-evicted-in-tanzania-for-ecotourism-and-land-conservation/.

Daily Nation. (July 12, 2015). *Every month 164 billion is transacted on M-PESA* (p. 13). Daily Nation.

Daily Nation Team. (July 19, 2020). *Some worshipers ignore State's COVID-19 rules as churches reopen.* Daily Nation. https://www.nation.co.ke/kenya/counties/embu/few-wear-masks-as-churches-reopen-1901388.

de Waal, A., & Richards, P. (April 1, 2020). *Coronavirus: Why lockdowns may not be the answer in Africa.* https://www.bbc.com/news/world-africa-52268320.

Elmendorp, R. (July 20, 2020). *COVID-19 diaries: Can Tanzania really be coronavirus-free.* VOA. http://voanews.com.

Emecheta, A. (July 23, 2020). *World's first commercial internet balloons gives rural Kenyans high-speed internet access.* African Vibes. https://www.africanvibes.com.

Farge, E. (May 12, 2020). *WHO sees 'potentially positive data' in treating coronavirus.* Reuters, World News. https://www.reuters.com/article/us-health-coronavirus-who/who-sees-potentially-positive-data-in-treating-coronavirus-idUSKBN22O1HG.

Gathara, P. (May 7, 2020). *Kenya is turning public health crisis into a law-and-order one.* The Washington Post. https://www.washingtonpost.com/.

Gathura, G. (September 5, 2020). *Pandemic peaks in Nairobi, Mombasa, and some towns.* The Standard. https://www.standardmedia.co.ke/the-standard-insider/article/2001385225/pandemic-peaks-in-nairobi-mombasa-and-some-towns.

Githaiga, P. (July 15, 2015). *Locals protest compensation, rights, they are shunned away* (p. 28). Standard.

Houreld, K., & Lewis, D. (July 8, 2020). *In Africa, lack of coronavirus data raises fears of 'silent epidemic'.* REUTERS. World News. Reuters.com.

International Monetary Fund (IMF). (May 6, 2020). *IMF executive board approves a US $ 739 million disbursement to Kenya to address the impact of the COVID-19 pandemic.* Press release No. 20/208. https://www.imf.org/en/News/Articles/2020/05/06/pr20208-kenya-imf-executive-board-approves-us-million-disbursement-address-impact-covid-19-pandemic

Jamah, A., & Kiplangat, J. (July 17, 2009). *Leaders blamed for receding forests* (p. 12). Daily Nation.

Jerving, S. (July 21, 2020). *Muddled messaging around COVID-19 complicates response in Tanzania.* Devex.com. https://www.devex.com/news/muddled-messaging-around-covid-19-complicates-response-in-tanzania-97590.

Joseph, S. (July 17, 2020). *Amref CEO to Kenyans: Stop wasting your time going to test for COVID-19.* News-af-fednews.com. https://ke.opera.news/ke/en/covid/e8d05a588a725a0828878b7418ac7e2e?news_entry_id=s1afa5134200717en_ke.

Kabale, N., & Ngotho, S. (May 13, 2020). *Foreigners testing positive to be turned back* (p. 5). Daily Nation.

Kegoro, G. (April 26, 2020). *Why state has failed to impress in managing COVID-19 pandemic* (p. 15). Standard. https://www.standardmedia.co.ke/article/2001369121/why-state-has-failed-to-impress-in-managing-covid-19-pandemic.

Kenyatta, U. President. (2020a, July 6). The 9th state address on the coronavirus pandemic, Nairobi, Harambee House.

Kenyatta, U. President. (2020b, July 27). The tenth (10th) presidential Address.on the COVID-19 pandemic. Nairobi: State House.

Kenyatta, U. President. (2021, March 12). The fourteenth (14th) presidential address on the coronavirus pandemic. State House, Nairobi.

Kernell, S. (1986). *Going public: New strategies of presidential leadership.* CQ Press.

Kinyua, J. K. (2020). *Compliance with workplace guidelines on COVID-19 pandemic. Office of the president. Head of public service.* Nairobi. Ref: OP/CAB.26/1/3A state house P.O. Box 40530-00100.

Kombe, C. (June 8, 2020). *Some Tanzanians resort to bogus steam treatment for Coronavirus.* VOA. Voanews.com. https://www.voanews.com/covid-19-pandemic/some-tanzanians-resort-bogus-steam-treatment-coronavirus.

Koskei, S. (July 14, 2018). *Revised marginalized areas to share Sh 14 b* (p. 21). The Standard.

Koskei, S., & Kipruto, N. (July 23, 2015). *State to abolish cash payments in ministries soon.* Says Ruto: Standard. https://www.standardmedia.co.ke/counties/article/2000170099/state-to-abolish-cash-payments-in-ministries-soon-says-dp-ruto.

Kwamboka, E., & Omondi, D. (April 23, 2020). *NIS gets billions to spy on patients, medics warn virus war is militarized* (p. 2). Standard.

Latoti, I. (May 22, 2020). *The many faces under COVID-19 mask wearers* (p. 12). People Daily.

Lee, J. (May 7, 2020). *Did Tanzania's president expose faulty testing COVID-19 testing by submitting non-human samples.* Snopes. https://www.snopes.com/fact-check/tanzania-president-covid-tests/.

Magufuli, J. President. (2020, June 5). Speech on earth day. World environment day. Chama Cha waalimu Tanzania, Dar es Salaam.

Marindany, K. (April 20, 2020). *Poor Masai to bear brunt of pandemic says senator Seneta* (p. 11). The Star.

Matara, E. (August 10, 2020). *Nakuru governor decries stigma on recovered COVID-19 patients, kin.* Daily Nation. http://www.nation.co.ke.

Mbilinyi, J. (April 30, 2020). *Akitema cheche bungeni, corona sio CHEDEMA.* TV. https://www.youtube.com/watch?v=erAxoc5ZrPY.

Mbula, R. (June 26, 2020). *Culture a major setback to Maasai women in politics, says Lydia Ntimama.* Daily Nation. https://www.nation.ke.co.

Muheria, A. A. (March 27, 2021). *Interfaith council for national response to corona virus pandemic.* Press Release. Nairobi, Kenya.

Muthetia, G. (May 1–3, 2020). *Fake sanitizer, cigarettes and alcohol impounded in Meru* (p. 22). The Star.

Mwai, P., & Giles, C. (June 19, 2020). *Coronavirus in Tanzania: What do we know? BBC reality check.* https://www.bbc.com/news/world-africa-52723594.

Mwaura, I. (May 1–3, 2020). *The making of a scandal: COVID-19 response cash already being stolen?* (p. 11) The Star.

Naingisa, S. F. (July 20, 2020). *Responses to the COVID-19 pandemic among the Maasai in Northern Tanzania.* Oxford, UK: InsightShare. https://insightshare.org/indigenous-maasai-response-covid-19-coronavirus-tanzania/.

Neustadt, R. (1980). *Presidential power.* John Wiley & Sons.

Ng, K. (May 6, 2020). *Tanzania coronavirus kits raise suspicion after goat, and paw paw test positive.* Independent. https://www.independent.co.uk/news/world/africa/coronavirus-tanzania-test-kits-suspicion-goat-pawpaw-positive-a9501291.html.

Ochami, D. (July 27, 2009). *Over 25 pc of Mau grabbed, report says. Mau Report* (p. 8). The Standard.

Ogemba, P. (May 13, 2020). *Government defends curfew rules* (p. 8). The Standard.

Oketch, A. (July 2, 2020). *COVID-19: A balanced diet key to fighting virus.* Daily Nation. July. https://www.nation.co.ke.

Oketch, A. (June 2, 2020). *Kemri is dead broke amid COVID-19 fight.* Daily Nation. https://www.nation.co.ke.

Okoth, D. (March 2, 2015). *Remote village residents risk their lives to make a phone call.* Standard. http://www.standardmedia.co.ke/lifestyle/article/2000153368/remote-village-where-residents-risk-their-lives-just-to-make-a-phone-call.

Ole, S. J. (May 17, 2020). *Archbishop of the Anglican Church of Kenya. 4th pastoral letter. All saints cathedral, Nairobi.* The Anglican Church of Kenya.

Olingo, A. (May 19, 2020). *Counties ramping up efforts to contain spread of corona virus* (p. 2). Daily Nation.

Ooro, V. (April 13, 2020). *New app saving Kibera residents from hunger. Daily Nation.* Daily Nation. http://www://www.nation.co.ke.

Otenyo, E. E. (2017). *Trade unions and the age of information and communication technologies in Kenya.* Lexington Books.

Otenyo, E. E. (2020). Africans combatting COVID-19 in song and music. *Practicing Anthropology, 42*(3), 12–15.

Otenyo, E., Harris, M., & Askew, K. (2020). *Where there is no formal social welfare system for an Indigenous people: Entrepreneurship, watchmen, and the reinvention of the Maasai Warrior.* Oxford University Press.

Otieno, K., & Olwenya, G. (July 17, 2009). *Major boost on fight against malaria: Vaccine tests to be done in seven African countries to check its effect in three years* (p. 10). Daily Nation.

People Daily. (May 17, 2020). *Governors call for border closures to combat COVID-19* (p. 8).

Pew Research Center. (June 7, 2018). *Weekly worship attendance is most common where life is shortest.* https://www.pewforum.org/2018/06/13/why-do-levels-of-religious-observance-vary-by-age-and-country/pf-06-13-18_religiouscommitment-01-01/.

Ray, S. G. (June 26, 2010). *The cervical cancer Bazaar.* Tehelka. http://old.tehelka.com/the-cervical-cancer-bazaar/.

Reporters Without Borders. (July 10, 2020). *Tanzania Suspends another media outlet over its COVID-19 Coverage.* RSF. https://rsf.org/en/news/tanzania-suspends-another-media-outlet-over-its-covid-19-coverage.

Republic of Kenya. (July 14, 2020). *Temporary scaling down of operations and services. Ministry of ICT, innovation, and Youth Affairs. Circular. Nairobi, Kenya.*

Rutten, M., & Mwangi, M. (2012). Mobile cash for nomadic livestock keepers: The impact of the mobile phone innovation (M-Pesa) on Maasai pastoralists in Kenya. In *Transforming innovations in Africa: Explorative studies on appropriation in edited by Jan–bart.*

Said, K. (May 29, 2020). *A problem of denial: Why Tanzania could lose the war against COVD-19.* The Elephant. http://www.theelephant.info.com.

Santos, R. (July 24, 2010). *The Maasai tribe goes mobile. 2010 Global Marketing.* http://2010globalmarketing.worldpress.com/2010/07/24/695.

Sayagie, G. (July 15, 2020a). *3,000 Maasai Mara workers to be tested for Covid-19.* Daily Nation. https://www.nation.co.ke/kenya/counties/narok/3-000-maasai-mara-workers-to-be-tested-for-covid-19-1799110.

Sayagie, G. (August 31, 2020b). *Outgoing Narok county commissioner hands over to achoki.* Daily Nation. https://nation.africa/kenya/counties/narok/narok-county-commissioner-samuel-kimiti-moves-1929730.

Sayagie, G. (March 26, 2021). *Narok's ICU beds empty as counties grapple with shortages.* Daily Nation. http://www.nation.co.ke.

Sharma, K. (April 17, 2010). *The other half: Uninformed consent.* The Hindi. https://www.thehindu.com/opinion/columns/Kalpana_Sharma/The-Other-Half-Uninformed-consent/article16123576.ece.

Standard. (April 23, 2020). *Punish rogue police officers who are undermining war on COVID-19* (p. 14). Standard.

The Star. (May 13, 2020). *Issue COVID-19 passports to distance truck drivers* (p. 20). EDITORIAL.

Tororey, T. (May 12, 2020). *Why Kagwe's COVID-19 briefings are not working* (p. 11). People Daily.

Voris, M. (May 20, 2020). *The vortex-arrest Bill Gates.* Church Militant. https://www.youtube.com/watch?v=uFzc_OgEDgI.

Wafula, P. (June 20, 2020). *Free bundles, airtime for 5,500 frontline health workers.* Daily Nation. https://www.nation.co.ke.

Wahinya, H. (July 16, 2009). *Wildebeest season begins…as wildlife populations decline in the Mara reserve* (p. 14). The People Daily.

Wakhisi, A., & Inyanji, J. (December 5, 2020). *Busia Children sue for injection gone terribly wrong.* The Standard. https://www.standardmedia.co.ke/health/article/2000225791/busia-children-sue-for-injection-gone-terribly-wrong.

Wangusi, F. W. (May 17, 2020). *Bridging digital divide through universal service fund (USF). Message from the director general, communications authority of Kenya* (p. 34). Daily Nation.

Wasuna, B., & Oketch, A. (June 20, 2020). *Scanty information on theft of donated COVID-19 equipment.* Daily Nation. http://www.nation.co.ke.

Wesangula, D., & Korir, M. (April 23, 2020). *It will cost you an arm and a leg in forced quarantine* (p. 4). Standard.

Wisner, B. (2013). Disaster risk and vulnerability reduction. In L. SygnaL, K. O'Brien, & J. Wolf (Eds.), *A changing environment human security: Transformative approaches to research, policy and action* (pp. 413−442). Routledge-Earthscan.

World Economic Forum. (July 6, 2016). *The global information technology report 2016. Geneva.* http://www3.weforum.org/docs/GITR2016/WEF_GITR_Full_Report.pdf.

CHAPTER 5

Managing the COVID-19 pandemic in India's Siddi zones

India in 2020 presents a particular challenge to all of us who have an interest in governance of a country with millions of poor people, whose social status is enshrined in the country's primary law. India regards large sections of its huge population as persons of a particular constitutional standing. We selected Siddi communities to investigate how information and communication technologies (ICTs) and government actions impact groups of Siddi people who occupy disadvantageous positions. Siddi people have achieved some notable positions of legal and political leadership over time including the election of a Siddi leader, Shantharam Budna Siddi, during the pandemic (Devaiah, 2020). However, Siddi people also continue to face multiple levels of discrimination and legal oppression. In the larger population where Siddi people live many leaders extend racism through the identification of dark skinned "untouchables," who have not benefitted from government investments in health care and education. COVID-19 became the occasion for another incursion of a disease. When India became one of the countries with the highest COVID-19 death rates in the world in the early part of the pandemic (and again over a year later) Siddi people like others living in systems of oppression suffered. There is still little discussion on how COVID-19 played out in these diverse communities.

Now, we have to state from the outset that it is improper to generalize the whole fabric of tribal life in India (Furer-Haimendorf, 1982). The various groups are diverse and rich in cultural outlooks. It is also ill advised to offer a visual picture of any of the tribes. What is apparent is that the social science penchant of classifying people forces us to stick to the age-old process of contextual labels embroiled in societal contestations concerning who any group of people present. Within different groups there are "multiple" diversities—"linguistic, religious, cultural, and regional." To add on that, India has a "strict caste hierarchy with its notions of purity and pollution" (Sharma, 1998, p. 214).

One of the important pieces of the puzzle of solving complex policy issues is to start by identification of case studies. India is a complex polity with overlapping cultures and subcultures, which, to a large extent, limit and confer benefits from the government bureaucracy. Siddi people number in the tens of thousands in population and have been in India for likely over 600 years. It is inescapable not to think about the impact of governance on marginalized communities of India. Through governance systems, needs of

The Inequality of COVID-19. https://doi.org/10.1016/B978-0-323-99867-3.00004-1

citizens are assessed and disbursed, sometimes based on political power calculations which are rooted in values and institutions that disregard the humanity of a people and yet endure as social norms.

Governing by scheduling and othering people

The social construction of identities is clearly visible in India's constitutional demarcation of Scheduled Tribes (STs) in its territories and union. Cristina-Ioana Dragomir states that "the Indian Constitution ensures certain protection for communities deemed as having Scheduled Tribe status. However, which groups should be accorded that status has been contentious" (Dragomir, 2017). Besides, the Indian constitution offers no clear criteria on which groups qualify to be on the ST list (Dragomir, 2017). In India's case, Kumar (1998, p. 223) said that the constitution of an ethnic identity, "is not only ambiguous but changes over a period of time. In other words, both horizontally and vertically it is fluid." India has more than 6000 castes and subcastes, 3743 of which are designated as "backward" based on social marginalization and lack of educational opportunities (New York Times (Online), 2007). Indeed, these characterizations are present in online journalism that originates from non-Siddi people who write about them in ways that reinforce stereotypical ideas.

The reinforcement of ongoing oppression is clear in the categorization of tribes through distinctive cultural traits and the fact that they often live on the fringes of so-called modern civilization (Negi, 1976). In fact, a Government of India (1952) Report of the Commissioner for Scheduled Castes and Scheduled Tribes gave currency to the "primitiveness" argument when they decided that the criteria for identification of members of the tribes on the schedule included a set of essentializing characteristics put together in a crude list that sounds like it was assigned to wildlife other than humans, a common way that powerful groups disenfranchise Indigenous people.

(a) Autochthony.
(b) Groupism or a very strong community fellowship, if not descent from a common ancestor or loyalty to a common headman or chief.
(c) A principal, if not an exclusive habitat.
(d) A distinctive way of life, or primitive or backward by modern standards and apart and aside from the main current of culture.
(e) Economic, political, and social backwardness.

These attributes, for the most part, were buttressed with cultural markers associated with groups wrongfully perceived as inferior on arbitrarily determined cultural criteria. This report is from the 1950s though to this day Siddi people must contend with social and political oppression based on these enduring ideas of being behind on a linear trajectory of progress. For popular culture commentators, to be in a scheduled tribe list has never been a badge of honor. Instead, it conjures an image of a people with low literacy achievements, tied to subsistence farming, and with low levels of technological literacy. The Karnataka's government's efforts, including prohibition of use of term "Dalit" to describe these tribes, have changed the enduring perspective of lower social status perceptions of these groups (Deccan Herald News, 2020).

In the state of Gujarat, and other places, references to STs as Scheduled Castes (SCs) cement this position. Granted that there is good faith to change the status of these groups by extending their reservation in the Lok Sabha and state assemblies, as witnessed by the Gujarat Assembly's ratification of a Constitutional bill to promote social equality and development of "backward communities," the political class still has much work to do (Times of India, 2020).

Although India is the world's largest democracy, it is also home to some of the poorest people on earth. Although as had been reported in Weiner and Osgood's (1975) important work, a number of these tribes retain a high degree of political cohesion, and their electoral strength remains perpetually weakened through enduring discriminatory structures. And a vast majority of those so described fall into statuses of "castes" and the STs' category (Oommen, 1977). These share similar indignation and economic marginalization and deprivation as Indigenous communities elsewhere.

STs are also settled in places that have common features with reservations, namely, geographical isolation. The vast majority "inhabit two distinct geographical areas—the Central India and North-Eastern Areas" (Government of India, 2013, p. 3). More than half of the ST population is concentrated in jurisdictions in central India, whereas Sharma (1998, p. 220) states, "the tribals are marginalized in their own homelands." There is a tendency to describe these groups as being outside of the mainstream social and economic systems. It is also widely believed that many STs have low and declining population. Further, most indigenous communities are scattered across many boundaries and are in overlapping jurisdictions. This is important for putting into context the existential threat of COVID-19 or any other pandemic.

Poverty and policy

Let us for a moment consider the economic matrices of the Siddi people living in Jambur village or Gir Forest in Gujarat, Ghats, and other enclaves of India. Programs such as the Integrated Rural Development Programme (IRDP) and Allied Programmes targeted STs through some assistance (Kapila, 2002). And even after India created Export Processing Zones to fully integrate in the global markets via the neoliberal economic policies of the 1990s, the standards of living in Siddi communities were only marginally impacted. If anything, India's economic prosperity is not shared by the vast majority of those in the SCs.

Although these groups have had improvements in their living conditions, they are still among the poorest communities in India. Granted that there is no agreed upon definition or yardstick to measure poverty in India, some observations cannot be ignored or wished away. Those living below poverty lines are predominantly agricultural laborers, small famers, disabled (handicapped) persons, and rural artisans (Sinha & Rana, 2001).

Thousands do not have running water and the availability of electricity is low. Generally, access of electric power is more prevalent in nearby urban centers but not in rural enclaves (Deaton, 2004). Although the federal government has tried to improve the health sector in tribal areas by providing health centers within reach of 10,000 people, the quality of facilities is relatively low and poorly staffed. And the average absence rates of health-care workers is close to that which prevails in underserved communities of India, at 35% (Choudhry et al., 2005; Das & Hammer, 2004; World Bank, 1990). These communities, like other Indian tribes also have poor sanitation infrastructures, spend high percentages of incomes on poor-quality food items, and experience malnutrition, which directly affects infant mortality (Cutler, Deaton, & Lleras-Muney, 2006; Wagstaff, 2003).

Many men work in low-wage jobs in nearby towns to supplement earnings from agricultural jobs. Women, many of whom have low literacy and educational attainment, in particular serve as domestic help in many urban centers. Hundreds of Siddi people live in large cities of Hyderabad and in Mumbai as well (Ansari, 2013). Those in Mumbai are concentrated in specific sectors of the city near Siddi dargahs (revered religious shrines)—in densely populated Dongri and Kurla, an important milling center with underserved neighborhoods.

Generally speaking, their wages are hardly above 2 US dollars per day, which is a measure that builds on the $1 per day marker or "line" of "extreme poverty" adopted by the World Bank conventions (1990). The poverty levels among Siddi people are also observable through low consumption of industrially produced goods, entertainment services, and absence of assets and saving bank accounts. Furthermore, very few Siddi people own computers or laptops, television sets, and mobile smartphones. Radios are available but have narrow FM bands. Many poor people prefer Indigenous entertainment services which are free, inexpensive, or cheaply priced.

By all accounts, and to quote an environmentalist, Dinesh Holla, the Siddi "have endured the worst of limitations among the tribes of Western Ghats" (Jain, 2020). The Siddi have shown much resilience and preserved their cultural identity amid all odds (Jain, 2020; United Nations, 2017).

However, these limitations have not stopped Siddi people from achieving higher education and participating in global networks and use of internet technologies during COVID-19.

Young people created online videos with songs about COVID-19 and created viral videos about mask wearing for prevention (see, for example, https://www.youtube.com/watch?v=QEIiZx4dS0s). Like other Indigenous people globally Siddi people continue to survive even in the midst of centuries of ongoing oppression.

COVID-19 and Indian federalism

Though India is a federation, the central government has responsibility in balancing and coordination development efforts through the large country that India is. Because of the complexities of devolution in India, managing COVID-19 required a form of creative centralization. The national government had to step up and to stop the COVID-19 pandemic from consuming the entire country. The national government, one would assume, had the role to protect the interests of citizens, regardless of state or tribal affiliations. India has a powerful central government, which recently guided the COVID-19 mitigation plans. The most far-reaching decisions were the lockdowns enacted immediately after India first reported cases of COVID-19 had been identified on travelers arriving from Europe, Dubai, Saudi Arabia, and other countries. The initial and immediate lockdown decisions, which are well documented, were made at both state and national government levels. Table 5.1 illustrates the different phases.

Somewhat, the pronouncements about COVID-19 from Prime Minister Narendra Modi were meant to accommodate the country's multilayered federal setup. His signaling messages had huge policy implications throughout the country. And the impressive nature of India's ICT infrastructure made this possible. Initially there was overwhelming public support for his actions, even after he ordered for national lockdowns.

Besides leveraging ICTs for communication, Modi's administration attempted to push through an agenda to employ digital contract tracing applications in lieu of pressure to reopen the economy after the lockdowns. The Indian government desired that its Aarogya Setu app be used throughout the territory (Chakravorti, 2020). The app did not

Table 5.1: National lockdown time frame.

Important date	Notes on lockdowns
March 25	Lockdown 1.0. Entire country 21 days; low-wage migrant workers, including Siddi, complained of shelter-in-place movement restrictions and lack of food supplies (Ganguly, 2020)
April 14	Lockdown 2.0. End of lockdown May 3, COVID-19 cases surge
May 4	Lockdown 3.0
May 18	Lockdown 4.0. Extended to May 31. Relaxations announced and new guidelines issued. In Karnataka, further restrictions announced
June 8	Areas unlocked. Central government directed opening of religious places, hotels, and malls

include robust security and privacy protocols. Its operations were based on use of GPS and Bluetooth technologies. Data were transferred to a government repository public GitHub, where security vulnerabilities were not assured. Initially, the government wanted the app to be mandatory for its workers, but opposition party leaders rejected the measure over privacy concerns and also because of allegations of corruption in procurement procedures (Chakravorti, 2020). In the end the requirement was not implemented but the government advised and recommended its use. These innovative possibilities did not occur in a vacuum and hence we next consider a brief reflection on the ecology of the ICTs in India.

Ecology of ICTs and other challenges

The transformative appeal of ICTs, through the internet, mobile phones, and other devices, have borne out the perspective the change is enduring. COVID-19 provided the environment to test out theories about e-Governance and change communication. The massive correspondences among citizens and government officials reflected the salience of horizontal networks. Although communications about COVID-19 remained fluid, the ability of the government in India to take stock of its e-governance infrastructure became much more a matter of national survival than pride, considering that the country has some of the most technologically savvy citizens anywhere in the world. India's ICT infrastructure exhibits core strengths in enabling laws and generally well-developed telecommunication infrastructures.

India's mass communication policy has always been part of the national development strategy. In the 1970s, the country utilized mass communication to promote education, skill development, and eradication of illiteracy. The latter is a constitutional mandate (Article 45 of the Indian Constitution). With a vast number of Indians residing in remote rural areas, development communications were directly linked to prosperity and societal transformations. Although the programs varied from state to state (and also union territories), the national government took measures to reduce the impact of the strong caste system and using communication and transport instruments sought to shrink the isolation embedded in the caste and class structures. New ideas were communicated through village leaders to promote social change.

For India, the integrated communication and media systems, including cinema, televisions, satellite television broadcasts, radio stations, newspapers, and magazines, were also an avenue for societal transformation. For decades, India had more newspapers than any other Asian country. And its cinema, through Bollywood productions, was more expansive than that of any country in the world. Thus, in terms of messaging on health challenges such as the COVID-19 pandemic, it was not so much an issue of avenues for outreach and public health education.

Besides, as a democracy, it is relatively open and many of its citizens enjoy freedoms to express ideas that would not be tolerated in dictatorships. In addition, India also has unlicensed news providers and documentary content creators whose work reaches both rural villages and urban areas. India's massive press provides demonstrably easy observations about the nature and scope of government responses to COVID-19.

During the COVID-19 pandemic, India witnessed tremendous growth in digital services. These services are the nervous system of modern India. Since the nationwide lockdown curtailed movements of people in all areas of India, people turned to digital services for education, telehealth, and entertainment. Individual states also employed tele-ICU approaches to support health care in rural enclaves. In Maharashtra, some doctors used ICTs to confer and consult with their colleagues in rural areas handling critical care cases (Chakraborti, 2020). Thus, the shift to digital services highlighted the unfinished agenda of strengthening the capacity of India to expand its digital infrastructures.

Due to the surge in demand of data doubling, the Ministry of Electronics and Information Technology charged the CSC e-Governance Services India Ltd., to oversee the implementation of common services and address the digital divide that was more pronounced in rural areas. According to the Economic Times (2020a) editorial,

> *There is a rural-urban component to the digital divide. Of the more than 60 crore internet users in India, 29 crore are in rural areas—internet density in rural areas accounting for 66% of population is around 25%, while among urban population (34%), internet density is almost 98% (Economic Times, 2020a).*

Prior to the pandemic, India invested heavily in the electronics sector to reflect its growing stature as one of the world's emerging political and economic powers. India capitalized on its long-standing relationships with global corporations to lay out foundations of informatization. Informatization refers to "the transition from an economy dominated by manufacturing to an 'information society' in which the occupational structure is dominated by information workers, and output is dominated by products and services related to communication and information" (Mueller & Tan, 1997, p. 13). As such, the links between information systems and central government planning were already at a mature level. Through creative tax policies and incentives, an information age system was already in place in the major hubs of the country, including in Hyderabad and New Delhi, among other centers. Although there remain unresolved issues, especially the digital divide issues in rural areas where the poor live, the institutions to solve the problems were available. And the government realized the importance of expanding the communications infrastructure and thus, funded additional Wi-Fi hubs to support the dissemination of COVID-19 relevant information.

State Ministries for Education moved learning programs to online platforms. Digital education became the norm during the lockdown periods. In Karnataka, the state Higher

Education Department recognized the digital inequalities. For example, the minister, Dr. C.N. Ashwah Narayan, observed:

> *Digital learning is a means to ensure quality education to all sections of the society without discrimination on the basis of caste, region, and religion. This is a solution to eliminate discrimination. Through digital learning, a student in a remote village will also get an opportunity to learn from an experienced and highly qualified teacher. In addition to this, students are getting study material in digital form. Keeping these important aspects in mind, all educational institutions should give stress to digital learning without wasting any time (Belur, 2020).*

To understand the claims of discrimination, we first have to acknowledge its various manifestations. In sociological terms, the notion of discrimination entails identification of categories and specific groups of persons who are treated differentially in many aspects of life. This does not mean that some Siddi persons as individuals "don't make it in life." They do, as did their Siddi ancestors.

Historically, the most well-known Siddi person was Malik Ambar (1548–1626), who ruled Ahmadnagar (1600–26). Although Malik, born in Ethiopia, sold by Arabs as a slave was a famous man of war, he also built several religious structures in the Deccan including the city of Khadki (Aurangabad) (Amarat, 2006, p. 69; Melwani, 2017). Hundreds other Siddi people also ruled in Janjira (1618–1948), becoming "generals, commanders, admirals, prime ministers, and rulers" in ancient India (Melwani, 2017; Rajan, 2013). And thousands owned fabulous business enterprises stretching back from the 17th century way until the independence of India in 1947.

As we understood, the Chief Minister of the State of Karnataka, Bookanakere S. Yeddyurappa nominated Shantharam Budna Siddi, from Uttara Kannada district as India's first Siddi legislator in the state body (Devaiah, 2020). The MLC post gave him a platform to highlight the grievances of the community and to support their human development efforts. Among his credentials was to help conserve trees in Western Ghats in Karnataka (Devaiah, 2020).

Shantharam, a university degree holder, and many others with college education have landed regular jobs. And like in other Indigenous communities, those lucky to have college education have opted to give back to their communities taking up jobs such as in social work and in the NGO sectors. For example, Shanta Pedru Dodmani, who earned a master's degree in Social Work from Rani Chennamma College, was employed with a number of NGOs including Sambudhi and CherYsh which empower women in the Siddi community for self-development (Ramdev, 2015). CherYsh provides food supplies to kitchens run by the Akshaya Patra Foundation. They also support education for girls and career counseling.

Although the Siddi people participate in activism for self-development, there is much to be done and their plight remains unknown to a vast majority of Indians (Harbi, 2020; Shroff, 2016; Vallangi, 2016). Their general status as a ST is one factor in the continued unequal treatment of Siddi people in many aspects of their economic and social lives. Many have been discriminated against when they look for jobs (Ketaki, 2014). While official policy is about "equality" the investments did not really match the pronouncements. The state government's policy initiative to provide 4G network across the state was still work in progress (Belur, 2020).

Communication and messaging

During the COVID-19 crisis, Siddi people continued to practice their cultural activities. For example, in Western Ghats, they celebrated "Siddi Habba." The vent involves dance forms like Pugudi Nrutya and Gamate Nrutya. The event was at a school in Bhagavati Halli of Haliyal taluk in Uttara Kannada district. Siddi's rich art form is celebrated in India. The Gamate Nrutya is performed by men, dressed in traditional attire who dance to music from the percussion instrument, gamate (Jain, 2020). The *gamate* is a clay pot which has an opening (mouth) sheathed with leather. The Siddi dance is often performed at all religious activities including Eid and Ramzan for Muslims, Christmas for Christians, and at Holi and Deepavali time as well. Siddi community embraces all faiths—Hindus, Muslims, and Christians (Jain, 2020).

Communication in different Siddi groups

One of the challenges in communication has to do with use of language that reaches out to the majority of people in a location. Siddi people have a mother tongue called Siddi Basha and many are fluent in Kannada. They also communicate in several other local languages like Urdhu (for those in Pakistan) and Hindi. These are languages through which the messages about COVID-19 were conveyed. Not their mother tongue, which has no mass communication imprints. Siddi people living in urban areas do have access to news channels. Many Siddi people tragically experienced COVID-19 through brutal government responses. In the state of Maharashtra, those living in Mumbai constantly heard speeches and comments, including address through social media platforms from the State Health Minister Rajesh Tope.

As the cases began to rise, Minister Tope called on Mumbai residents to stay calm and not panic (Chavan, 2020). The Republicworld.com's Varsha Chavan quoted him:

> *In an online address to people, Tope said, "There is no need to panic over the central government's team projections that Mumbai might have 42,604 cases by April 30 and 6.50 lakh patients by May 15. It is a mathematical model and there is a science about it. However, things here will not be static because every possible effort is being made to reduce the threat and improve the situation arising out of Covid-19. The death rate has decreased, and the recovery rate is 13 percent. Even the hotspots have decreased from 14 to five" (Quoted in Chavan, 2020).*

This was after the national government projected a surge in COVID-19 cases in Mumbai. The state had constituted a task force and collaborated with an Inter-Ministerial Central Team to oversee the COVID-19 pandemic mitigation measures. The city was among the hot spots alongside with Mumbai Metropolitan Region, Pune, Nagpur, and Nashik. The region benefitted from testing capacity at the National Institute of Virology testing center in Pune.

The media widely reported that tests were ongoing and that it was essential to continue following social distance and covering face with mask. The Minister of Health of the state of Maharashtra also commented on policies of reopening the economy and easing up restrictions that had been announced at the national level. Maharashtra announced lockdown on April 22, 2 days before the Prime Minister Modi of India made a nationwide lockdown. Minister Tope, for example, allowed alcohol to be sold in the state but with a caveat, that social distancing be applied at point of sale. In April, some states had begun allowing wholesales of liquor. The unfolding events had been prompted by pressure from the Confederation of Indian Alcoholic Beverage Companies (CIABC) which wanted businesses to resume (Jain, 2020).

To manage the different levels of virus loads, patients were separated into three different categories, the most critical were confined to designated hospitals. Next, the mild and moderate patients were in another category. The last group, the asymptomatic cases, were separated into different hospitals. The state government also purchased personal protection equipment (PPE) including face masks for frontline workers in the government-run hospitals. The government also investigated private hospitals that were rumored to be charging high fees for COVID-19 treatments and care. Amid complaints from the public, the state slashed fees for testing in private labs (Economic Times, 2020b).

In the state of Karnataka, especially in the north, where the vast majority of Siddi villagers reside, they reported several cases of COVID-19 among the Siddi population. Karnataka is also the first state to list Siddi as a ST to allow them access to numerous government programs.

As for Andhra Pradesh and Gujarat, the respective governments ensured that additional beds were provided in government hospitals. Media gave daily updates on cases, recoveries, and deaths from COVID-19. Cases in Andhra Pradesh crossed the 1.5 lakh mark on August 1. Official government bulletin reported that the state tested over 20,12,573, with a return of a positivity rate of 7.46% (Express Web Desk, 2020).

The state of Gujarat, which together with Maharashtra had the highest concentration of COVID-19 cases, is home to thousands of Siddi people. The hot spots from the Ministry of Health maps showed a disproportionate burden of COVID-19 in areas where Siddi people live. The city of Ahmedabad, Gujarat, had high cases of COVID-19 as did Mumbai, and Pune in Maharashtra, areas where many Siddi people work as domestic workers.

Siddi villages like Amreli, Junagadh, Surendranagar, and Bhavnagar, the last two which are roughly 91 miles away from Ahmedabad, were among those with reported highest infection cases. In many of these villages, hospitals and facilities run by the Vanvasi Kalyan Ashram, a Hindu social welfare nonprofit group, were equally overstretched with tribal patients.

In Gujarat, the total number of COVID-19 deaths surpassed 2509, as of August 3, 2020. The data showed a fatality rate of 3.88%, which was high in spite of state government actions. In Gujarat, the government closed schools, colleges, shopping malls, and social centers including swimming pools on March 15, way before Prime Minister Modi's national lockdown. This was followed by other closures of wildlife sanctuaries and parks, close by to some Siddi homes. The state government also prohibited railway transportation and joined in the national curfew efforts to stem the spread of COVID-19. Government announced it would provide relief food to members of the STs and other less fortunate and marginalized communities. Further, the migrant workers were provided temporary accommodation during lockdowns. Many Siddi workers benefitted from these gestures. In addition, the Gujarat government unrolled an app to track down those quarantined to their homes.

International and nationwide messages

Although all news media reports spoke about the origin of the pandemic as being China, people in the medical community suspected that the specific variant of coronavirus that was dominant in India may have originated via European travelers. This gave the hope that finding a cure was feasible. Scientists believed that the national lockdowns announced in March and April of 2020 were effective in reducing the spread of COVID-19. In addition, India's ban on international flights and also intra-Indian air travel helped stem the rise in SARS-COV-2 cases (Express Web Desk, 2020).

TV Stations were quick to set agenda against Chinese companies who were reportedly selling masks and faulty PPE (including Rapid Test Kits) that were ineffective. By some estimates more than 90% of the kits were faulty, which rendered them useless. The Indian Council of Medical Research also voiced its concern about the testing kits and communicated its frustrations on TV Stations. The companies mentioned for producing poor-quality PPE were: Wondfo Biotech Co., Ltd, Zhuhai Livzon Diagnostics, Inc., and MGI Shenzen. These companies sold product to states such as Rajasthan. Their products, including N95 masks, had also been questioned by authorities in the Netherlands, the United Kingdom, Slovakia, and Spain.

The Indian Ambassador Vikram Misri reported that the Indian government was working with Chinese companies to ensure that only bona fide ventilators were exported to India. India imported ventilators from China.

Citizen reactions and policy feedback

One of the key issues that emerged from studies on COVID-19 was the preliminary assessment and confirmation of cases. For purpose of managing the pandemic, it was necessary to determine how many persons were infected and those who succumbed to COVID-19. The case fatality rate (CFR), in Israel was 0.58 in contrast to a high 12.3% in Italy. This number is a function of variable such as age, underlying population, and system response, guided by national communication efforts.

The problem in India was that computing the CFR number as a means for determination of the devastation caused by COVID-19 was affected by weaknesses in registration of deaths (Gupta, 2020). In other words, the data collection mechanisms made the calculation of the CFR unreliable. India's Vital Statistics Database, available on the Office of the Registrar General & Census Commissioner (Birth and Death Registration), required user login credentials and, therefore, not available to majority of the Siddi community tribe.

Since the institutions in tribal society lacked much broadband capacity, there were lags in registration of deaths as prescribed in the Births and Deaths Act of 1969. India as a whole had weaknesses in its Integrated Disease Surveillance Program (IDSP) of the Government of India. Its outreach was more robust in communities with greater broadband capabilities and reliable vital registration systems. The Siddi regions, such as Siddi (Nayaka) in Goa and Siddi in Daman and Diu territory, Siddi (Badshan) in Amreli, Bhavnagar, Jamnagar, Junagadh, Rajkot, and Surendranagar districts in the state of Gujarat, Siddi (in Uttara Kannada districts, mainly concentrated in Haliyal, Yellapura, Sirsi, Mundagoda, and Karwar taluks) of Karnataka State, are all listed as STs. And the problem in these areas was complicated by the absence of tracking of deaths outside hospitals. State governments downplayed the actual impact of COVID-19 on poor Siddi households.

More specifically, in Siddi villages, the deaths that may have been captured through the IDSP system were largely, missed out. The prevalence of mobile phone–based death notification systems in Siddi villages also did not capture the magnitude of the problem.

Reaction in urban settings

Data on COVID-19 in urban centers such as Mumbai were well publicized. In Mumbai, the capital city of Maharashtra and home to a sizable number of Siddi, the death toll of all residents was a high 6395 according to figures released end of July (Mumbai Mirror, 2020). These were not disaggregated into ethnicities. The city had among the highest density of cases in the entire country. At the same time, the state had recorded over 2,66,883 recovered patients (Mumbai Mirror, 2020). The state also had the highest numbers of COVID-19 cases in the country (Upadhyay, 2020).

In the city of Bengaluru, in Karnataka, after restrictions were announced, there was opposition from MLA (Economic Times, 2020c). They did not favor lockdown in the city to control the spread of COVID-19 citing the need for government to plan on providing resources to the poor. The Karnataka government was advised to give those under lockdown food. Government, then, released Rs. 25 *lakh* for each ward to provide food and medicines for residents in sealed-off areas (Economic Times, 2020c). The Government also offered transportation for patients and increased beds in government hospitals.

Stigmatization beyond caste

In most large cities, there was also concern that those who had recovered from COVID-19 reported getting into a state of depression and paranoia. This experience was consistent with findings in other parts of the world (Chakraborti, 2020). Clinicians also reported an increase in posttraumatic stress disorder, fatigue, and delirium, a confused state of mind. Many of the COVID-19 patients reported lack of sleep and poor eating habits after treatment.

There were also reported cases of stigmatization of those who had recovered from the disease. This behavior was experienced in several communities. This was very much associated with the general fear surrounding the COVID-19 pandemic. Fear was manifest in several forms, including fear of those who recovered feeling they would pass on the disease to loved ones, even after being cleared of the disease. This fear, perhaps, also explains the reported cases of increased paranoia. Many of those in the Mumbai areas often showed obsessive compulsive disorder (OCD) and continued to constantly clean their hands with sanitizer and wear masks at home.

Cases of police beating migrant workers during lockdowns were reported as well (Ganguly, 2020). These instances of police brutality frequently target the poor who include Siddi workers in large cities of Hyderabad and Mumbai, where they are forced to show Identification cards because of many are misconstrued for being foreigners. One of the unintended consequences of the lockdowns was to expose the poor labor conditions of migrant workers. Most need health care and labor protections at work.

As in other societies, the stigmatization of COVID-19 patients was also reported in India, and especially among the poor members of the society. This fear of others was heightened by an already existing climate of hate mongering by Hindu nationalist members of the ruling Bharatiya Janata Party (BJP), under Narendra Modi. After Modi took power in 2014, there was a remarkable rise in attacks on students, Muslim cattle traders, and "others." This hatred was reflected in the reported incidence where BJP supporters were quick to blame Muslims for COVID-19 (Ganguly, 2020). A vast majority of Siddi people are Muslims and were not treated kindly for the same reasons. Health workers were reportedly denied entry into some neighborhoods which complicated the process of checking the spread of the pandemic.

Economic self-reliance and bureaucratic corruption

Prime Minister Narendra Modi was among the members of the ruling elite to call for self-reliance in production of essential medical supplies. He and many Indian policy influencers were concerned the COVID-19 revealed India's vulnerable position in terms of manufacturing essential medicines and health-care supplies. COVID-19 is, therefore, a wakeup call for national renewal, especially in the health sectors.

The distribution of health-care resources also revealed the corrupt nature of some elements in the Indian national bureaucracy who provided benefits based on corruption and favoritism (Ganguly, 2020). There were concerns that the poor were often denied services from state institutions which made them more vulnerable to the worst consequences of COVID-19. It is for this reason that in some states, authorities raised concern over private hospitals raising fees for treating COVID-19 patients and in Karnataka, prompting the Chief Minister's intervention to support vulnerable groups, including Siddi people.

Students of government know that transparency is a key value of public administration. In the context of COVID-19, it is about the free flow of information on the modes of transmission of the coronavirus and best ways to protect the citizens. Where transparency of operations exists, the chances are that uncertainty is reduced and rules and regulations are likely to be followed. Without transparency in management of public health challenges, there is a very strong likelihood that mobilization of resources to combat the pandemic is hampered.

Issues of transparency became front page news items in India's richest state of Maharashtra. Although there had been claims that government was not transparent in releasing information about COVID-19. The leader of the Opposition and BJP Maharashtra Legislative Assembly (MLA) Devendra G. Fadnavis had claimed that Chief Minister Uddhav Thackeray's government was hiding deaths to show a lower tally. There was claim that the mortality rate in the state was 4.2% but had risen to 5.2% in Mumbai—this was well above the Indian nation's average death rate of 3.3%. This was initially denied by officials who instead issued a clarification noting they had corrected the tallying errors (Suryawanshi, 2020).

There were also reports that government had released fake news on Coronavirus in Telangana state, which to the north borders Maharashtra and Karnataka, to the west. It is also home to the major technology industrial city of Hyderabad. This prompted a court case to be filed against Telangana Government (V6 News Telgu, 2020).

In Karnataka, the degree of transparency was evidenced through its daily bulletin, available on YouTube (ABP News, 2020). Besides, the whole state was informed of the hospitalization of state Chief Minister Bookanakere S. Yediyurappa after he tested positive for COVID-19 (Kulkarni, 2020). In addition, there was also a robust COVID-91 tracking system, based on Ministry of Health data on all major news outlets in the country. For instance, Karnataka had 129,287 cases, Gujarat 62,463, and Maharashtra 33,479 as of August 2 (Economic Times, 2020d). However, the claims to government transparency fade as a resurgence of COVID-19 emerged toward the end of 2020. By early 2021, India became an epicenter of the pandemic, and most epidemiologists and public health journalists openly denounced the national government's failure to be transparent in COVID-19 efforts. The numbers of those who died of COVID-19, including members of the Siddi, were, for the most part, undercounted. In the state of Gujarat, the opposition demanded proper disclosures and accounting of the deaths (Desai, 2021).

Feedback: Communicating Grievance and NGO-driven self-reliance efforts

Communicating Grievance of Neglect: Changing focus—rather than COVID-19 messaging—shift to Air grievance on development gaps—especially the Poor Health Care Ecosystem Siddi areas are resource constrained and do not have high-quality health care. There is an acute shortage of doctors in the village centers. In India, 2.4 million die each year due to lack of access to health care or poor-quality health care (Kruk et al., 2018). Journalist Kalyan Ray (2018) citing the Kruk and other data in the Lancet pointed out that in Indian hospitals, 4300 deaths are recorded every day. A vast majority of these deaths are caused by poor diagnosis, wrong treatment, and underfunded public health facilities. A number of these deaths are in scheduled tribal communities such as Siddi, Any Naga, Gond, Saur, the Gonds of the Adilabad District (*Adivasi* or tribal people), and the socially marginalized groups known collectively as India's Dalits (untouchables).

Sustainable Development Goals established benchmarks for universal health coverage across the nations. India too sought to improve its health-care services. But mere expansion of health-care services did not prove to be a game changer. In India, a program, called Janani Suraksha Yojana was set up in 2005 to provide cash incentives for women to deliver their children in health facilities (Powell–Jackson et al., 2015). The program increased coverage but not desired outcomes, namely improved maternal or newborn survival. This program was not impactful among Siddi communities.

Within Siddi communities are several NGOs, including Vanavasi Kalyan Ashram. There were concerns that some NGOs in Indian states witnessed drying up of funds as the COVID-19 pandemic continued to take its toll on the country's health and economic well-being. Many of the NGOs supported work in areas of public health and education. India had the third highest numbers of cases after the United States and Brazil. There were allegations that funds raised by the central authorities, in centralized funds, were handled in an opaque manner. This allegation was repeated for the United States and Kenya too (Byatnal, 2020; Delevingne, 2020; Mwaura, 2020).

In the case of India, the disbursement of the Prime Minister's Citizen Assistance and Relief in Emergency Situations Fund (PM CARES Fund) was established in March at the beginning of the COVID-19 crisis. An estimated $1 billion had been raised, which ironically denied monies to NGOs, which are dependent on donations from some of the donors, which include international agencies and business concerns. There was concern that the Prime Minister's office pressurized organizations to make donations to the PM CARES Fund as a part of their social responsibility mandate (Byatnal, 2020).

Reactions: use of traditional religious beliefs and reliance on patron Saint Bava Gor

Bava Gor (Baba Ghor) is said to have come to India from Africa as a trader. He specialized in agate (a semiprecious stone) trade, establishing a base in Gujarat. Bava Gor practiced mystical forms of Sufi Islam. His tomb is at agate mines at Ratanpur. According to experts on his religious practices, the tomb remains a holy shrine and Africans and Muslim devotees still make pilgrimages to the site (Amarat, 2006, p. 19).

During the pandemic, Siddi Muslim *pirs* (holymen) in Gujarat led prayers of small groups of worshippers to pray for the community. During the Idd ceremonies, which occurred amid the pandemic, drums beats were heard across the towns. When religious gatherings were allowed, a few Siddi held devotional Zikr dances invoking the healing help of their religious leaders. Many of the Siddi felt very strong that they would rely on the patron Saint Bava Gor to protect them from COVID-19. Visits to his shrine in Ratanpur, Gujarat, were notably higher during the pandemic. There was widespread belief that the shrine, like Bava Gor himself, cured demonic spirits, like COVID-19.

COVID-19 and India's health-care innovations

There was some hope that many Siddi people would also benefit from the vaccine developed over 100 years ago to prevent the spread of tuberculosis (Redelman-Sidi, 2020). Bacillus Calmette-Guerin (BCG) has been used extensively in Indian communities. It was touted as a possible line of defense for the Siddi and other populations. These claims were not verified or promoted by competent health-care authorities. BCG has been used for protecting children (infants and neonates) and there was some hope that it could offer help to COVID-19 patients.

If there are technology innovations occasioned by the COVID-19 crisis, then scientists and entrepreneurs were quite busy providing new solutions to a host of emerging COVID-19-related problems. Globally devices emerged including cellphone tracking apps, robots, and drones to deliver medicines and food, and variations of thermal imaging COVID-19 screening systems just to name a few were all part of the new world of public health infrastructure. And Indian scientists and entrepreneurs were not left behind, they too joined in the effort.

For example, Dr. Ajay Bhandarwar came up with an innovative idea of transporting COVID-19 patients. His device, the "Kavach," which in Sanskrit means an armor was tried in Mumbai (Hindustantimes, 2020). It was designed as a chamber for contactless movement within the hospital premises. Because many health workers were contacting the disease, it became necessary to move them in a way that minimized infecting health-care workers. The device, a glass transport trolley fitted with HEPA filter at the end served to insulate the patients. Patients could also be incubated in the trolley and also receive sonography tests while in the chamber. These and other innovations became part of the COVID-19 story.

India's innovative capacity was also exhibited as it joined the world in searching for COVID-19 vaccinations. In July, The Indian Council of Medical Research (ICMR) (Department of Health Research) informed the public that the ICMR had partnered with Bharat Biotech International Limited (BBIL) to fast-track clinical trials of the indigenous COVID-19 vaccine (BBV152 COVID Vaccine). This vaccine was the first indigenous vaccine with support from the Government of India. According to a letter from Professor Balram Bhargava, Director General ICRM, the vaccine was derived from a strain of SARS-COV-2 isolated by ICMR-National.

India's daily tracking of morbidity and mortality rates was well reported by both government departments and the local and international media. The earlier reported cases in February and March showed lower morbidity within Indian populations, prompting self-praise from the country's leadership. In April all indications were that India, a country with over 1.3 billion people, was controlling the spread of the coronavirus. After Prime Minister Modi had announced lockdown in March 24, it looked like the projections that the health system would collapse were alarming. Modi extended heightened nationwide lockdown on April 14 (Hollingsworth, 2020). However, numbers picked up in May, and in June (as seen in Table 5.2) through August, the country, including members of its STs, fared poorly. There was no longer room for self-praise and congratulatory statements.

Table 5.2: Example of daily tracking cases June 3–4.

	June 04	June 03
Active Cases	1,10,157	1,05,021
Discharged	1,106,923	1,03.3382
Deaths	6,249	6,012

Live data: Sourced from Deccan Herald, D. H. (June 4, 2020). Daily Covid-19 bulletin. Available on Facebook.com. The coverage includes patient ages and other related health data of those who died from the virus.

According to the Johns Hopkins University (2020) coronavirus tracking database, India was in the company of Brazil and the United States as the countries with most confirmed cases of coronavirus infections. By mid-August 2020, India had 2,647,663 cases and 50,921 deaths. At the same time, the United States reported 5,403,213 cases and 170, 052 deaths. The numbers for Brazil were 3,340,197 cases and 107,852 deaths, respectively. These high morbidity cases were worse than in China, which nationalist Indians and Republican Party adherents in the United States blamed on China. The social media amplified the public and private discontent and regional TV station links embedded within Facebook proved to be extremely useful in bringing live coverage of the unfolding virus-related events.

States such as Karnataka and Maharashtra, which have large numbers of Siddi people, were among the places with high infection rates. Nationwide, there were concerns that the social repercussions of weak health-care infrastructures begged national attention and investments. Although it was not enough to blame China, social media mockery and criticism of China-made COVID-19 PPE continued. Artists had a field day communicating messages about the defects of the Chinese PPE products were and the immediate negative effect on India's own industrial capacity. COVID-19 brought about increased conversations about economic nationalism.

Conclusion

In conclusion, while it is true that India's strong leader Modi framed COVID-19 as a war, the vision of a nationwide defense strategy did not work well because members of the STs like the Siddi people did not receive any special federal resources to cure the perennial problems within tribal health-care systems. Whether the fight against COVID-19 was conducted as a means toward preventing a nationwide crisis, the efforts temporarily provided a sense of unifying message to defeat the enemy. However, the enemy within, which is marginalization of Siddi people, requires more than rhetoric and symbolic provision of water stations in rural villages. COVID-19 required large portions of national resources to systematically plan, and direct mitigation efforts aimed at reducing health-care inequalities. COVID-19 demands the focus on poverty within scheduled communities in rural and urban India. Not as entities for vaccine experimentation but as persons who, if given equal chances will be part of the discovery of new medicines to cure future diseases that cause major public health crises.

References

ABP News. (2020). *Health bulletin*. https://www.youtube.com/watch?v=zFD3pG5OVt4.

Amarat, H. (2006). *African elites in India*. Mapin Publishing.

Ansari, H. (March 24, 2013). *Invisible Indians*. Hindustan times. https://www.hindustantimes.com/mumbai/invisible-indians/story-sHA1BAtR99UPmgJwPJ8fjP.html.

Belur, R. (June 13, 2020). *Amid debate, higher education minister stress for digital edu*. Deccan Herald. https://www.deccanherald.com.

Byatnal, A. (July 16, 2020). *In India, prime minister's relief fund for COVID-19 jeopardizes NGO sector.* Devex.com. https://www.devex.com/news/in-india-prime-minister-s-relief-fund-for-covid-19-jeopardizes-ngo-sector-97706.

Chakravorti, B. (July 10, 2020). *Digital contact tracing's mixed record abroad spells trouble for U.S. efforts to rein in COVID 19.* Government Executive. https://www.govexec.com/technology/2020/07/.

Chavan, V. (April 23, 2020). *Don't panic over centre's COVID-19 projections for Mumbai: Maha health minister Rajesh Tope.* Republicanworld. https://www.republicworld.com/india-news/general-news/dont-panic-over-centres-covid-19-projections-for-mumbai-rajesh-tope.html.

Choudhry, N., Hammer, J., Kremer, M., Muralidharan, K., & Rogers, F. H. (April-May). Teacher absence in India: A snapshot. *Journal of the European Economic Association, 3*(2).

Cutler, D., Deaton, A., & Lleras-Muney, A. (2006). The determinants of mortality. *Journal of Economic Perspectives, 20*(3), 97−120. https://doi.org/10.1257/jep.20.3.97

Das, J., & Hammer, J. (2004). *Stained Mercy: The Quality of Medical Care in Delhi. Policy Research Working Pare, No. 3228.* Washington, D.C: World Bank. https://openknowledge.worldbank.org/handle/10986/14725.

Deaton, A. (2004). Measuring poverty. In A. Banerjee, R. Benabou, & D. Mookhejee (Eds.), *Understanding poverty.* Oxford University Press.

Deccan Herald DH. (June 4, 2020). *Daily COVID-19 bulletin.* Facebook.com.

Delevingne, L. (July 13, 2020). *Amid pandemic, wealthy U.S. families approved for government loans.* Reuters. https://www.reuters.com/article/us-health-coronavirus-ppp-familyoffice/amid-pandemic-wealthy-u-s-families-approved-for-government-loans-idUSKCN24E130.

Desai, D. (May 21, 2021). *61K Covid deaths not counted in Gujarat: Report.* Hindustan Times. https://www.hindustantimes.com/india-news/61k-covid-deaths-not-counted-in-gujarat-report-101621027267608.html.

Devaiah, D. (July 24, 2020). *Shantharam is India's first legislator from African -origin Siddi community in Karnataka.* Bengaluru: Indian Express. https://indianexpress.com/article/cities/bangalore/shantharam-is-indias-first-legislator-from-african-origin-siddi-community-in-karnataka-6521265/.

Dragomir, C. (March 13, 2017). *Scheduled tribe status: The need for clarification.* Center for the Advanced Study of India. (CASI). India in Transition. University of Pennsylvania. http://www.casi.sas.upenn.edu.

Economic Times. (April 9, 2020a). *Let lockdown not grow digital divide.* https://economictimes.indiatimes.com/blogs/et-editorials/let-lockdown-not-grow-digital-divide/.

Economic Times. (June 13, 2020b). *COVID-19 testing: Maharashtra govt slashes charges for coronavirus tests by private labs.* https://economictimes.indiatimes.com/news/politics-and-nation/dont-crave-for-any-post-party-gives-us-post-and-can-take-it-back-sachin-pilot/videoshow/77469723.cms.

Economic Times. (June 27, 2020c). *Majority of MLAs against lockdown in Bengaluru: Karnataka govt.* https://economictimes.indiatimes.com/news/politics-and-nation/majority-of-mlas-against-lockdown-in-bengaluru-karnataka-govt/articleshow/76655956.cms.

Economic Times. (August 2, 2020d). *ET Covid-19 state spread tracker.* https://economictimes.indiatimes.com/news/politics-and-nation/test-et-covid-19-tracker/articleshow/74750649.cms?from=homepage.

Express Web Desk. (August 2, 2020). *Hyderabad. Andhra Pradesh, Telangana coronavirus live updates: Telangana reports over 1800 cases in a day; Andhra tally touches 1.5 lakh mark.* The Indian Express. https://indianexpress.com/article/cities/hyderabad/andhra-pradesh-telangana-coronavirus-live-updates-6535098/.

Furer-Haimendorf, C. (1982). *Tribes of India: The struggle for survival.* University of California Press.

Ganguly, M. (May 26, 2020). *India cannot expect to beat COVID-19 without tackling hate and corruption.* https://www.hrw.org/news/2020/05/26/india-cannot-expect-beat-covid-19-without-tackling-hate-and-corruption#.

Government of India. (1952). *Report of the commissioner for scheduled castes and scheduled tribes.* New Delhi: Government Printer.

Government of India. (2013). *Statistical profile of scheduled tribes of India 2013.* Statistics Division of the Ministry of Tribal Affairs. Shastri Bhavan. http://www.tribal.nic.in/ST.

Gupta, A. (April 13, 2020). *COVID-19 and the importance of improving civil registration in India*. Center for the Advanced Study of India. (CASI). India in Transition. University of Pennsylvania. http://www.casi.sas.upenn.edu.

Harbi, S. (2020). Travelling across "AfroRoutes". TECHVANGART. https://techvangart.com/2020/08/20/travelling-across-afroroutes/.

Hindustantimes. (August 1, 2020). *Mumbai hospital devices "kavach" to move COVID-19 patients*. https://www.hindustantimes.com/mumbai-news/mumbai-hospital-devises-kavach-to-move-covid-19-patients/story-KFzJZJztf6t0Bi6dvBlFCL.html.

Hollingsworth, J. (April 29, 2020). *How does India, a country of 1.3 billion people, have around 1,000 coronavirus deaths?* CNN. https://www.cnn.com/2020/04/28/india/india-coronavirus-outbreak-explained-intl-hnk/index.html.

Jain, A. (March 14, 2020). *On a cultural drawbridge: The Siddi Tribe*. Decca Herald. DH. https://www.Deccanherald.com.

Jain, Y. (April 21, 2020). *Maharashtra minister says 'there should not be a ban on liquor stores' if distancing norms followed* (Vol. 18). CNBC TV. https://www.cnbctv18.com/healthcare/alcohol-sale-to-be-allowed-in-maharashtra-amid-coronavirus-lockdown-5734701.htm.

Johns Hopkins University. (2020). *Coronavirus resource center. Tracking cases and trends*. https://coronavirus.jhu.edu.

Kapila, U. (2002). *Indian economy: Issues in development and planning and sectoral aspects*. Academic Foundation.

Ketaki, S. (2014). *A certain grace: The Sidi, Indians of African descent*. https://www.youtube.com/watch?v=s5ikcz125qE.

Kruk, M. E., Gage, A. D., Joseph, N. T., Danaei, G., García-Saisó, S., & Salomon, J. A. (November 17, 2018). Mortality due to low-quality health systems in the universal health coverage era: A systematic analysis of amenable deaths in 137 countries. *Lancet, 392*(10160), 2203–2212. https://doi.org/10.1016/S0140-6736(18)31668-4. Epub 2018 Sep. 5.

Kulkarni, T. (August 2, 2020). *Coronavirus: Karnataka chief minister B.S. Yediyurappa tests positive for COVID-19*. Bengaluru: The Hindu. https://www.thehindu.com/news/national/karnataka/karnataka-chief-minister-bs-yediyurappa-tests-positive-for-covid-19/article32255466.ece.

Kumar, P. (1998). Indian Federalism: Issues and challenges in the context of demands for new states (222-232). In K. J. Azam (Ed.), *Federalism and good governance: Issues across cultures*. South Asian Publishers.

Melwani, L. (August 2017). *Africans in India: From slaves to sultans*. Khabar Magazine. http://www.khabar.com/magazine/cover-story/africans-in-india-from-slaves-to-sultans.

Mueller, M., & Tan, Z. (1997). *China in the information age: Telecommunications and the dilemmas of reform*. Greenwood Publishing.

Mumbai Mirror. (August 2, 2020). *Covid-19 live updates: Maharashtra reports 9,601 cases, Mumbai's death toll climbs to 6,395*. https://mumbaimirror.indiatimes.com/coronavirus/news/covid-19-live-updates-unlock-3-begins-mumbai-maharashtra-thane-lockdown-mira-bhayandar-kalyan-dombivali-containment-zones-international-flights/liveblog/77296219.cms.

Mwaura, W. (May 7, 2020). *Coronavirus in Kenya: Fearing 'money heists' amid pandemic*. BBE. https://www.bbc.com/news/world-africa-52540076.

Negi, T. S. (1976). *Scheduled tribes of Himachal Pradesh: A profile*. Raj printers.

New York Times. (June 3, 2007). *Indian shepherds stoop to conquer caste system*. https://www.naytimes.com.

Oommen, T. K. (1977). Scheduled castes and scheduled tribes. In S. C. Dube (Ed.), *India since independence* (pp. 156–157). Vikas Publishing House Pvt.

Powell-Jackson, T., Mazumdar, S., & Mills, A. (2015). Financial incentives in health: New evidence from India's Janani Suraksha Yojana. *Journal of Health Economics, 43*, 154–169.

Rajan, S. (January 29, 2013). *Nawabs of Janjira and Sachin were African slaves*. AB Wire. The American Bazaar https://www.americanbazaaronline.com/2013/01/29/nawabs-of-janjira-and-sachin-were-african-slaves/.

Ramdev, D. (May 13, 2015). *Racing to the finish: Siddi tribal women in TCS world 10 Km run*. Nation current affairs. Deccan Chronicle http://www.deccanchronicle.com/150513/nation-current-affairs/article/racing-finish-siddi-tribal-woman-tcs-world-10k-run.

Ray, K. (September 10, 2018). *4,300 Indians die daily due to poor hospital care*. DH Deccan Herald. https://www.deccanherald.com.

Redelman-Sidi, G. (April 27, 2020). Could BCG be used to protect against COVID-19? *Nature Reviews Urology, 17*, 316–317. https://doi.org/10.1038/s41585-020-0325-9

Sharma, T. R. (1998). The federal and the marginal: The shifting styles of interface between nationalism, sub-nationalism and sub-sub-nationalism in India. In K. J. Azam (Ed.), *Federalism and good governance: Issues across cultures* (pp. 214–221). South Asian Publishers.

Shroff, B. (2017, March). India's Forgotten Africans. *Documentary. Down to Earth*. https://www.youtube.com/watch?v=2I4vUBTHc_4.

Sinha, P. B., & Rana, A. (2001). *Concise Handbook of Indian economy: A Guide to the Quintessence of Indian economy*. M'n' M series. Jawahar Publishers.

Suryawanshi, S. (June 17, 2020). *Maharashtra finally owns up to 1328 COVID-19 backlog deaths but denies concealing data*. Express News Service. https://www.newindianexpress.com/nation/2020/jun/17/maharashtra-finally-owns-up-to-1328-covid-19-backlog-deaths-but-denies-concealing-data-2157847.html.

Times of India. (January 10, 2020). *Gujarat house passes bill on SC/ST quota in Lok Sabha, assemblies*. https://timesofindia.indiatimes.com/city/ahmedabad/.

United Nations. (2017, February 21). The Siddis: India's Forgotten Africans. Documentary. https://www.youtube.com/watch?v=Q4SNsgwXsys.

Upadhyay, A. (May 1, 2020). *With the highest COVID-19 cases in the country, Maharashtra's health minister Rajesh Tope talks about the fight against coronavirus*. Banega Swasth India. https://swachhindia.ndtv.com/with-the-highest-covid-19-cases-in-the-country-maharashtras-health-minister-rajesh-tope-talks-about-the-fight-against-coronavirus-44256/.

V6 News Telgu. (May 28, 2020). *Telangana govt release fake health bulletin on corona cases*. Petitioner Varun Rao.

Vallangi, N. (August 20, 2016). *India's forgotten African Tribe*. BBC Travel. https://www.asiaone.com/travel/indias-forgotten-african-tribe.

Wagstaff, A. (2003). Child health on one dollar a day. *Social Science & Medicine, 57*(9), 1529–1538.

Weiner, M., & Osgood, J. (1975). *Electoral politics in the Indian states* (Vol. II). Manohar Book Service.

World Bank. (1990). *World development report*. World Bank.

CHAPTER 6

COVID-19 communication and Indigenous Australians

When we turn our attention to political communication in developed countries, perhaps the starting point is to note that top-level bureaucrats and heads of administrative agencies have long traditions of distinguishing active political actions from rule-based managerial practices. In that sense, the expectation is that on January 25, 2020, after COVID-19 appeared in Australia, the country's political leaders throughout the subcontinent would discharge duties in the most rational manner. And to a large extent, they treated the issue as a national emergency and crisis. An ecological approach to understanding policy management during crisis situations requires we scan physical, geographical, social and technological spaces within which COVID-19 was managed.

Governance framework and public health policy

As a federal system, with six states and two territories, Australia is home to many governments including those of First Nation people who were subjected to colonialism and settler economic structures that have persisted. Of Australia's 25 million people, about 3% are Indigenous. Australian is an island continent, in the middle of the Pacific Ocean, which is by itself an ideal isolation location. As of May 17, 2020, Australia had recorded 7045 cases of COVID-19 and 98 deaths. By end of July 2020, the numbers of Australian deaths from COVID-19 were 313. A total of 21,397 cases had been positively identified (Australian Government, 2020). Thus far, the greatest number of confirmed cases were in Victoria (VIC) (14,957, with 322 deaths), New South Wales (NSW) (3875), and Queensland (QLD) (1089) (Australian Government, 2020). According to the Commonwealth of Australia's Department of Health (2021), by the end of March 2021, VIC still had the highest deaths at 820, NSW 54, Western Australia (WA) 9, and South Australia (SA) 4, respectively, confirming the relative policy stability in mitigation processes throughout the territory.

Perhaps it is best to start by acknowledging that historically, the largest populations of Aboriginal and Torres Strait Islander Australians lived in NSW, QLD, and WA. This brings us to reflect on just how the Indigenous populations relate, in a technical sense to Australia's federalism. The Australian states include Australian Capital Authority (ACT), New South Wales (NSW), Northern Territory (NT), Queensland QLD, South Australia

The Inequality of COVID-19. https://doi.org/10.1016/B978-0-323-99867-3.00001-6

(SA), Tasmania (TAS), Victoria (VIC), and Western Australia (WA). Australia also administers several external island territories.

Originally, home to First Nations people, homeless immigrants from Britain and Ireland arriving after World War II participated in dispossession. There have also been immigrants from Southeast Asia. The core organizing colonial principle has always been to segregate Fist Nations communities and preventing them from joining white communities. It is the unification of the colonies after Federal Council of Australia Act of 1885 and the 1900 federal constitution that gave form to current cooperative federalist streak.

In this system, the Governor General represents the monarch as titular head of the federal and state governments, but daily government operations are managed by the Prime Minister. Australia has a collective body of ministers who, with the Prime Minister, are members of the Federal Executive Council. The colonial structures attempted to reduce Indigenous people to a dependency and semisovereign scheme that, in itself, conferred tasks or in sociological terms, roles of managing public health challenges to the different jurisdictions.

As in the British system, the cabinet makes policy decisions, which, in the case of Public Health, is under the purview of the Department of Health (DOH), which through the Australian Health Protection Principal Committee (AHPPC) coordinates efforts and ensures that pandemic management is integrated and consistent across the country. The national response is summarized in the Australian Health Sector Emergency Response Plan for Novel Coronavirus (COVID-19 Plan) (Australian Government, 2020b). And since Australian states and territories have their own devolved governments, they too have health departments that oversee the local response to COVID-19. This, they do by working together with the Australian government and share information on logistics.

Between the collaborative approaches among different Australian states and territories are features for managing the pandemic among the Indigenous communities and also persons with disabilities. There were specific guidelines to manage the COVID-19 crisis for the Aboriginal and Torres Strait Islander populations. Published in July 2020, the plan incorporated values of the Indigenous societies (Commonwealth of Australia, Department of Health, 2020).

Arguably, because of the existence of what Cousins (2020) refers to as "a neglected health system" these populations were considered high risk and also required aggressive teams that would engage the communities and provide information on COVID-19. This assertion is based on well-known research on social determinants of health and well-being. Most significantly, the places where people are born and live are related to life expectancy and ability to cope with consequences of illness, including those brought about by viruses. Research continues to support the thesis that First Nations people have a high burden of diabetes, respiratory ailments, and heart disease, which were considered to be underlying risk conditions with COVID-19.

Cousins (2020) reported that Aborigines and Torres Strait Islander people suffered deaths and hospitalizations at a higher rate than the general population during the previous H1N1 pandemic in 2009. Thus, efforts would also be made to empower Indigenous citizens tackle the COVID-19 pandemic. The plan envisioned development of a culturally appropriate, respect for human rights, data-driven, and safe risk reduction strategy. This entailed adherence to principles of culture, country, family, and community in shared decision-making and disease control management.

At its core, the plan was collaborative and involved Indigenous communities and other stakeholders. The first approach was to support local operational plans and then target areas of concern. The targeted action plans were in different stages and involved containing community transmissions and public education and awareness campaigns in case when outbreaks were reported. The plan also included a "stand down and evaluation" stage in which the effectiveness of the responses would be evaluated, and lessons learned.

How did it go?

For the first 6 months of the pandemic, it was evident that the primary voice on the management of the pandemic came from the Federal Prime Minister Scott Morrison. He, on behalf of the federal government, announced in February that any person arriving from countries with high rates of infection would quarantine for 14 days in a third country before they would enter Australia. The government effectively closed the borders after the World Health Organization (WHO) declared COVID-19 a global pandemic. The executive actions were deemed necessary after the death toll in Australia rose in mid-March. The Australian press covered numerous stories of panic buying, toilet paper shortages, and other behaviors associated with COVID-19 events and activities.

The government took to reacting to the ensuing barrage of information. The Prime Minister gave daily press conferences and delegated medical and public health decisions to health providers. Thus, those with leeway to speak on the technical issues regarding the COVID-19 pandemic were not politically motivated individuals. Health experts like the Chief Medical Officer of Australia took lead as the Prime Minister stepped back and cheered the experts avoiding politicizing the messages.

Perhaps the most visible department that led the coordination efforts was the DOH. The department constituted an AHPPC—a subcommittee of the Australian Health Ministers' Advisory Council which was part of the Council of Australian Governments to coordinate mitigation efforts between federal and state governments. The advisory council was constituted to make government more responsive and collaborative. The leadership ensured that stakeholders and access to vital information and that in the interest of fairness, special interests were not given unlimited access to decision-makers. The advisory council was also quite representative and took a balanced view on important issues concerning COVID-19 in

Australia and its territories. Although there were challenges in consensus building, disputes were resolved quickly and did not become a front-page news item as was in the case of the United States where political heads openly disagreed with members of their own government agencies (see, for example, Baker, 2020; Higgins-Dunn, 2020).

The Australian Department of Health posted COVID-19-related materials on the government websites on a daily basis. Officials also held meetings several times daily with regional governments and provided information on various government platforms. Announcing "hot spots" and other global progression of the COVID-19 pandemic ensured that the public was aware of what was going on globally.

Australian authorities had done advanced planning and positioned the country on a trajectory that would serve the public interest. The planning process was national, and it took persuasion to bring the state governments on board. The states seemed to adhere to the tenets of "cooperative federalism" and agreed to work together during the pandemic. According to Dr. Lesley Russell, Associate professor at the Menzies Centre for Health Policy, Australia empowered health agencies to make professional decisions avoiding the more political campaign style processes. The country conducted several tests and regularly updates the citizens on actions being taken to check the spread of COVID-19.

Testing and tracing

As the days passed, activities took a more nuanced regional approach but with strict enforcement of national guidelines. Among the approaches adopted was institution of a robust tracing regime.

Australia developed a nationwide tracing app and had a surplus of test kits. Australia invested in testing early enough including facilitating citizens to download the contract tracing app throughout the country. Australia also avoided bureaucratic debates over faulty test kits and had a decent test results reporting system.

Health organizations helped recruit contract tracers who put into place a system that enabled authorities to know where the disease density was highest and its trajectory. Australia did a good job at establishing the risks it faced and acted quickly whenever the data showed areas of concern. The costs of testing was also relatively inexpensive and available to larger sections of the community. It helped that the publicly funded Australian public health-care system is widely popular and covered significant COVID-19 cases. Those who needed care and testing were not denied on the basis of lack of insurance. The health-care system had a better track record of improving health literacy within the commonwealth. This encouraged many of the lower income Australians.

Logistic arrangements made it possible for Australia's less empowered communities to effectively handle the COVID-19 isolation plans. For example, those working did not fear to take leave because of unsupportive employment law. In Australia, the nervousness associated with losing a job due to staying away during the pandemic was largely minimized through a nationwide policy. Full and part time employees have 4 weeks of paid leave and 10 days of paid sick leave. Such arrangements favored the lower-level essential service workers who had borne the brunt of the pandemic in the United States.

The summer months

By late summer, with less than 300 cases in early July, Australia had brought its numbers largely under control (Packham, 2020a). At the end of July, the National Indigenous Television reported that 60 members of the Aboriginal or Torres Strait Islander people had been confirmed as COVID-19 cases. Of these, there were no reported deaths or even intensive care hospitalization cases. The country as a whole had recorded approximately 10,000 cases.

This was a remarkable success. However, the country did not achieve the same level of success as neighboring New Zealand where the Prime Minister Arden called two cases "unacceptable" signifying her administration aggressive mobilization of resources to check the spread of COVID-19 in the region.

Notwithstanding the region's relative success there were exceptions as some eruptions emerged in the state of VIC, prompting confinements and reintroduction of mitigation measures. Toward the end of June, the state government of VIC deployed military operations to manage the quarantine. It also increased the levels of testing in the worst affected suburbs. Although each state determined how the easing would be organized, the country's borders were still closed. And Australia sought an exemption from a requirement that its citizens be quarantined for 14 days upon arrival in the United Kingdom (Packham, 2020b).

Realities of COVID-19 prevention initiatives and solutions among Aboriginal and Torres Strait Islanders

Although the federal government had established a national Aboriginal and Torres Strait Islander advisory committee (group) on management of COVID-19 to move forward an agenda for preventing outbreaks of COVID-19 within Indigenous societies, the motivation to do so was also largely organic. It should be emphasized that at the outset that the bulk of the success of the lower rates of infection of COVID-19 among Indigenous communities is their own resilience and ability to understand and improve their defenses against the coronavirus pandemic. Many community leaders and First Nations controlled health sectors working with state health professionals and experts in communicable diseases were already consulting widely with the National Indigenous Australians Agency and the Australian Indigenous Doctors' Association over strategies to adopt and prevent the spread of the coronavirus in their native lands. These groups shared information and agreed on principles that would be applied within their societies.

Australian Indigenous communities demonstrated that in spite of the limitations they face, the collective will to institute mitigation systems was extraordinarily successful. As Madeleine Keck (2020) reported, the Indigenous people used unique methods to save lives. She argues that Aboriginal-led responses to the COVID-19 crisis "invert the belief that First Nation peoples are at a higher risk from the virus."

Evidence suggests that the expected health gap between Indigenous and non-Indigenous Australians did not produce the devastation predicted as were the cases in Brazil and the Navajo nation in the United States. The Australian Indigenous communities narrowed this gap by deployment of the Aboriginal-controlled health services for health literacy campaigns and training. Indigenous-tailored messaging which adhered to culturally appropriate wellness promotional communications seemed to work very well. Thus, deservedly, according to Professor Fiona Stanley, an epidemiologist, the outcomes from managing COVID-19 were handled much "better than any other Indigenous nation internationally" (Quoted in Keck, 2020).

One of the most profound strategies within the Aboriginal-controlled health services was the deployment of social media platforms to convey Indigenous-tailored health messaging. Keck (2020) explained: "The Aboriginal Health Council of Western Australia uploaded animated videos to Facebook to convey messaging on social distancing and the importance of maintaining regular check-ups using telehealth services." This was hardly surprising. Following the great work done by Carlson (2015) and others, we know that social media presents an efficacious platform for development of culturally appropriate wellness promotion and mental health-care programming. Carlson and Frazer (2018) identify that Facebook was most popular app for family and community messaging. Facebook's appeals were probably enhanced because it also embedded COVID-19 posts and broadcasted video clips from other media outlets. Carlson and Frazer (2018) add that while the usefulness of Twitter was mainly in the realm of politics, Indigenous Australian youth preferred to use Snapchat and Instagram for their social interactions while connecting to their elders through Facebook. Using social media for informing Aboriginal people about prevention of COVID-19 was, therefore, not unexpected. It comes from a space of proven efficacy. And it worked to make Indigenous Australians among the best global performers in using social media to communicate mitigation strategies. Journalist Julie Power writing for *The Sydney Morning Herald* reported,

> *But a rapid response by Aboriginal leaders to stop the virus worked. Of the 907 deaths and 27,698 cases in Australia, including 685 deaths in aged care, there have only been 149 cases among Indigenous Australians. There were no cases in remote communities and not one death of any Indigenous person. Only 13 Indigenous people were hospitalized (Power, 2020).*

There was also the ability to use other ways for effective communication, especially in local Indigenous languages (Cousins, 2020). Data show that 60% of people in the NT communicate in Native languages and not English (Ralph et al., 2020). This barrier could only be broken by either using interpreters and/or local speakers to explain the COVID-19 mitigation protocols. Thus, for example, the health-care professional teams at various Aboriginal Health Service Centers were regularly featured on community radio stations. Keck (2020) provides an example of the Derbari Health Service, which provided appropriate hygiene information on the Perth, WA-based Noongar Radio Station, 100.9 FM. This award-winning station, which was licensed in 2008, hosted several panel discussions about the coronavirus pandemic and provided advice on how to prevent the spread of the disease. The messages conveyed on radio bulletins were consistent and clear for the Indigenous listeners.

As was in cases elsewhere, radio stations utilized the services of local or Indigenous celebrity sports personalities and other influencers to persuade the populations to adhere to health protocols such as washing hands, wearing face masks, social distancing, and avoiding crowds. Additionally, government provided community radio (WAVE) service announcement platforms to help Aboriginal and Torres Strait Islander peoples look after elders and older people during the pandemic. The information was posted on a national ministry of health websites (Australian Government, 2020c). Audio functionality was available and frequently updated.

The messages, mostly in English language, advised on keeping away from each other and staying in contact to keep elders strong. The government advised that members of the Indigenous communities continue to set up time to connect, offer to run errands for elders, including buying groceries and placing at doors and to chat over the fence (social distancing). The information was changed at regular intervals to respond to the conditions created by the novel coronavirus. The government maintained a live call Coronavirus helpline.

At state levels, messages on giving hope to elders were embedded on Facebook platforms as well. For example, the "Help keep Aboriginal elders and communities safe from Coronavirus (COVID-19)" message was posted by the Department of Communities and Justice NSW. It reiterated the need for communities to stay together.

Another reason for the positive outcomes was that Indigenous communities throughout Australia did not repeat mistakes made by mainstream health-care systems and providers. The Aboriginal health-care services were quick at learning about failures elsewhere and achieved tremendous success in organizational self-correction. It is also imperative to note that many communities in remote areas of the country moved quickly on their own discretion, to prevent visitors into their communities by closing their territorial borders. This was done ahead of the federal government's requirements and advice.

For instance, the Anangu Pitjantjatjara Yankunytjatjara Lands in remote SA and Cape York communities in Far North Queensland were among some of the Indigenous communities to take a proactive step and lead the nation in implementing self-governance strategies for checking the spread of localized coronavirus outbreaks (Cousins, 2020).

As alluded to, huge populations of First Nations people live in urban settings as well. It is a region where 30% of the population identifies as Aboriginal or Torres Island. So, the numbers of those with inadequate shelter are relatively higher. Homelessness is not just about shelter, but it affects a person's ability to enjoy basic freedoms and rights, including health care. An inquiry on Youth homelessness showed that many young homeless Australians were also at risk of mental illness (Australian Human Rights Commission, 1993, 2006).

Many First Nations people reside in overcrowded dwellings and were poorly positioned to prevent outbreaks, if these occurred in city dwellings. It was observed that many Aboriginal and Torres Strait Islander people in urban settings returned home to their native lands to isolate. One must point out that the remote places do not necessarily offer advanced standards of living. There are still areas in remote communities where overcrowding in houses was observable. And in those areas, plans were arranged for setting up tents to ensure that social distancing was possible. Others opted to camp out in bushes and protect their elders for purposes of preserving traditional knowledge (Cousins, 2020). This strategy was a rational calculation to preserve community interests and save lives.

Although variations in Aboriginal cultural values are well documented, the overriding principles adopted by community included the recognition that elders were to be protected at all costs. Not only for their services as custodians of Indigenous knowledge, but as persons who hold the narratives that support emotional health. Communities held the views that the self is relational and place enabled culture to flourish. Thus, in cases where people returned from urban living to rural homes to isolate, the overriding cultural meaning was that identity is defined by place.

As members of the communities cooperated to do what the health experts recommended, there was a sense of togetherness that was both spiritual and culturally defined. The common fear of loss was unmistakably recognized in the actions taken. It was observed that the meaning of the threat of COVID-19 was shared, and Indigenous people were quick to recognize the connections between their own individual well-being and that of the community. In other words, collective despair and cultural well-being of the whole community were part of the same wellness equation. There was, therefore, a strong need for community control of COVID-19 mitigation efforts. This motivation meant that community also managed the stigma arising from the virus. Indigenous communities internalized the principle that health was beyond the notion of absence of disease but required emotional connectiveness and solidarity of people. It is, therefore, family support and community empowerment that proved an important piece of the holistic healing witnessed in the Indigenous communities. These values were realized in several ceremonies that brought everyone together, even when practiced within the framework of social distancing.

Viral misinformation

Perhaps, one of the challenges that developed countries like the United States and Australia faced was that of COVID-19 misinformation along with blaming and stigmatization of groups of people. While some of the misinformation came from bad journalism, media overload, and careless talk from political leaders, others creeped in (Savage, 2020). Misinformation created rumors, and in the case of Australia, there were reported instances where Chinese Australians were subjected to anti-Chinese language and abuse online and on the streets (Lee, 2020). Debbie Bargallie Sulistyo (2020), one of the leading authorities on race relations in Australia, called out this form of racism and advised that the country required to invest in racial literacy.

In tandem with the internal frustrations, at the state level new frontlines emerged, partly fueled by misinformation campaigns. Diplomatic dispute emerged after Australia joined in the call for an international investigation into the source of COVID-19 and its spread (Mercer, 2020a). China had rejected this move. Rumors were not just associated with mainstream Australian press houses. Within the ranks of Indigenous Australians, rumors took a different form. For example, in the NT, myths were circulated to the effect that First Nations peoples were immune to COVID-19. And that "heat kills the virus so people in Australia's tropical north are immune" (Kerrigan et al., 2020). Another rumor peddled around was that the disease had been unleashed to kill Aboriginal peoples. However, the data did not support that assertion and in fact (Kerrigan et al., 2020).

During the pandemic there was a clear rise of misleading viral information on Twitter, Facebook, Instagram, and other social media platforms (Samios & McCauley, 2020). UNICEF office in Australia warned that "misinformation during times of a health crisis can spread paranoia, fear, and stigmatization. It can also result in people being left unprotected or more vulnerable to the virus" (UNICEF, 2020).

To ensure against misinformation, the government employed a clear messaging strategy and quickly countered any distortions concerning the disease. Speaking of the misinformation, communications Minister Paul Fletcher urged Australians to be aware of information that was misleading. The minister called for use of fact-checkers on social media platforms to remove misleading dangerous information. In addition to several well-known and trusted media organizations, the federal government's DOH (health.gov.au) website was the recommended website for the most up-to-date and accurate information concerning COVID-19 (Samios & McCauley, 2020).

An Australia-based independent technology organization concerned with ethical practices, Responsible Technology Australia, called upon social media corporations to create "a list of the top 100 viral pieces of information about Coronavirus … with demographic context about where it is being shared and also flag if this content is coming from bad or foreign actors" (Singh, 2020). As time went by, corporations like Google and Facebook flagged down misleading information on COVID-19. Likewise, Twitter removed posts that were false.

Prime Minister Scott Morrison's administration was given high marks for the manner in which it provided transparent information to Australian citizens. According to polls by the Clemenger Group's Australia NOW research, 70% of Australians believed the government was giving the right information about the impact of coronavirus pandemic (McIlroy, 2020). The government of Australia regarded misinformation as an affront to democracy and took steps to enhance cybersecurity after warning of Russia and China's disinformation campaigns (McGuirk, 2020). Australian officials accused Russia and China of "using the anxiety around the Coronavirus pandemic to undermine Western democracies by spreading disinformation online" (McGuirk, 2020; Murphy, 2020). Although no examples were offered, the government posited that disinformation contributed to a climate of division and fear.

American intelligence agencies and the State Department's Global Engagement Center reported the existence of several English-language websites working together to project Kremlin talking points during the COVID-19 pandemic (Tucker, 2020). The disinformation campaigns were quite active in the early months of February and March and continued into April. The disinformation campaigns promoted the conspiracy theory that the virus was a bioweapon deployed against China. The disinformation networks worked to undermine Western democracies including Australia and normalize undemocratic decision-making processes which would, in turn, undermine efforts to prevent the spread of COVID-19.

Information overload

The COVID-19 epidemic inundated government agencies with data of all kinds, from information around patient symptoms, and relief fund disbursement to economic recovery and unemployment claims. Indeed, the Clemenger Group's Australia NOW research poll indicated that 17% of the respondents were concerned about receiving too much information from the government (McIlroy, 2020). This concern was therefore not just about private citizens but of significant policy implications for the government. Government required data that were accurate in order to make the best decisions. The need for elimination of poor-quality information was thus a key responsibility of government agencies across the globe.

Struggling with information overload was reported as one of the unintended consequences of the coronavirus control and management regime. As Australian workers moved to shelter-in-place and work from their homes, levels of stress appeared to increase. Research from Open Text revealed that one in three Australian office workers felt they did not have digital tools to work remotely and were inundated with information overload. Stress levels increased, in part because of challenges of unplugging from various devices including mobile phones and computer apps. IT Brief contributor Shannon Williams (2020) reported that digital services were widely employed for social communication, health-care information, and also ecommerce including buying groceries.

Stress was also associated with inability to make sense of the information to make health-enabling decisions. Here is where the issue of managing misinformation again became a key part of the COVID-19 management portfolio. Journalist Graham Readfearn (2020) amplified the view from experts that deciding what was good information and what wasn't was an important aspect in managing the pandemic. And especially, if the information was going to be spread through social media. This behooves societies to consider the sources of the information and rely on expert epidemiologists and other health-care experts. Quoting Lyndal Byford of the Australian Science Media Center, the main thing is to consider "who was making the claim" and who funded the various online outlets (Readfearn, 2020). In general, most health experts from Australia voiced the opinion that the WHO and national science academies were reliable sources of information on COVID-19.

The Open Text research found that, "two-fifths (41%) of Australian respondents agree that the number of information sources email, news feeds, diaries, social media sites, company drive, shared drive, etc. they check each day has increased in the last five years. On average, almost two in ten (17%) Aussie workers typically spend more than a minute searching for a specific file or piece of information for work purposes. Only a quarter (26%) can usually find the file they require in less than ten seconds" (Williams, 2020).

In sum, Brad Creighton, a Portfolio strategist, Dynamics Markets, Sydney, Australia, advised that in the age of information, it was important to separate noise from the regular flow of information. More important, while the mainstream media emphasized the mortality rates and the undeniable human tragedy there was also merit in providing context and information that measured the level of disruption to the economy and denouncing "wild conspiracy theories" which caused much fear. Thus, framing was an important piece of the COVID-19 mitigation strategy (Creighton, 2020).

Ecosystem of ICTs and other lingering challenges

Australia's information and communication technology (ICT) ecosystem is mature and when cases of COVID-19 were first reported, the entire country was instantaneously alerted. Australia is among the 20 best performers on the Global Information Technology report of 2016; its readiness to apply ICTs to all human development challenges has accelerated and covered wide sections of the population (Knoema, 2016; World Economic Forum, 2016, p. 16). This means that the widespread and systematic usage of ICT products and e-government positively impacts societal transformations in all areas of the economy, including health care and education.

Australia ranks far behind Finland, Switzerland, Sweden, Canada, Japan, the United Kingdom, Singapore, the Netherlands, and the United States which by 2016 were the countries with the heaviest investments and innovation for generation of digital products and services. These were also countries where affordability, skills, and infrastructure placed them in good standing to bridge the digital divide for social impact. Australia's digital divide remains a work in progress and is the key concern for the gaps in telehealth and e-learning within First Nations groups.

In previous work, various scholars have shown the negative outcomes of the problem and especially its lopsided nature, disadvantaging rural communities and people who face cultural and language barriers (Black & Atkinson, 2007; Erdiaw-Kwasie & Alam, 2016; Gibson, 2003; Lloyd & Hellwig, 2000; Thomas et al., 2016). Even among the Indigenous people, the disadvantages for women, disabled persons, and elderly were quite pronounced (McLaren & Zappala, 2002). Most significantly, the digital divide in Australia exacerbates the existing inequalities. Unequal access to the internet and computers also lowers productivity for those who are unable to have continuous access. The effects were visible for Indigenous Australians working from home where computing power was limited.

Digital divide concerns

Thanks to the efforts by the Australian Communications Media Authority, internet usage in Indigenous communities has improved. It took much government effort to narrow the layout of the infrastructure for the information society that Australia has become (Rooksby et al., 2002). Some of the better-known responses included the Networking the Nation Program, which funded initiatives around the country. Others included the CTC@NSW program, the Reach for the Clouds wired housing scheme, Access@ Schools program, and numerous localized training programs across the different states (Curtin, 2001; Meredyth & Thomas, 2000; Samaras, 2005; Turk, 2003; Willis &Tranter, 2006). The Australian Bureau of Statistics (ABS) report (2018) on Household Use of Information and Technology (HUIT) reveals that access in remote Indigenous communities is still problematic but satellite connections and power provided through generators have eased the hardships associated with lack of infrastructures.

The HUIT survey collected the Multipurpose Household Survey for 2016–17 and confirmed the existence of information inequalities. For instance, the proportion of households with access to the internet at home has been steadily climbing since 2004–05 but remained constant between 2014–15 and 2016–17 at 86%. Devices used to access the internet at home were mainly desktop or laptop computers. These are used by 91% of connected households (Australian Bureau of Statistics, 2018, pp. 2016–2017).

Similarly, mobile or smartphones are also used by 91% of connected households. The survey revealed that tablets were the next most popular type of device used by 66% of households to connect to the internet. Internet connected TVs had the largest increase in usage by households, from 27% in 2014–15 to 42% in 2016–17. Internet connected music or video player devices showed a drop-in usage from 26% in 2014 to 19% in 2016–17. The three most popular online activities were entertainment, social networking, and banking (all 80%) (Australian Bureau of Statistics, 2018, pp. 2016–2017). This is consistent with studies done on Indigenous usage of online platforms for social messaging.

From a health policy perspective, there is still much to be done. The proportion of internet users accessing the internet health services of health research increased from 22% of internet users in 2014–15 to 46% in 2016–17, which means that telemedicine is yet to be mainstreamed. However, overall usage had been growing steadily, especially driven by the internet enabled mobile phone handsets. The term "digital divide" in Australia can, therefore, be linked more to the speed, hardware, and education elements rather than a lack of government effort to provide basic infrastructures. The federal government had established internet platforms such as myGov and aggressively promoted an ICT Sustainability Plan to benefit all its people. The plan involved development of the National Broadband Network (NBN) which provided high-speed broadband to cover the country.

NBN supported connections to provide education and training to community members within Indigenous societies. One of the criticisms leveled against Australia is the expensive character of fixed broadband subscriptions (US $ 46.7 PPP per month, ranked 100th worldwide) but it has ranked 10 for highest penetration of mobile broadband (World Economic Forum, 2016, p. 27). Still, the country is a leader in providing online services and allowing citizens' participation generally.

There were other supplementary efforts that included the National Centre of Indigenous Excellence with the Telstra Foundation's Indigenous Digital X partnerships that strengthened the shift toward digital economies. Further, geographical challenges have made it difficult for all Indigenous people to be guaranteed of uninterrupted online services.

Policy feedback: reactive communication and justice messaging

Since Australians speak over Indigenous people, the stories about COVID-19 from mainstream media were not necessarily accurate descriptions. However, going by social media accounts and feeds, the messaging from Indigenous influencers was quite clear.

Underlying the threat of COVID-19 is an injustice that the ruling elite is unwilling to confront. Thus, the emphasis on low numbers of positive cases among First Nations people and directly impacted by the pandemic paled when weighed against the injustice of massive incarceration Indigenous people in Australian jails. In fact, it was in jail and prison environments where the density of COVID-19 cases was located. This fact has not been given much attention in the media, partly because "the settler Australian public simply does not see Indigenous deaths in custody as an act of violence, but as a co-morbidity" (Whittaker, 2020).

This injustice is part of the COVID-19 narrative because there were real fears that prisons would be hot spots forcing the Australian government to issue statements concerning that possibility. Since 1991, over 432 Indigenous people have died in custody, of these only 2 cases made it to serious but unsuccessful court indictments. In addition, families of victims are systematically silenced (Whittaker, 2020). Aboriginal and Torres Strait Islander people have employed social media to bring attention to these human rights concerns, but regular media continues to reinforce the existing perceived normality of the colonial-leaning prison and law enforcement situation in Australia. The choice of language and media narratives frame Aboriginal communities as "lawless" and "dysfunctional" (Cunneen & Porter, 2017; Porter, 2012, 2016).

Carlson and Frazer (2020) point to use of social media for strengthening the power of the oppressed. They too have a platform to reject the various forms of violence meted against them. Worldwide, that was what COVID-19 unearthed, the persisting inequalities that forced the ruling elite to see the failure to provide health care to disenfranchised and marginalized people. In Australia, the government's inability to adequately fund health-care services for Indigenous people was linked to the fact that the political class, in 2019 and always, was looking after interests of the ruling elite. And it was all about winning elections and channeling resources to gain political mileage (Seccombe, 2020).

Professor Bronwyn Carlson of the Department of Indigenous Studies at Macquarie University has done much groundwork on the topic of social media and political advocacy within indigenous societies in Australia. She points out that social media "amplifies Indigenous voices and provides necessary information for non-Indigenous people to learn more about Indigenous people, histories and cultures" (Carlson, 2020). In her view, Twitter has a role to play in political teaching and learning. And it does this so well because messages can be retweeted several times. Since Twitter was prominent in microblogging for political voice, it is noteworthy that citizen reactions to political aspects of COVID-19 and the overall health care for Indigenous people underscored the narrative that more needed to be done for the welfare of all Australians. *The Conversation* provided an excellent expose of the 10 Twitter accounts that offered feedback on policies affecting Indigenous Australians. Table 6.1 sums up the accounts.

Table 6.1: 10 Twitter accounts you should be following if you want to listen to Indigenous Australians and learn.

Twitter handle	Description
IndigenousX @IndigenousX	Created in 2012 by Gamilaroi man, Luke Pearson. Wakka Wakka woman, activist, and student Bizzi Lavelle are hosts.
Lynda-June Coe @LYndaJune1	She is a PhD candidate, a Wiradjuri and Badu Island Yinaa from Erambie, Cowra, NSW, women. Demands justice for Aboriginal people.
Dr. Debbie Bargallie @debbiebargallie	Senior research fellow at Griffith University, Recipient of 2019 WEH Stanner Award for best doctoral thesis by an Indigenous person. She is a descendant of the Kamilaroi and Wonnarua peoples of NSW.
Nessa Turnbull-Roberts @TurnbullVanessa	A Bundjalung activist, winner of Australian Human Rights Commission 2019 Young People's Medal.
Hayden Moon @hayden_seek94	A Wiradjuri Brotherboy and activist for LGBTQi + persons with disabilities.
Aboriginal Health in Aboriginal Hands @NACCHOAustralia	The National Aboriginal Community Controlled Health Organization. The handle covered COVID-19 pandemic. Advice on safe practices at protests held during the pandemic.
Secretariat of National Aboriginal and Islander Child Care @SNAICC	SNAICC is a nonprofit organization that offers family support and early childhood education. The voice of children and COVID-19 resources to families affected by the pandemic.
Celeste Liddle @Utopiana	Celeste is an Arrernte feminist and union organizer. Her contributions are on *Growing Up Aboriginal in Australia* and *Mothers and Others.*
Amy McQuire @amymcquire	A Darumbal and South Sea Islander person from Rockhampton. Currently PhD Candidate at Queensland University. Served as journalist for Buzzfeed, *The Guardian*, and *The New York Times*. Podcaseis Curtain.
Dr. Sandy O'Sullivan @sandyosullivan	A Wiradjuri based at the University of the Sunshine Coast. Researcher on First Nations' Communities.

Based on Carlson, B. (June 10, 2020). Ten Twitter accounts you should be following if you want to listen to Indigenous Australians and learn: The Conversation. *https://theconversation.com/ten-twitter-accounts-you-should-be-following-if-you-want-to-listen-to-indigenous-australians-and-learn-140353.*

Following some of these Twitter accounts, we note the formation of a narrative that praises Indigenous actions to combat COVID-19 pandemic, while calling for more action to eliminate social injustices. Since COVID-19 pandemic coincided with the Black Lives Matter global social movement, there were several interlocking messages that highlighted for the Australian government to end its brutality against Indigenous Australians, who like African American are disproportionately represented in their countries' prison system. The unmistakable links between poverty, poor health, and racism were major features in hundreds of tweets conveyed by prominent Australian microblogs.

As protests against racism occurred in NT and other areas, the activists advised on safe practices at protests held during the pandemic. One may make the case that the impact of COVID-19 on Indigenous societies in Brazil, the United States, and elsewhere added to the activism against racism. This prompted Australia's minister of Indigenous Australia to revisit some of the assumptions in existing policies that endangered the lives of Aboriginal people in the Pacific region. One may assume Scott Morrison's government summer 2020 pledge to appoint more Indigenous Australians in top positions was such a possible consequence.

Professor Bronwyn Carlson had commented that three in every four Australians held negative views about Aboriginal and Torres Strait Islander people (Carlson, 2020). Following Bronwyn Carlson's (2020) observations, it is clear that the trust levels between the Aboriginal people and governments were a source of concern. This, in itself provided motivation for Indigenous organizations to use all communication tools to educate their citizens about COVID-19. As a follow-up, communication policy decisions were made through leadership of Indigenous public health experts. Again, this affirms Carlson's (2020) that Indigenous people were best served when they spoke for themselves.

Perhaps, we must mention the importance of the National Aboriginal Community Controlled Health Organization (NACCHO). The organization's Twitter handle covered COVID-19 pandemic in a compressive way. NACCHO's official website provided COVID-19 updates and information that supplemented efforts of the Australian Federal Government DOH. One of the links provided educational materials on caring for people with COVID-19. This resource detailed the various therapeutics in use to minimize the effects of coronavirus.

The overall handling of COVID-19 pandemic was praised in numerous letters. We offer the example in Table 6.2 to illustrate the importance of feedback processes in policy implementation in Aboriginal communities.

Health communication on video clips

Perhaps, Kerrigan et al. (2020) provide the best example on health and wellness communication messaging in the NT of Australia that targeted Aboriginal people. They write, a Darwin-based health communicator collaborated with Aboriginal leaders to produce messages tailored to local communities urging them to make choices that would enhance their health and not make it easier for COVID-19 to undermine their health outcomes. For instance, the messages urged Indigenous people to stop smoking, wash hands, "stay strong" by taking their prescribed medicines as needed.

Furthermore, there were short health promotional video clips provided to cultural educators, politicians, and health professionals that used English and Indigenous languages including Kriol, Warlpiri, Tiwi, and Djambarrpuyngu—some of the largest populations

Table 6.2: Excerpt of letter in praise of indigenous efforts to reduce spread of the coronavirus.

Letters to the editor
In addition to the issues raised in Patricia Turner's timely and important article, something seems to have gone unnoticed. Since the start of the epidemic the total cases of coronavirus in Aboriginal and Torres Strait Islander peoples is 59 nationwide. This represents only 0.7 percent of all cases, a considerable under-representation, as First Nations peoples comprises 3 percent of our population. Most cases have been in urban areas and none in remote communities. Only 13 percent needed hospitalization, none needed intensive care and there were no deaths. Given the high-risk characteristics of this population and the predictions of how devastating the pandemic could have been (and has been overseas), this is a remarkable outcome. Pat and her team at NACCHO, in co-operation with all the Aboriginal community-controlled health services, and other groups, effectively lobbied governments to close remote communities early. These leaders also acted quickly to prevent the pandemic in urban and regional centers by lobbying for personal protective equipment, enhancing capacity for testing and contact tracing, enabling better housing to manage social distancing and developing a brilliant set of health information videos that were clear and aimed at local groups. Most of this activity, which has prevented major illness and death and avoided costly care and anguish, has been done by these groups themselves, galvanizing state, territory and federal government departments to participate appropriately. This is clear evidence of how effective it is to give power to First Nations populations. In addition to the excellent recommendations in Pat Turners' article, we should all support the implementation of the Uluru Statement from the Heart. Fiona Stanley, Perth, WA.

From Stanley, F. (June 6—12, 2020). Indigenous leadership on Covid-19 *(304th ed.): The Saturday Paper. https://www.thesaturdaypaper.com.au/2020/06/06/aclands-destruction-not-the-last/15913656009944.*

in NT. These videos explained the importance of maintaining healthy practices including attending to routine care like dialysis and taking medicines for other ailments as a means to mitigate the coronavirus effects.

Many more of these health-care-related messages were rebroadcast in local radio, TV networks, Facebook, Twitter, and on smartphone apps. The Menzies School of Health Research embedded some of the messages on their website (Menzies, 2020). Kerrigan et al. (2020) observe that successful messaging was not just confined to translating COVID-19 related information, but it included words meant to inspire resistance among Aboriginal and Torres Strait Islander peoples.

Arts and music also helped. Aboriginal artists locked down in COVID-19 biosecurity zones performed to online audiences (Mercer, 2020b). Musicians from northeast Arnhem Land in the NT joined the worldwide trend of artists in lockdown performing from home. They gave free weekend performances online, which helped to sustain them during the pandemic. The virtual shows helped open up their music to the world (Mercer, 2020b).

What we learned

Although COVID-19 affected all lives globally, there were unique and nuanced impacts specific to Indigenous people. Evidence of the deleterious psychological effects of

COVID-19 is beginning to emerge. There was already a recognition that Indigenous communities were disadvantaged in terms of having underlying health conditions. And high incidence of mental health issues including alcoholism. Being exposed to new health challenges in addition to having considerable life experience disadvantages is, therefore, a line of inquiry that will have to be explored in more systematic inquiries. At the preliminary level, we can observe the emergence of such concerns among those infected by the disease.

Data shared in informal social media spaces reveal increased risks of anxiety and depression. In addition, other disorders were reported but to a lesser extent. COVID-19 was a traumatic event because it threatened the very existence of those who held Indigenous cultural memories, the elders. Thus, any reporting a case of COVID-19 infection of an Indigenous elder on social media increased fear for those who had no symptoms of the disease.

There were also adverse effects on those who were separated from their families during the quarantine involving those who tested positive and were required to isolate. The separation was equally a traumatic event as seen in the commentary by members of the communities. Qualitative methods that tracked comments on tweeter and other social media reveal that COVID-19 created stressful situations within the Aboriginal communities, which forced adaptation mechanisms. Yet, the lack of sufficient mental health professionals in those communities complicated the healing and recovery processes.

Still, survivance among Indigenous people shined through. Comments about overcoming past epidemics were reported in newspaper outlets and blogs. As Carlson theorized, there was increased use of social media for empathy and grief in Aboriginal communities. Social media served as a tool to bring together friends of a person, regardless of their ethnicity who succumbed to COVID-19. In some ways, social media provided a platform to build resilience and to boost coping strategies.

Social trauma

Because the COVID-19 impact is an evolving area, we have to make assumptions about the range of variables that can explain the coping strategies. What we can extrapolate is that the pandemic exacerbated the social trauma of daily living for all communities including Indigenous peoples. And the severity was noted because of the unique difficulties of poverty.

From comments available in documentary data, we can discern the importance of family support as a coping strategy. Aborigines fully employed social support from family and friends as a means to cope with the pandemic. The poorest family members received emotional support and supplies from their kin. The large well-developed family networks proved to be a strong support system for the communities at the time of national lockdown and quarantine. Community members also participated in online chat groups as a means to discuss problem with government handling the COVID-19 crisis.

Conclusion

Understanding how different advanced countries managed the global pandemic is an important part of organizational learning. How Australia was able to leverage Indigenous initiatives to combat the spread of COVID-19 within Aboriginal lands is a lesson that can show policy experts the merits of different cultural managerial styles and effects of different interpretations of what constitutes empowerment and sovereignty.

Australian Indigenous communities certainly did not face the delayed disbursements in federal aid as was experienced in the case of the Navajo nation in the United States. By including the case of the Aboriginal people, we have identified how national oversight and communications can offer leadership signals that enable lower levels of government to be more successful in combatting pandemics. Certainly, there were operating characteristics that are unique to each Indigenous society. However, realistic managerial challenges from the governance networks made a difference.

References

Australian Bureau of Statistics. (ABS). (2018). *Household use of information technology, Australia.* Canberra https://www.abs.gov.au/ausstats/abs@.nsf/mf/8146.0.

Australian Government. (2020). *Coronavirus (COVID-19) current situation and case numbers*. Department of Health. https://www.health.gov.au/news/health-alerts/novel-coronavirus-2019-ncov-health-alert/coronavirus-covid-19-current-situation-and-case-numbers.

Australian Government. (March 23, 2020b). *Australian health sector emergency response plan for novel coronavirus (COVID-19) short form.* https://www.health.gov.au/sites/default/files/documents/2020/03/australian-health-sector-emergency-response-plan-for-novel-coronavirus-covid-19-short-form-the-australian-health-sector-emergency-response-plan-for-novel-coronavirus-short-form.pdf.

Australian Government, Department of Health. (May 2, 2020c). *Coronavirus (COVID-19)-radio-for Aboriginal and Torres Strait Islander peoples-looking after elders and older people.* https://www.health.gov.au/resources/audio/coronavirus-covid-19-radio-for-aboriginal-and-torres-strait-islander-peoples-looking-after-elders-and-older-people.

Australian Human Rights Commission. (October 20, 1993). *Human rights and mental illness: Report of the national inquiry into the human rights of people with mental illness.* https://humanrights.gov.au/our-work/rights-and-freedoms/publications/human-rights-and-mental-illness.

Australian Human Rights Commission. (August 2006). *A statistical overview of Aboriginal and Torres Strait islander peoples in Australia.* https://humanrights.gov.au/our-work/statistical-overview-aboriginal-and-torres-strait-islander-peoples-australia.

Baker, P. (April 12, 2020). *Trump lashes out at Fauci amid criticism of slow virus response*. New York Times. https://www.nytimes.com/2020/04/12/us/politics/trump-fauci-coronavirus.html.

Black, R., & Atkinson, J. (2007). *Addressing the digital divide in rural Australia. In AusWeb: Sic erat in fatis (so it was fated)* (pp. 45–62). Southern Cross University.

Carlson, B. (June 10, 2020). *Ten Twitter accounts you should be following if you want to listen to Indigenous Australians and learn*. The Conversation. https://theconversation.com/ten-twitter-accounts-you-should-be-following-if-you-want-to-listen-to-indigenous-australians-and-learn-140353.

Carlson, B., Farrelly, T., Frazer, R., & Borthwick, F. (2015). Mediating tragedy: Facebook, Aboriginal peoples and suicide. *Australasian Journal of Information Systems, 19*, 1–15. https://journal.acs.org.au/index.php/ajis/article/view/1174/687.

Carlson, B., & Frazer, R. (2018). *Social media mob being Indigenous online*. Report. Sydney Australia: Macquarie University. https://researchers.mq.edu.au/en/publications/social-media-mob-being-indigenous-online.

Carlson, B., & Frazer, R. (June 10, 2020). *"They got filters": Indigenous social media, the settle gaze, and a politics of hope*. Social Media + Society. https://doi.org/10.1177/2056305120925261

Commonwealth of Australia, Department of Health. (July 2020). Australian Health Sector Emergency response plan for novel coronavirus (COVID-19). Publications number: 12750. In *Management plan for Aboriginal and Torres Strait Islander populations operational plan for Aboriginal and Torres Strait Islander populations*. https://www.health.gov.au/sites/default/files/documents/2020/07/management-plan-for-aboriginal-and-torres-strait-islander-populations.pdf.

Commonwealth of Australia, Department of Health. (March 21, 2021). *Coronavirus (COVID-19) at a glance*. https://www.health.gov.au.

Cousins, S. (May 19, 2020). *Indigenous Australians avert an outbreak -for now*. Foreign policy (FP) dispatch. https://foreignpolicy.com/2020/05/19/indigenous-australians-avert-coronavirus-outbreak-for-now-aboriginal/.

Creighton, B. (March 2, 2020). *Information overload: An update and insight on the coronavirus epidemic*. AMP Capital. https://www.ampcapital.com/au/en/insights-hub/articles/2020/february/information-overload-an-update-and-insight-on-the-coronavirus-epidemic.

Cunneen, C., & Porter, A. (2017). Indigenous peoples and criminal justice in Australia. In *The palgrave handbook of Australian and New Zealand criminology, crime, and justice* (pp. 667–682). Palgrave Macmillan.

Curtin, J. (2001). *A digital divide in rural and regional Australis*. Canberra: Department of Parliamentary Library.

Erdiaw-Kwasie, M. O., & Alam, K. (2016). Toward understanding digital divide in rural partnerships and development: A framework and evidence from rural Australia. *Journal of Rural Studies, 43*, 214–224. https://doi.org/10.1016/j.jrurstud.2015.12.002

Gibson, C. (2003). Digital divides in New South Wales: A research note on socio-spatial inequality using 2001 census data on computer and Internet technology. *Australian Geographer, 34*(2), 239–257. https://doi.org/10.1080/00049180301734.

Higgins-Dunn, N. (July 10, 2020). *Trump's rift with White House health advisor Fauci widens as coronavirus cases continue to hit new records*. CNBC. https://www.cnbc.com/2020/07/10/coronavirus-trumps-rift-with-white-house-health-advisor-fauci-widens-as-cases-hit-new-records.html.

Keck, M. (July 17, 2020). *COVID-19 rates among Indigenous Australians are low thanks to "extraordinary" aboriginal-led response*. Global citizen. https://www.globalcitizen.org/en/content/covid-19-rates-indigenous-australians/.

Kerrigan, V., Lee, A. M., Ralph, A. P., & Lawton, P. D. (May 30, 2020). Stay strong: Aboriginal leaders deliver COVID-19 health messages. *Health Promotion Journal of Australia*. https://doi.org/10.1002/hpja.364

Knoema. (May 14, 2016). *The global information technology report: Country profile*. World Economic Forum. https://knoema.com/infographics/ljisicg/the-global-information-technology-report-country-profiles.

Lee, C. (February 12, 2020). *We're in danger of drowning in a coronavirus "infodemic". Here is how we can cut through the noise*. The Conversation. https://theconversation.com/were-in-danger-of-drowning-in-a-coronavirus-infodemic-heres-how-we-can-cut-through-the-noise-131303.

Lloyd, R., & Hellwig, O. (2000). *Barriers to the take–up of new technology*. Canberra.

McGuirk, R. (June 17, 2020). *Australia accuses China and Russia of virus disinformation*. AP. Federal News Network. https://federalnewsnetwork.com/world-news/2020/06/australia-accuses-china-and-russia-of-virus-disinformation/.

McIlroy, T. (April 9, 2020). *Australians risk coronavirus 'information overload' financial review*. https://www.afr.com/politics/federal/australians-risk-coronavirus-information-overload-20200408-p54iaq.

Menzies. (2020). *Resources*. Menzies School of Health Research. http://www.menzies.edu.au/resources/?keywords=coronavirus.

Mercer, P. (May 2, 2020a). *Aboriginal Australians take music to the world during COVID-19 lockdown*. VOA. https://www.voanews.com/covid-19-pandemic/aboriginal-australians-take-music-world-during-covid-19-lockdown.

Mercer, P. (April 30, 2020b). *Australia-China tensions over call for global COVID-19 probe*. VOA. https://www.voanews.com/covid-19-pandemic/australia-china-tensions-over-call-global-covid-19-probe.

McLaren, J., & Zappala, G. (2002). *The digital divide among financially disadvantaged families in Australia* (Vol. 7). First Monday. https://doi.org/10.5210/fm.v7i11.1003 (11).

Meredyth, D., & Thomas, J. (2000). Virtually no policy: Modulating the digital divide. *The Southern Review, 33*(2).

Murphy, K. (June 11, 2020). *Home affairs flags steps to help Australians identify fake news by foreign powers*. The Guardian. https://www.theguardian.com/australia-news/2020/jun/11/home-affairs-flags-steps-to-help-australians-identify-fake-news-by-foreign-powers.

Packham, C. (June 24, 2020a). *Australia, once nearly free of coronavirus, reports worst case surge in 2 months*. Global News. Globalnews.ca.

Packham, C. (May 21, 2020b). *With low coronavirus figure, Australia wants exemption to British quarantine rule*. Global News. https://globalnews.ca/news/6972039/australia-coronavirus-uk-quarantine/.

Porter, A. (2012). Aboriginal sovereignty, "crime" and criminology. *Current Issues in Criminal Justice, 31*(1), 122–142.

Porter, A. (2016). Decolonizing policing: Indigenous patrols, counter–policing and safety. *Theoretical Criminology, 20*(4), 548–565.

Power, J. (November 13, 2020). *When it came to COVID-19, Indigenous Australians sent it packing*. The Sydney Morning Herald. https://www.smh.com.au/national/when-it-came-to-covid-19-indigenous-australians-sent-it-packing-20201112-p56e5u.html.

Ralph, A. P., et al. (2020). Improving communication with Aboriginal hospital inpatients: A quasi-experimental interventional study. The communicate study group. *Medical Journal of Australia, 213*(4), 180–181. https://doi.org/10.5694/mja2.50700

Readfearn, G. (March 21, 2020). *Coronavirus overload: Five ways to fight misinformation and fear*. The Guardian. https://www.theguardian.com/world/2020/mar/22/coronavirus-overload-five-ways-to-fight-misinformation-and-fear.

Rooksby, E., Weckert, J., & Lucas, R. (2002). The rural digital divide: The rural society. *Rural Society, 12*(3), 197–210.

Samaras, K. (2005). Indigenous Australians and the digital divide. *Libri, 55*(2–3). https://doi.org/10.1515/LIBR.2005.84

Samios, Z., & McCauley, D. (January 28, 2020). *Minister urges skepticism as fake virus news spreads*. The Sydney Morning Herald. https://www.smh.com.au/business/companies/minister-urges-scepticism-as-fake-virus-news-spreads-20200128-p53vjn.html.

Savage, M. (May 6, 2020). *Coronavirus: How much news is too much?* BBC. https://www.bbc.com/worklife/article/20200505-coronavirus-how-much-news-is-too-much.

Seccombe, M. (June 13–19, 2020). *Who is Morrison looking after*. The Saturday Paper. htps://thesaturdaypaper.com.

Singh, A. (February 5, 2020). Responsible technology Australia. In *Social media sites urged to flags coronavirus misinformation*. https://www.technologydecisions.com.au/content/it-management/news/social-media-sites-urged-to-flag-coronavirus-misinformation-1081630878.

Sulistyo, B. D. (July 26, 2020). *Telling the China story in Australia why we need racial literacy*. http://www.thechinastory.org.

Thomas, J., Barraket, J., Ewing, S., MacDonald, T., Mundell, M., & Tucker, L. (2016). *Measuring Australia's digital divide: The Australian digital inclusion index*. Telstra. https://doi.org/10.4225/50/57A7D17127384

Tucker, P. (August 6, 2020). *Russian disinformation networks detailed in new state department report*. Defense One. https://www.defenseone.com/technology/2020/08/russian-disinformation-networks-detailed-new-state-department-report/167537/.

Turk, A. (2003). Digital divide remediation: An Australian Indigenous community case study. *Southern Review, 36*(1), 48–63.

UNICEF Australia. (March 7, 2020). *Misinformation during times of a health crisis can spread paranoia, fear, and stigmatization, says UNICEF Deputy Executive Director for partnerships, on coronavirus*. Australia: UNICEF. https://www.unicef.org.au/about-us/media/march-2020/coronavirus-misinformation.

Whittaker, A. (June 3, 2020). *Despite 432 indigenous deaths in custody since 1991, no one has ever been convicted. Racist silence and complicity are to blame*. The Conversation. https://theconversation.com.

Williams, S. (July 9, 2020). *Australians struggling with information overload as they move to remote working in wake of COVID-19*. IT Brief. https://itbrief.com.au/story/australians-struggling-with-information-overload-as-they-move-to-remote-working-in-wake-of-covid-19.

Willis, S., & Tranter, B. (2006). Beyond the "digital divide": Internet diffusion and inequality in Australia. *Journal of Sociology, 42*(1). https://doi.org/10.1177/1440783306061352

World Economic Forum. (July 6, 2016). *The global information technology report 2016*. Geneva. http://www3.weforum.org/docs/GITR2016/WEF_GITR_Full_Report.pdf.

CHAPTER 7

An overview of pandemic response

Global health systems and COVID-19 preparedness

In the previous chapters, we have highlighted the pandemic responses in four regions using a variety of social science frames to explore what happened in the immediate onset of the COVID-19 pandemic. The preceding chapters have drawn attention to specific actions taken by both local communities, local and national governments to mitigate the spread of the virus. Most of those efforts offer lessons for understanding best practices in the management of evolving pandemics. A holistic overview of policy actions and management of pandemics places this matter theoretical lens useful for refining how policy practitioners and citizens think about public health systems in regions railing from colonially induced marginalization. In this chapter, we tease out the pertinent issues of communications and government handling of the COVID-19 crisis as applied to selected Indigenous societies. In the end, the discussion sums of the lessons learned from the experience of Indigenous societies, which are structurally located within nation-states. We recognize the importance of playing a small part to increase representation of Indigenous communities in the historical narrative of the COVID-19 crisis.

The purpose, thereof, is not just to present the story of government handling of the COVID-19 pandemic with reference to selected marginalized communities but to offer an overview of the global handling of the pandemic. This allows us to learn more about the impact of policies in which wellness and health outcomes of Indigenous people across the continents are debated. While many organizations worked within an ICT-enabled governance network to prevent the spread of coronavirus, the outcomes differed across the different jurisdiction. The nuances explaining these differences are worth our analysis. By shedding light on only a few cases, we continue a conversation about the most pronounced crisis of our lifetime. We are fully aware that COVID-19 is not the last pandemic in human experience and lessons learned will be helpful in future public health crisis management efforts.

For the sake of argument, going back to 2017, we recognize that public health systems in different countries were at different levels of development and readiness in terms of tackling the vast array of challenges. Perhaps, we all assumed that the public health system was ready to address challenges brought about by diseases such as COVID-19. After all, Johns Hopkins had ranked US facilities as the best in the world. This would

The Inequality of COVID-19. https://doi.org/10.1016/B978-0-323-99867-3.00008-9

imply that there would be some trickle-down effect on communities such as those on the Navajo Nation and other Native Americans. We were wrong. Sarah Hatcher et al. (2020) in Centers for Disease Control and Prevention Study (CDC) revealed that the incidence of COVID-19 among Native Americans was 3.5 times that of white counterparts. Among Native Americans and Alaska Natives, there were 594 COVID-19 cases per 100,000 people. For white counterparts, the number was 169 cases per 100,000 signifying massive disparities in the health-care system.

Thus, in 2020, the American public health system showed us that was expensive, fragmented, individualistic, and designed to cater to wealthier members of society. And the national ecosystem was awfully prepared to manage a devastating disease that COVID-19 is. The Navajo Nation among others lacked adequate testing and protective equipment. In addition, their capacity to track infected individuals was severely hampered by logistical issues.

And there is more to these inequalities. In most of the United States, workplaces lacked benefits like paid leave, which would have enabled sick employees to stay home and reduce infections. In addition, there was no national system to provide employees income replacements when they stayed at home as a strategy to prevent community spread. Worse still, the Republican Party—led government caused disruptions in the health system by undermining efforts to establish affordable health care for the most vulnerable members of society. Simply put, when COVID-19 arrived on American shores, there were far too many leaking spots in the health-care system.

There was also the expectation that communities such as those of Maasai people, in Africa, would have no defenses against viral attacks. After all, the public health systems in developing countries have periodically demonstrated a lack of capacity to handle simple challenges that include malaria, typhoid, cholera, and a host of challenges. Speaking of the public health situation in Kenya, where Maasai people reside, Professor Khama Rogo wondered aloud "when coronavirus arrives, what will be our reaction?" (Rogo, 2020). He argued that Kenya's health-care sector was more important than the economy and should never have been neglected. His remarks on the health sector are important and need to be given more emphasis:

> *Many have said, ignore this sector at your own peril. This sector is more important than the economy because it can bring the economy down, it is more important than the military, it can bring the military down. In Iran mosques of Friday are empty. ... When people fall in, where do they go to? Even hospitals get overwhelmed. People will not go to hospitals because it will be a dangerous place. Epidemics reduce the number or health workers*
>
> **(Rogo, 2020).**

Professor Rogo was right. COVID-19 proved beyond a doubt that it could bring down economies across the globe. Professor had observed the early implications of COVID-19 on northern Italy and come to a conclusion that if Italy, an advanced country struggled to fight COVID-19, the challenges for developing countries would be much more unsurmountable. And for a country like Kenya where the health-care system was corrupt, underfunded, characterized by low productivity, and lacking critical health workers, a disaster was in the making. Many instances of hospitals procuring fake COVID-19 tests were reported in the media. For example, Langa'ata Hospital Laboratory was shut down for malfeasance and fraudulent activities (Kimuyu, 2020). And a nurse at Kenyatta National Hospital was charged for making fake COVID-19 certificates (Citizen, 2020; Munguti, 2020). The issue of corruption in Kenya's health sector needs to be underlined because, as we shall show, it directly affected the COVID-19 polymerase chain reaction (PCR) testing protocols along the border of Kenyan and Tanzania where the Maasai reside.

Besides, the lack of equipment and overpriced medicines made the management of COVID-19 an unpromising endeavor. Although the clouds were dark for a country like Kenya, those fears were, as time progressed, lacking in nuanced information resources. The many unknowns included the density of the COVID-19 strains and the African communities' abilities to self-govern outside of the formal structures and communication systems.

Remarks on the unpreparedness of the public health system in India were also observed and noted in several studies. The World Bank (2012) has noted tribal people in India have far worse health indicators than the general population. In addition, Scheduled Tribes (STs) lack sanitation systems which made them vulnerable to health challenges. The World Bank supported health-care systems in Rajasthan, Karnataka, and other areas where Siddi people live. In addition, these area-specific interventions included partnerships between public and private providers. Still, the overall system did not adequately serve Siddi people, whose poverty levels are high and cannot allow them acquire ICT hardware.

Significantly, these economically marginalized societies face a lack of awareness about measures to take in efforts to combat diseases such as COVID-19. According to the World Bank,

> *Without awareness of health issues, most tribal populations tend to fall ill more frequently and wait too long before seeking medical help or are referred too late by untrained village practitioners. In the past, most health awareness campaigns, which need significant investments over long periods of time for noticeable impact, were planned by the medical community instead of by communication experts. The form and content of health messages was not pre-tested to ensure proper comprehension and absorption by target groups. Moreover, the campaigns' meager effect was easily nullified by the tribal population's poor experience with health workers (World Bank, 2012).*

Broadly, these information inequalities are a wider component of systemic failures. Thus, there was genuine fear that COVID-19 would have negative impact on many of the STs.

There was also an element of systemic failure on the part of global institutions. Some countries opted to choose nationalist isolation over global solidarity. United Nations Secretary General Antonio Guterres said on CNN Fareed Zakaria show (June 28), COVID-19 attacked the global community when nations were more inward-looking than focused on solving global problems together (Zakaria, 2020a). The nationalist streak in public policy management meant countries, especially major powers like the United States, Russia, and China were expected to join efforts to jointly combat COVID-19. The lack of sufficient sharing of information resources negatively affected the ability of countries and communities to effectively combat COVID-19 pandemic.

As Herbert Simon (1976) showed us, that, in government studies, information is key to making rational decisions. If governments in 2017 had known that COVID-19 would result in thousands of deaths and the huge losses in productivity that was experienced in 2020 the vast majority would have made every different decision. Or if they only realized how unprepared they were at managing a pandemic, this world would be different.

The point is, governments across the globe did not have full information about COVID-19, its manifestations, and consequences in the social and economic realms. If that were the case, perhaps, we would be communicating differently all together. However, things turned out differently in all government sectors, and the realities of COVID-19 in different Indigenous societies were complex and often painful. Thousands of lives were lost, and economies devastated. And perceptions about the pandemic became political (Hardy, 2020, pp. 3–4). With the passage of time, we are now in a better position to take stock of the scourge and its different dynamics. For example, scientists now recognize that the previous focus on numbers of those infected and killed by COVID-19 was an incomplete picture of the devastation from the calamity.

Julie Steenhuysen (2020) of Reuters captures this sentiment in her well-written piece that cited research from Oxford University. Scientists noted that the effects of COVID-19 are enormous and can last for several years. Patients experience health challenges which go beyond the respiratory systems and affect the pancreas, brain, kidney, and other organs. COVID-19 also caused neurological disorders and blood clots disorders which could result in strokes. In many instances, costs of rehabilitation of patients were significant. Thus, the total costs of treatment go beyond what was discerned in the first 6 months of the pandemic. These new findings were not captured in any of the early narratives and communications concerning COVID-19.

Government leaders and policy experts now know that, with passage of time, some information was incomplete, misinterpreted, or not considered in decision that were put in place to combat COVID-19 pandemic. By tracing our steps backwards, we desire to learn the mistakes that were made using the available information and how it was presented to the different publics. The perspectives of COVID-19 within and around Indigenous societies offer us lens through which we can decipher a full range of how different communication styles and technologies are relevant in managing pandemics. This is important for future decisions and approaches to managing pandemics and other large-scale public health crisis. For information about COVID-19 to be fully and effectively communicated, it must be acquired and processed within Indigenous communities in ways that do not cause confusion and cause further injustices and devastation within those societies.

Data on COVID-19 messaging and social media

The setting in which data on COVID-19 was acquired within government jurisdictions was drawn from institutions that are largely controlled by international bodies such as the United Nations and its affiliate, World Health Organization (WHO), the national governments within which sovereign First Nations and Indigenous societies are domiciled, as dependent nations, and of course, the local Indigenous authorities. These are policy venues that made decisions which ultimately affected people residing in all the communities that we discussed. These decision-making entities have various capabilities including resources on various diseases and how to combat those particular diseases, including the novel COVID-19.

Notice that the various agencies gathered scientific data on how to combat COVID-19. Indigenous community leaders also gathered data on how to prepare the people to avoid catching the diseases and infecting others. More accurately, Indigenous community leaders employed data for saving their communities from consequences of poor public health policies which had over generations worsened the health outcomes in those societies. Even when tribal members had to physically stand outside of homes these communications methods sometimes worked more effectively than those in areas with high connectivity. Communication was a major aspect of both the immediate prevention of further negative outcomes and prevention of adverse effects on those societies and required local and creative negotiations of existing resources.

Data generated by various institutions including universities and research centers, funded by both private and public sectors, aided in making management of the COVID-19 pandemic possible. At one level, Indigenous public health providers utilized Indigenous approaches to managing wellness within the various communities. Still, national and international organizations also took the responsibility to support these societies in monitoring and limiting the spread of the highly contagious disease to adjacent neighbors. That is, organizations like the WHO and CDC served as custodians of useful scientific data that would benefit decision-making in the COVID-19 global pandemic. In the process of interacting with national government leaders and also health-care experts drawn from different organizations, Indigenous providers acquired valuable information and also shared their own experiences with experts from other communities.

Data from Indigenous societies told the same stories as in other areas. The numbers of those infected and recovered or passed on spoke to whether the battle against COVID-19 was being lost or won. Moreover, information tools used by caseworkers and community members to narrate their experiences revealed the characteristics of the pandemic and its progression in different populations. This is valuable in global public health management.

When governments collected data on who was dying of COVID-19, the purpose was not just to understand the vulnerability of the health systems. The data conveyed in print and electronic media also served to remind us that Indigenous societies required the ability to develop and direct their own programs and enact policy interventions to change health-care outcomes. When data on who was dying and at what rate were analyzed in different media, it communicated a powerful message on government structures, cultures, and that human societies needed to do better in handling future pandemics.

For tribal governments, there was a consistent narrative that COVID-19 created a situation in which the very important knowledge base of those societies was at risk, if numerous elders were not treated and prevented from the fatalities associated with COVID-19. The acquisition of data on which age sets were most adversely affected by the disease was an important point in making decisions on combatting the disease. Health-care workers and political leaders were, thus, guided by the data in their policy decisions.

This, of course, meant that the information on COVID-19 in Indigenous societies was processed with the goal to prevent the scourge from ravaging communities that were already underdeveloped by capitalist systems, which, for the most part, left these communities in weaker political power arrangements. Thus, information gathered and processed about COVID-19 had to be culturally relevant and accurate. As we have described, the signals coming from some of the national leaders did not serve the Indigenous communities well enough to avert the devastation on their economies, health, and wellness. The Indigenous societies in counties like the United States suffered from a poor national strategy, which was largely manifest in disjointed communications and ill-preparation on the part of the national bureaucracy. The positive can be said of Indigenous societies in Australia, which, to a large extent, benefited from a more unified communication strategy.

Even though communication errors were reported in all societies, there is evidence that different Indigenous communities internalized the information on COVID-19 in ways that they understood. Thus, community members transmitted messages that were relevant to their cultures and history. For example, the errors represented in rumors and misinformation on social media were interpreted in historical terms of marginalization and oppressions. This was particularly true of communities such as those among Maasai people, whose social media influencers shared information about diseases brought into African villages by Europeans prior to and within the colonial lens. And there was also the strong widely shared narrative on the need to fall back on African herbs as remedy instead of going for western medications. Thus, the importance of combining oral communications and new social medias in storytelling about COVID-19 will remain a key lesson for the future.

Taken together, the transformation of data gathered from various media and combining it with lived experiences of Indigenous peoples is now a part of the COVID-19 communication system. It is the reality of the lived experience of those who occupy Indigenous territories. It worked just well enough because all Indigenous societies now have computer-based data archival systems that are easily available through mobile telephones. The outreach of mobile telephony has made it possible for Indigenous societies to embrace a form of consciousness within existing digital communities.

To elaborate, the countries covered in this work the United States, Australia, India, and East Africa are at different levels of economic and social development. This implies that their digital maturity levels are also quite diverse, with the African societies lacking the fastest reliable broadband connectivity. Still, we cannot claim that the digital divide and all its access manifestations have been resolved in developed societies. Quite the contrary, the digital divide, which many believe cannot be affectively measured, remains a problem even among Navajo people.

Still, the disparate spread of ICT infrastructures explains some differences in the COVID-19 management problematic. But it is not everything that shaped the outcomes of the pandemic. Indeed, even messaging on the now nonexistent analogue TV screens could have been a factor in how COVID-19 messaging unfolded. For example, when Donald Trump spoke about using disinfectants for "washing away" the virus, media outlets across Indigenous worlds quickly reacted and warned society members not to inject such chemicals or other dangerous substances (Rogers et al., 2020).

As evidence from the city of New York suggests, Trump's comments may have led to isolated incidents of ingestions of disinfectants and other poisonous substances. The New York City Department of Health and Mental Hygiene's Poison Control unit linked at least 30 cases to the president's comments (Slotkin, 2020). The CDC also reported cases of deaths directly linked to ingestion of hand sanitizers (Haglage, 2020). Social media memes in most national languages made Donald Trump on object of laughter (Applebaum, 2020).

These memes were accessible to Indigenous people through platforms such as WhatsApp and Facebook. Staff writer at *The Atlantic* and Senior fellow of the Agora Institute at Johns Hopkins University, Anne Applebaum (2020) observed, "quotations from the president's astonishing April 23, 2020, press conference have appeared on every continent, via countless television channels, radio stations, magazines, and websites, in hundreds of thousands of variations and dozens of languages—often accompanied by warnings, in case someone was fooled, not to drink disinfectant or bleach" (Applebaum, 2020).

Trump's messaging, which he later claimed was "sarcastic," was a disaster in messaging that will be one of the defining blunders in pandemic communication. While the president's thinking is a function of his political judgments, there are other aspects of poor messaging that were less political in impact. As the leader of the world's most advanced society, Donald Trump goofed and sent wrong communication signals about COVID-19 and the danger it posed on human societies regardless of the country. Applebaum (2020) posited that the:

> *American president's simultaneously catastrophic and ridiculous failure to cope with it. … The tone of news headlines ranges from straight-faced in Kompas, a major Indonesian news outlet—"Trump Usulkan Suntik Disinfektan dan Sinar UV untuk Obati COVID-19," or "Trump Proposes Disinfectant Injection and UV Rays to Treat COVID-19"—to snide, from Le Monde in France—"Les élucubrations du « docteur » Trump,"or "The Rantings of 'Doctor' Trump." The incredulous first paragraph of an article in Sowetan, from South Africa, declares that "US President Donald Trump has again left people stunned and confused with his bizarre suggestion that disinfectant and ultraviolet light could possibly be used to treat COVID-19." El Comercio, a distinguished Peruvian newspaper, treated its readers to photographs of Deborah Birx, the White House coronavirus -response coordinator, grimacing as the president asked her whether the injection of disinfectant might be a cure (Applebaum, 2020).*

Applebaum was not alone in commenting about Trump's poor handling of the COVID-19 crisis. Writing for *The Irish Times*, Fintan O'Toole (2020) was even more blunt. He observed that essentially Trump's actions were a part of his malignant narcissism, which amplified its lethality through his refusal to save lives of those who lost their lives through COVID-19. O'Toole further stated that for generations, America was a country that was admired, hated, and envied. Under Trump, America was now a country to be subject of pity, thanks to his handling of the COVID-19 crisis (O'Toole, 2020). The observation that Trump destroyed the country he promised to "make great again" was essentially fueled by his inability to act strategically in the interest of the country and communicate well about combatting a dangerous pandemic. Anthropologist Wade Davis (2020) commented,

> *In a dark season of pestilence, COVID-19 has reduced to tatters the illusion of American exceptionalism. At the height of the crisis, with more than 2000 dying each day, Americans found themselves members of a failed state, ruled by a dysfunctional and*

> *incompetent government largely responsible for death rates that added a tragic coda to America's claim to supremacy in the world. For the first time, the international community felt compelled to send disaster relief to Washington (Davis, 2020).*

One other example of the possibility of wrong messaging was evident in the confusion brought about by the recommendations for use of alcohol sanitizers as remedy for preventing spread of the coronavirus. It has been documented that after the pandemic broke out, some businesses sold alcohol-based hand sanitizers in stores across Arizona and in the Navajo Nation. In fact, the demand for sanitizers and cleaning product was worldwide. However, with the rise in demand, the unintended consequence was the incidence where persons were misinformed and ingested products known as methyl alcohol or wood alcohol, which were also offered as sanitizers on the Navajo Nation (Jensen, 2020). The consumption of these methanol products resulted in hospitalizations and deaths on the reservation and elsewhere.

In June 2020, the Food and Drug Administration (FDA) issued public notifications to advice against consumption of these products. A recall of some of the products, mainly imported from Mexico, was also initiated. Most had been sold under different product names, causing much confusion (Jensen, 2020).

The fact that critics, especially, from the French media regarded President Trump's unscientific ideas about treating COVID-19 patients amounted to emergence of a dangerous policy narrative, that is, an emergence of an antiscience value system. This is in complete contradiction to the ideas of the founding fathers of policy sciences including Harold Lasswell, who had provided the vision of using scientific knowledge to solve public policy problems (Lasswell, 1951). As President Donald Trump spewed solutions that were unscientific, Europe's leading politician, Dr. Angela Merkel, a Ph.D. holder in the field of quantum chemistry, also led the world in upholding the virtues of the emergent "Government by Scientist," which laid scientific plans to combat COVID-19 pandemic (Ricard, 2020). And countries that followed science performed better in reducing the deaths caused by the coronavirus.

Immediate government communication

Local radio stations, social media, and other internet-based platforms exhibited much innovation. Different video clips in numerous Indigenous languages communicated information on living in the era of COVID-19. TikTok added much humor to an otherwise emotionally draining curfew and lockdown climate, which characterized the various government response regimes. The new medias also spread lots of inaccuracies at a cost. Governments had to respond, and correct information peddled throughout the ordeal.

And it was obvious that internet-based town halls were a convenient method of reaching out to the savvier netizens. For example, President Jonathan Nez of the Navajo Nation held a series on live Facebook Town Halls in which he communicated directly to members of the Native American nation. And public officials in various Indigenous health centers also effectively applied new technologies to answer questions about COVID-19. Many of the video clips from these communications were duplicated and shared over and over again.

Looking at the rare mirror, we have tried to understand whether government communications during COVID-19 captured their intensions. Clearly, communication is an important part in management of pandemics and other major crises. The question to raise is, were the various communication strategies effective and efficient, especially for Indigenous societies? Few people would argue that government bureaucrats serving Indigenous communities employed error-free communication styles and processes. This is not even about languages employed in the various communications and press briefings. If anything, national governments across the board have not necessarily invested enough to promote Indigenous languages. There are also programs that are designed to promote certain languages and to frustrate others, at the very least. The COVID-19 official communications from national languages in the United States, India, Australia, and Kenya and Tanzania lacked depth in simplification of the technical terms employed during the epidemic. Terms such as "bending the curve" were technical and meaningless in all the Indigenous languages captured in this analysis.

Many examples of poor communicators and misinformation quickly found themselves being ridiculed by reporters. Since we noted that fighting COVID-19 was framed as military endeavor, there are plenty of examples to show that some of the government spokespersons used command, rather than persuasion to initiate behavior change within different societies.

Take the case of President John P. Magufuli, he, like President Trump, was not factual in his claims that Tanzania was free of coronavirus. Magufuli claimed there was no one in Tanzania with corona, which was false. Against advice from health-care experts Magufuli initially told Tanzanians not to wear masks. "I want to urge you Tanzanians not to accept donations of masks, instead tell the donors to go and use them with their wives and children" (BBC, 2020). While giving a roadside address in Mkuranga he told his audience that prayers healed Tanzanians from COVID-19. "There was no one covering their mouth and that people should continue to rely on God." He strangely praised atheists, Muslims, and Christians. Magufuli noted, much like many global leaders, that the biggest fear was to be afraid. These remarks were covered live by AZAM TV and repeated the same message in other rallies (BBC, 2020).

Perhaps, of all world leaders with jurisdiction over Indigenous communities, Trump's comments on managing coronavirus pandemic were most problematic in terms of impact. He, after all, was the leader of the nation with the most resources to lead the world in the global fight against the spread of coronavirus. As Ed Yong *The Atlantic*'s science writer emphasized, Trump ignored all the warnings and, with impunity, asserted that "the coronavirus is very much under control" and "like a miracle, it would disappear" (Yong, 2020). He further stated that the cases would get to zero. Instead, millions were infected.

Perhaps, as expected, Indigenous society activists and leaders used the COVID-19 pandemic to remind the world of the unending pandemic of racism. This was the case in the more advanced countries, Australia and the United States. For instance, many of President Jonathan Nez's widely televised messages on local and international TV stations carried important reminders of historical injustices and betrayals of unfulfiled promises made to tribal societies. Obviously a reminder that American federalism misses one point—tribal nations signed treaties with the national government and are older institutions, which means their issues are, and should be prioritized as ancient treaties required. COVID-19 offered opportunities for sagacity, as is evidenced in the quotes sampled in Table 7.1. Indigenous voices were abundantly evident from many leaders' comments on COVID-19.

Unintended consequences

Attacks on the WHO and weakening of support of its programs by the US administration ushered in an era of uncertainty in the UN system. The fracture of the WHO's philosophy of leading efforts to combat global pandemics, as one human race, was shattered after

Table 7.1: Indigenous thought and sagacity: Quotable quotes.

"Some individuals think we're using scare tactics or extreme measures ... but we are losing lives here on the Navajo Nation, and I'm going to do everything I can to help save lives." Jonathan Nez, President, Navajo Nation. Remarks on measures taken to prevent spread of COVID-19 on the reservation. "We're barely getting bits and pieces." ... "You have counties, municipalities, already taking advantage of these funds, and tribes are over here writing our applications and turning it in and waiting weeks to get what we need." Jonathan Nez, President, Navajo Nation. Commenting on the slow response of the Federal government to provide emergency funds to support Native Americans combat COVID-19. Source: Scher, I. (April 20, 2020). *The Navajo Nation's coronavirus infection and death rates are 10 times higher than the neighboring state of Arizona*. Business Insider. https://www.businessinsider.com.
"We're in a unique time," ... "All eyes are on Navajo right now. Sadly, it took a virus for people to see the issues we're facing here, but I'm not above using it to advocate for our people. Why shouldn't we use it for all its worth?" Remarks by Vice President Myron Lazio, Navajo Nation. On why he had to attend Trump Rally in Phoenix, Arizona, to lobby for Navajo Nation's issues. Source: Yurth, C. (June 25, 2020). *Lizer defends attendance at Trump event on lobbying*. Navajo Times. https://navajotimes.com/reznews/lizer-defends-attendance-at-trump-event-as-lobbying/.
"There are many people calling on the hataaliis for help but many of them don't have the resources to have a ceremony or to pay the hataalii. ..." "We want to pay our hataaliis so they can continue to heal our nation." Michelle Kahn-John, Secretary of the Diné hataalii Association of the Navajo Nation. Remarks to 24th Navajo Nation Council Health, Education & Human Services Committee (HEHSC) Members. Committee chair, Hon. Daniel E. Tso. Vice chair. Hon. Carl Roessel Slater. Source: Navajo Times. (June 25, 2020). *'We must return to the teachings of Hozho': Hataalii Assoc. Poised to receive $ 1M for coronavirus response*. Navajo Times. https://www.navajotimes.com.

President Trump's administration decision to cut off US support for the organization. This approach was completely in sync to the neonationalist turn in American politics. Trump, a self-proclaimed nationalist, campaigned on a promise to make America First. Thus, the lowered engagement with multilateral institutions was not at all surprising. The WHO Director-General, Tedros Adhanom Ghebreyesus, in an emotional response to Trump's notification to leave the WHO, warned that humanity was failing because of lack of leadership and solidarity. This was a veiled indictment of President Trump's policies. Dr. Tedros Ghebreyesus's emotional statement included the words:

> *The COVID-19 pandemic is a test of global solidarity and global leadership. The virus thrives on division, but it is thwarted when we unite. How is it difficult for humans to unite and fight a common enemy that is killing people indiscriminately? Are we unable to distinguish or identify the common enemy? Can't we understand that divisions and the cracks between us are an advantage for the virus. I think I do not need to remind you because we all know that these are the basics (Ghebreyesus, 2020).*

In psychological terms, America's continued aloofness had already impacted WHO operations. And it affected the UN body's apparent bid to salvage its reputation as a global entity. Already, messaging on COVID-19 reflected a lack of cohesion when some statements attributed to WHO officials appeared contradictory. Dr. Maria Van Kerkhove, an infectious disease epidemiologist and member of WHO's COVID-19 Technical team, reported that some data suggested asymptomatic person was not necessarily infectious. This headline was captured in numerous news organizations and also broadcast widely on WhatsApp and other social media applications. For example, Greg Kelly of Newsmax TV called this finding very good news. He invited Dr. David Samadi, Director of Men's Health at St. Francis Hospital, to verify the statement that asymptomatic persons did not infect others, which he did noting that the WHO had made 360° turn (Kelly, 2020).

This idea was quickly refuted by fact-checkers consulted with the WHO. Dr. Fauci also refuted the claims on ABC News *Good Morning America* Show (Lovelace, Jr., 2020). The WHO issued a statement on June 9, 2020, to clarify the remarks by Dr. Van Kerkhove. The organization noted that she had referred to unpublished studies that followed asymptomatic cases involving limited contract cases. The Logical Indian Fact check organization quoting WHO's Dr. Kerkhove noted that:

> *Modelling studies have not put the extent of asymptomatic transmission to about 40%. There have been two-three studies that followed asymptomatic cases over time and looked at all contacts and came to the conclusion that there was no spread but that is a very small subset of studies. In trying to articulate what we know, I used the words "very rare" and there was a misunderstanding ... what I didn't report yesterday because this is a major unknown is that there are some modeling studies that estimate that 40% transmission may be due to asymptomatic models. It is a modeling study, so I didn't include it in my answer yesterday (Logical Indian Fact Check, quoting Kerkhove, 2020).*

The apparent backtracking by WHO dented its reputation for issuing clear statements on pandemic data problems. Because of the viral spread of the earlier comments, statements from WHO were now a new data point in the proliferation of misinformation about COVID-19. In fact, clips circulated on social media platforms announced erroneously that COVID-19 was no longer an infectious disease.

Elevation of public health-care professionals in policy processes

There is a silent question that must be added to the discussion on governance networks, just what did the crisis bring to the policy process table especially as concerns the emergence of specialized policy venues for decision-making on preventing the spread of COVID-19? What are the consequences of establishing specialized COVID-19 Task forces? In the short term, there appears to be definite influences on policy-making and governance in all the countries covered. One, they elevated a few key individuals to national prominence. The task forces offered opportunities for health professionals to take center stage in national conversations. This includes the elevation of health officials in Indigenous communities.

Next, there was a degree of increased transparency—even where countries had been in denial. Citizen journalism helped capture various dynamics of the disease, including corruption of officials conducting PCR testing in Kenya. The task forces also reduced the costs associated with communication with the broad publics. Communication from central authorities trickled down to devolved units run by tribal authorities. The hope was that the policy choices made at the top would be congruent with those made at various tribal authorities. That may not necessarily have worked out well in some communities where national and state governments frustrated local Indigenous governments from doing what was in the best interest of their tribes.

A different angle to the question at hand is the elevation of discourses that link public health to development as a whole. The emergence of COVID-19 as an existential threat to tribal communities forced policymakers to revisit the need for completing the unfinished projects in areas such as education, housing, and water development in marginalized communities. Therefore, the notion of institutional specialization was rendered illogical by the salience of policy-making within teams managed by the task forces. The task forces, therefore, highlighted the need for contextualizing problems beyond narrow confines of single governmental structures.

Power of government surveillance and the question of democratic governance

Perhaps an unintended consequence of the COVID-19 era was the proliferation of the surveillance society concepts. This really amounted to escalation of the surveillance society or in more dramatic terms, surveillance on steroids. Mobile phone companies across the board added features for contact tracing persons with COVID-19. One would contend that big brother came back big time in all governments. Thus, the main concerns expressed have been in the area of privacy and ethics around the use of the data and how it would be used. Contact tracing devices include mobile apps, apps on smart watches, and health tracking wearable smart watches. In its most intrusive form, contract tracing apps can share locations of infected people with media personalities and issue text alerts.

In South Korea, where the apps were used, government was alerted to the misuse of the app to call out members of the LGBTQ + community. It turns out that some bigoted persons misused the apps to target clubs and bars visited by members of the LGBTQ + communities. Surveillance technology was not just about contact tracing but expanded into the realm of monitoring people's movements. South Korea also extended its surveillance capabilities through use of CCTV images of infected persons (Chakravorti, 2020). Thus, these new technologies are now capable of ensuring that citizens comply with guidelines on movements (Harari, 2020).

The troubling unintended consequence of democracies seizing the moment to extend surveillance systems was that there were no sunset laws to end the practices after the COVID-19 pandemic was over. Also, the technology companies did not add features that would enable the devices to self-destruct at the end of the pandemic. Democratic countries such as India, Israel, and South Korea took advantage of these tools to control the spread of COVID-19; there was potential that governments would be reluctant to let go their newfound technical ways of monitoring their citizens at the expense of privacy rights. Thus, debates over surrender of privacy and freedom from state surveillance persisted during the crisis.

COVID-19 enabled governments to not only utilize the surveillance systems available through ICTs but also to threaten use of other laws to enforce compliance with strategies to combat the spread of COVID-19. For example, the government in the United States was legally capable of forcing students and citizens to get COVID-19 vaccinations, if available (Rim, 2020). According to legal experts, states can force vaccinations on individual to prevent the spread of contagious diseases. The law supports vaccinations when the disease threatens societies.

A 1906 case, Jacobsen v. Massachusetts ruled that each state had police power to enact health-related laws and regulations to protect its citizens. These include vaccination and quarantine mandates. Articles in the American Medical Association's *Journal of Ethics* had revisited the question of enforceability of vaccines during the SARS pandemic of 2006. At that time, the consensus was that states had obligations to ensure public health. Thus, the quarantines and other measures taken during the shelter-in-place regulations were all supported by existing legal frameworks (Rim, 2020).

Internationally, COVID-19 provided an excuse for some governments to resort to antidemocratic practices ostensibly to combat the pandemic. This thought came in different forms. In some instances, the overreach of governments was packaged as an affront on democracy and privacy. In some instances, it dovetailed the misinformation agenda and associated with ways in which governments pushed for societies to embrace nonpharmaceutical interventions like wearing masks. This was manifest by those who spread the idea that governments were not really advising or requiring people to wear masks to prevent the spread of COVID-19, but as prelude to pending infringement on rights and also an aspect of corporate "scamdemic"—to make money off selling face masks (Branson-Potts, 2020). This idea, although not widely spread, suggested that "mandate son wearing masks were a communistic move by government to see how much people are willing to give up their freedom" (Branson-Potts, 2020).

In some countries, like Kenya, members of the corrupt police force departments used the wear mask mandate to attack and arrest citizens on the flimsy ground of not wearing masks, even in their own private vehicle! As Larry Diamond (2020) posited, many countries that had previously mad efforts at building democratic institutions were now clearly backtracking and utilizing fear of the pandemic as a tool to stifle legitimate contrary voices on governance. In his work published by Stanford University's Hoover Institution, Professor Diamond articulated the idea that countries like China were not transparent with its own dealing on the death rates of COVID-19. In addition, some of its material offered to other countries to support efforts to contain the pandemic were of remarkably inferior quality. Democracies, on the other hand, provided numbers on impact on human societies. More open and democratic countries like Israel, Germany, Taiwan, South Korea, and a few Scandinavian nations initially appeared to have better and more open strategies for managing COVID-19. Their responses were above board and numbers of cases were widely shared and used big data to track the pattern of the virus's trajectory. They were also supported by strong public health systems, which allowed experts to be the voice for combatting the pandemic.

Aeyal Gross (2020) reported that Israel assigned contact tracing to the General Security Services, which notoriously applied tactics employed for controlling Palestinians. Israel initially established a military command supported by ministry of health bureaucrats. In fact, the Defense Minister Naftali Bennett also took an active role in messaging mitigation strategies. He called upon Israeli to use Skype and WhatsApp and other social media to keep in touch with the elderly. In his view, Israel's best plan was to move toward herd immunity but protect the most elderly. However, Siegal Sadetzki, the leading epidemiologist and director of the Israeli Health Ministry's Public Health Services, resigned from her position in early July amid surges in COVID-19 cases (Estrin, 2020).

She claimed the government had lost direction by opening up some economic sectors earlier than her recommendations. Still, by early July 2020, Israel had 338 deaths and 12,700 active cases, which was still a lower better per capita rate than that prevailing in the United States (Estrin, 2020). Deepening the democratic recession democratic regimes became more authoritarian, for example, Russian President Vladimir Putin eliminated democratic institutions and became president for life after a referendum was carried out in the middle of a pandemic. Viktor Orban, a right-wing populist leader of Hungary, was named first dictator in the European Union after passing sweeping emergency measures that allowed him to rule by decree amid rising COVID-19 cases (Dettmer, 2020).

India, the world's largest democracy, saw a rise in an authoritarian practice after Prime Minister Narendra Modi with only 4 h' notice shut down government operations on March 24, 2020. He made full use of the Disaster Management Act to curtail movements across the union and imposed criminal penalties on those refusing to adhere to the imposed guidelines to control the spread of the coronavirus. State and regional governments also joined in to enforce lockdown measures under the Epidemic Diseases Act (1897). The government, without hesitation, used drones and other invasive technologies to monitor quarantined individuals (Grover, 2020).

There was more to the story. Modi reportedly used colonial era draconian laws to pressurize journalists covering the pandemic (Freedom House, 2021; Rezaian, 2020). In the first 3 months of the pandemic, small media outlets were financially crippled by restrictions on movement to prevent stories being covered. In addition, social media was used extensively to attack journalists. In Uttar Pradesh and other states where Modi's nationalist Bharatiya Janata Party (BJP) controlled levers of power witnessed much intimidation and oppression (Majumder, 2020). Attacks on the media were carried out across the country including regions where STs resided. Even before the pandemic, he had previously unsuccessfully attempted to fight "fake news," which many read as a declaration of war on journalists (Safi, 2018). Many feared that India was narrowing freedoms and drifting toward an illiberal democracy in the wake of the pandemic.

Also, in question was the ability of nations to conduct elections during the pandemic. In America, 2020 was a year of elections and some primaries had to take place within the uncertain health emergency climate. Initially, debate over use of mail in ballot papers was presented as a challenge, even though it has traditionally been used in several states. And many military personnel overseas vote that way regardless of who the candidates for the president are. In 2020, the situation seemed different as President Trump publicly joined the debate.

The idea that elections would take place during pandemics, although seen as a willingness to extend political freedoms in countries such as Tanzania where President Magufuli won the October 2020 in a landslide (over 84%) victory, will be a debated issue since many meetings and rallies occurred without respect for COVID-19 mitigation protocols. The strong state projection of normalcy suggests weaknesses in governance networks. From this perspective, our view is that electioneering practices done at the expense of public health protocols may have led to harmful outcomes. This was clear, at the White House, where Trump and many of his allies contracted COVID-19 in the course of campaigns.

We will never really know of the implications in other countries where elections took place amid emergence of the pandemic. For instance, in central east Africa, Malawi physically conducted elections during the public health emergency. Opposition candidate Dr. Lazarus Chakwera defeated incumbent Peter Mutharika of the Democratic Progressive Party (DPP). Chakwera referred to his win as a "victory for democracy and justice" (Lusaka Times, 2020). Previously, Mutarika had been declared winner in an election held in 2019. But after a court nullified the election, a repeat election was ordered and held June 2020. The opposition party, Malawi Congress Party (MCP) won by a 58.57% vote. The Malawi Electoral Commission (MEC) reported that 4.4 million people had voted out of 6.8 million registered (Masina, 2020). Although there were no international election observers due to COVID-19 restrictions, the Malawi Human Rights Commission declared the election reflected the will of the people of Malawi and was widely hailed as a fair process.

From a politics standpoint, elections provided room for recalibration of COVID-19 mitigation strategies. However, in terms of governance, the pandemic weakened the economic base of many civil society organizations, especially those that were active in recruitment of Indigenous people for purposes of participation in self-governance. In addition, the imposition of social-distancing rules worked against effective participation in civic affairs. The problem was more severe where key resource persons were negatively impacted by the disease. As was described, many societies lack reliable Internet services and hence the possibility of Internet voting, which would require paper trails was not a viable option in any of the societies covered.

Supporting the community empowerment effort and return to war imagery

Perhaps, a more enduring image of the management of the coronavirus pandemic was the increased government and society appreciation of frontline health-care workers. In India, the mainstream press referred to and hailed the medical professionals as "frontline COVID-19 warriors" (Garari, 2020a). This language was also used by President Jonathan Nez of the Navajo Nation when he made remarks in reference to disputes with the Council over expenditure of monies allocated for combatting the COVID-19 pandemic. President Nez had stated that certain members of Council "put politics over the needs of our frontline warriors and the Navajo people" (Krisst, 2020).

Since, the pandemic had been framed as a war effort, some of the salutations were expressed in military terms. In Arizona, the state Air National Guard flew over the Navajo Nation on May 6 to show support for frontline workers. The Arizona National Guard teamed up with the US Air Force for a multiple-plane flyover from Phoenix into Northern Arizona cities of Flagstaff and Sedona, and the Navajo Nation areas of the capital at Window Rock, Fort Defiance, Chinle, Kayenta, and Tuba City. The event featured a KC-135, and three F-16 Fighting Falcons launched from a base in Tucson, Arizona (Baker, 2020; Curtis, 2020).

In a press release, US Air Force Maj. Gen. Howard Purcell, the Arizona National Guard Air Component Commander, stated that, “We support the community during Covid-19 by showing goodwill, and our hope is to work together to respond and grow from this event.” … “We want to show our support and we stand behind our community. We will move forward together” (Curtis, 2020). One notable outcome of increased rates of infection was an apparent recalibration of the war imaginary and its replacement by an unclear policy agenda. This criticism applies mostly to how the national government changed its approach to handling COVID-19 in August. There was strong evidence that the CDC was politicized and no longer at the helm of the epidemic control process (Varmus & Shah, 2020). Furthermore, at one point, the opposition Democratic Party, through its presidential candidate Joseph Biden, announced that President Trump had surrendered to the coronavirus (Stevens & Kaplan, 2020). In graphic terms, Biden accused Trump of having “waved the white flag and left the field of battle” (Wilkie, 2020). Late in August, the organizations that had been instituted to confront the coronavirus crisis were suggestively seen to be moving toward a “herd mentality strategy,” which was not a military approach that had been touted in early 2020 (Abutaleb & Dawsey, 2020). The Trump administration lacked a clear strategy or any cohesive war planning and operations. Different jurisdictions developed their own localized and community-driven anti-COVID-19 campaign strategies.

Community-based informal organizations spread throughout the Indigenous enclaves. Unfortunately, there is no dependable database to assess the roles and structures of these organizations. Therefore, we cannot generalize about the community-based organizations that were established to mitigate impacts of COVID-19 in Indigenous societies. Nevertheless, sufficient information is provided in numerous accounts of group leaders and members. For instance, we know that community groups organized in the Navajo Nation were self-organized and designed their strategies according to their intimate knowledge of the organization and values of their own communities. Most, if not all, established their own mandates and organizational structures. The vast majority were never really completely independent of external support, whether through publicity or financial aid. And most focused on relief and specific patient and family supplies needs, including water, sanitation, food, and companionship and mental health support.

The community-based COVID-19 pandemic relief groups relied on consensus building as a means for deliberation and decision-making. And a vast majority felt accountable to Indigenous people and Indigenous leadership from within.

The resurgence of global groupthink?

If there is anything to learn from the management styles of public officials during the COVID-19 pandemic, it is the proliferation of groupthink mentality. As novel oncornavirus, the dangers of COVID-19 infections invoked memories of previous flu pandemics that killed millions of people. Previous experiences offered valued lessons of history, especially the ability of bureaucracies to "speak with one voice" on matters of preventing the spread of the disease. However, speaking with one voice also conjured the images of groupthink decision-making within the administrative communities. Groupthink, as we know, is about people collectively agreeing about a critical decision when on their own, they might individually be in a different place (Janis, 1971, 2005). Groupthink implies that dissent is suppressed, and officials may privately have different views but publicly support a certain line of thinking and sets of decisions. The pressure to conform to the official script on how to manage COVID-19 pandemic existed in all communities, albeit at different degrees.

A US military veteran, and tech entrepreneur, Yinon Weiss makes the case that lockdowns were not commensurate with the threat posed by the coronavirus (Weiss, 2020). Weiss (2020) contends that "academic epidemiologists presented widely-off-the-mark doomsday predictions." He refers mostly to Dr. Neil Ferguson of Imperial College who had made inaccurate predictions on the pending casualties from COVID-19 pandemic. The models that informed the predictions contributed to a domino effect and contributed much of the groupthink decisions adopted by governments to manage COVID-19. While this simplistic argument was meant to demonstrate that countries should have not been quick to order lockdowns and travel bans because a vast majority of the population were never really at significant risk. The author provides skewed data that hint at morbidity before and during the COVID-19 crisis.

Furthermore, the decisions to require face mask covers or shields and cancellation of large gatherings were a judgment call. Although science had put the issue of masks to rest, doubts persisted (Walton, 2020). Thus, for Weiss (2020) to claim that decision-makers had information to make more accurate policy decisions on COVID-19 management is an inaccurate statement. Most medical and biological science experts shared the view that far too many unknown variables characterized the COVID-19 pandemic. There was consensus that knowledge about the novel coronavirus was in a state of flux.

Still, knowledge, which changed almost daily, was often disseminated through preliminary scientific reports and required additional evidence for validation (Cereceda, 2020; Frenk, 2020; New York Times, 2020). Field surveys in China, where the first cases were reported, there appeared to be preliminary results on questions such as the incubation period from SARS-CoV-2 infection to onset of symptoms. Work done in Hubei province revealed that the median incubation period was about 5 days and that 97.5% of infected patients exhibited symptoms by 11.5 days (Lauer et al., 2020). Such important published work helped to inform decisions about monitoring and quarantine for 14 days, which became a standard globally. Scientific work that appeared in *Nature* also clarified the key role of presymptomatic or asymptomatic phases in transmission of COVID-19, which was helpful in designing post-lockdown control policy strategies (Lavezzo, Franchin, & Ciavarella).

However, the unknowns were arguably more problematic from policy-making perspectives. Among these unknowns are the linkages to childhood maladies like Kawasaki disease (Elkind, 2020; Woodyatt & Howard, 2020). Dr. Erin Bromage, a biology professor at the University of Massachusetts, Dartmouth, shared similar views on CNN and other news outlets (Curnow, 2020; Merrill, 2020). Dr. Bromage's blog and explainer kept millions of readers abreast with new trends in prevention of the disease. In addition, he highlighted the impact of COVID-19 on various organs, including the heart, lungs, and intestines (Curnow, 2020; Puntmann et al., 2020). The association between COVID-19 and vascular conditions is now becoming a key hypothesis in understanding the full manifestations of the disease. Dr. Daniel Jacobson and his colleagues' work published in *eLife* disclosed that COVID-19 had bizarre symptoms because it is primarily a vascular disease, rather than a respiratory one (Smith, 2020). There were also unverified studies on links between blood groups and COVID-19 infections (Clark, 2020). Of significant policy implication is the debate over the efficacy of vaccines, which prompted two leading scholars to write an opinion piece assuring the public that vaccines work (Akiko & Medzhitov, 2020).

COVID-19 was also mutating, which complicated the full range of its epidemiological effects. Data from Italy also showed that the virus was systemic and attacked all organs and creating brain fog (Couzin-Frankel, 2020; Ramsay, 2020). Scientists across the globe were confused by claims that the virus was mutating and changing its genomic structure. New symptoms such as diarrhea, vomiting, headache was found in some patients thereby complicating COVID-19 care protocols (Garari, 2020b). With time, there were additional expert reports concerning public health consequences including "coronasomnia," a rise in chronic insomniacs, decline in productivity, hypertension, and risks of depression among the patient populations (Brulliard & Wan, 2020).

Reactions: are we in this together?

A popular perspective and rallying point of community and world leaders was that COVID-19 was a world problem and that the human race would find common solutions together, as one. The narrative that "we are in this together" was overplayed across the countries. On its surface, it implied that COVID-19 gave us a chance to see our humanity as members of the earth, our planet home. And in spite of some symbolic gestures that communities faced the challenge together, there were also instances when communities were apart, which made managing COVID-19 a more daunting task. Among the best gestures expressing the togetherness dimension in the management of COVID-19 was the gesture by about 24,000 donors from Ireland who gave over $820,000 in an online fundraiser (GoFundMe) to support Hopi and Navajo Indigenous communities to buy food and supplies during the pandemic. The fundraiser conducted by Native Americans as the pandemic ravaged through the reservations. Various members of the electronic press reported that money donated by the Irish was part of $ 3.6 million raised to support the Navajo and Hopi people (Hedgpeth, 2020). Most of the donations were in small amounts—$10, $20, $30 some exceeding $100 (Carroll, 2020). According to various accounts, the gesture was a returned favor that rekindled the times when the Choctaw Nation of Oklahoma gave food to the Irish during the famine of 1847.

History records that during the Potato Famine, as millions of Irish people starved, news reached the Choctaw leaders through a letter read at a meeting of the Memphis Irish Relief Committee. The message was also portrayed in the *Arkansas Intelligencer* newspaper of April 3, 1847, which he spoke to the condition of Ireland. The support accorded was sent to the Society of Friends in Dublin, Ireland. *The Washington Post*'s journalist Dana Hedgpeth (2020) description of events amplifies the persisting stereotypes:

The *Arkansas Intelligencer* article said the Choctaws "knew nothing more, cared for nothing more, than the fact, that across the Big Water, there were thousands human beings starving to death." It went on to say, "Is not this a sublime spectacle? The Red man of the New, bestowing alms upon the people of the Old world!"

Clearly, as explained by Hedgpeth's piece (2020), the gesture from the "red brethren of the Choctaw nation" who were also "separated by miles of land and an ocean's breadth" was not the first time that Indigenous societies supported settlers. As *The Guardian* reported, the Choctaw nation set aside its own impoverishment and suffering from the Trail of Tears relocation and empathized with Irish people who were starving 4000 miles away (Carroll, 2020). Considering that Former President Mary Robinson visited Oklahoma in 1995 to thank the Choctaws, even after 6 months of traumatizing COVID-19 experience, no cabinet-level member of the Trump administration visited the Navajo Nation at its greatest hour of need.

This disconnect and appearance of lack of empathy resulted in much social media displeasure after Navajo Vice President Myron Lizer, a Republican, was reportedly present at a "Students for Trump" rally in late June 2020. Lizer defended himself by suggesting that he used the event to meet some key congress people including Senator Martha

McSally, Rep. Paul Goser, and Rep. Debbie Lesko (Yurth, 2020). Lizer told Tséyi' Bureau reporter that he wanted to lobby for an extension of the deadline for spending money received from the federal government for coronavirus relief (Yurth, 2020). And he added that he and the Navajo president, Nez visited the hot spots all over the Navajo Nation delivering supplies.

To demonstrate the apparent, disconnect between the Trump and administration and Native American communities, there was no policy announcement concerning Indigenous societies tackling COVID-19 at Vice President Mike Pence's press briefing when he visited the state of Arizona on July 1, 2020, following a surge in numbers of coronavirus cases in the state (Sanchez & Oxford, 2020). Although Sanchez and Oxford (2020) highlighted Pence's comments that "help is on the way and we're going to spare no expense to provide the kind of reinforcements that you need across the state," the vice president and head of the White House Coronavirus Task Force gave no specifics on aid to Native American populations. And when the Governor Ducey added that the state was going to protect the "most vulnerable," the target groups were not articulated. Arizona's most vulnerable include immigrants in Immigration and Customs Enforcement (ICE) holdings, prisoners in Florence and other prison centers, and thousands of Indigenous on reservations. Governor Ducey did not specify how those specific groups would be supported. There was also criticism that Ducey had reopened the state to business without aggressive mitigation measures, which partly explained the upsurge in COVID-19 cases.

There are more stories of foreign groups and governments supporting the Navajo Nation's battle against COVID-19. The South Korean government also delivered 10,000 N 95 masks and other Personal Protective Equipment (PPE) to the Indigenous communities of Arizona (mainly the larger groups, the Navajo and Hopi) to honor members of the Navajo community who had served as "code talkers" during the Korean War (1950–53). Code talkers were Navajo speakers who helped the American army deliver coded radio messages that were concealed from the enemies. The gesture was coordinated by the Korean Ministry of Patriots and Veteran Affairs in consultation with the University of Arizona's College of Medicine and US Rep. Greg Stanton with the Arizona Korean Association (Scoop Upworthy, 2020). Over 800 Navajo fought in the Korea war helping to defend the South from communist rule. It begs the questions that why did private European citizens appear to show more compassion and empathy than American federal government officials, who by all accounts delayed in disbursing resources to combat the scourge on Navajo and Hopi lands?

An unusual remdesivir purchase and selfishness

We could argue about the emotions of suffering that linked all human beings. However, the nationalist streak and selfishness tampered the otherwise goodwill which was widely desired. One evidence of this assertion is President Trump's administration decision to buy all the global supplies manufactured by Gilead Sciences (Shivji, 2020). Initial reports claimed the medicines would be supplied to Veteran Affairs and Department of Defense hospitals, Indian Health Services, and the Coast Guard among other American health facilities. The unit price was about $ 390 per vial for the US government. The discounted price was not meant for Medicare or Medicaid. According to some estimates, a full-five-day treatment would cost about $ 2340 (CNN, 2020). This action denied any other nations or Indigenous societies outside of the United States to access the medication for 3 months. This measure to buy global supply of a drug which had been proved to lower death rates among the most critical COVID-19 patients was considered a selfish act. Dr. Anthony Fauci, Chief medical Advisor to the president, considered such an action as not in the interest of promotion of "equitable distribution of scarce vital resources" (Werman, 2020). Even if the medication was available on the market, it was out of the reach of the world's poor communities.

Observers also argued that the United States had adopted an aggressive posture by seizing PPE to prevent other countries from access to the vital supplies. The attempt to protect and control existing treatments did not augur well for the narrative that nations were "in this together," instead it was a manifestation of the global power inequalities, which fueled inequalities in global health outcomes. A related dynamic of the inequalities and global racism was the effort by Western corporations to test vaccines on relatively poor African communities. In particular, the Oxford University Vaccine Group rushed to conduct trials in Africa which Dr. Adrian Hill excused with the pretext that "there was only a 50% chance of the Oxford coronavirus vaccine working because cases in the UK are declining so fast" (Sky News, 2020). The point Africans were raising was that there were other places for the trials to be conducted. Many Africans were not ready to watch their people used as Guinea pigs by Western medical researchers. Among the more affluent Maasai, the projections about vaccine development conjured the images of the Tuskegee Syphilis racially motivated and ill-informed experiments. In the case of Tuskegee, African Americans were made experimental subjects at the hands of influential members of the American medical fraternities. Many of the findings were misrepresented to aid agents of eugenics and social Darwinian theories that depicted black people in negative terms.

Massive WhatsApp campaigns, by more educated Africans, were launched in Ghana, Nigeria, and Kenya to speak against testing vaccines for COVID-19 on Africans, who at the early stages of the pandemic did not have as high rates of the disease as did people in China and Europe. Dr. Harriet Washington, professor of bioethics at Columbia University and author of *Medical Apartheid: The Dark History of Medical Experimentation on Black Americans from Colonial Times to the Present*, opined that the vaccination tests planned and executed by Western institutions, including universities and pharmaceutical corporations, did not seek informed consent approvals to test on African people as they would in their own home countries. Here is how she explained her case:

> *These are Western universities, pharmaceutical companies, institutions, and so Western oversight is necessary. The FDA and other groups should be providing oversight. They should be checking documents, making sure there's informed consent, making sure that everything happens as it should happen. When you're not doing that, you're leaving the field wide-open for abuse (Harriet Washington, in Pfeiffer, 2020).*

Dr. Washington, therefore, on national radio expressed solidarity with marginalized groups who include Indigenous people not treated equally and fairly by testing authorities (Pfeiffer, 2020). One area of reform that was needed is, for governments to reform patent laws and also recognize the importance of Indigenous knowledge included in some of the research outcomes. Evidence suggested that these communities contributed to finding solutions to killer diseases but did not necessarily benefit from discounted costs of medication that emerged from the research projects involving them. This was a morally wrong position for pharmaceutical corporations to take.

When asked during a live TV Show what the impact of coronavirus is, French philosopher Bernard-Henri Levy, remarked that the pandemic had revealed some aspects of generalized selfishness. To Levy, selfishness increased instead of acts of humanity (Zakaria, 2020b). He referred to the loss of America's global leadership and its replacement with an "America First" agenda that led to more suffering not only in the United States but globally.

In his book, Levy (2020) comments that in the age of the coronavirus, the actions of some of the major powers had reversed the existing progressive order. As such, countries like the United States appeared to be turning back to the pre–Christopher Columbus era, it is as if America did not exist. For the most part, the COVID-19 pandemic triggered a sense of darkness and denial of reality and an attack on liberty. This absence of America led to some global leaders turning to draconian ruling styles.

In addition, the selfishness was manifest in President Trump and other like-minded leaders' denial of the coronavirus disastrous effects. Philosopher Levy posited that those who denied the virus were, in effect, selfish and didn't seem to care about the thousands of deaths caused by COVID-19 infections. Further, he noted, Trump's refusal to wear face mask amounted to an act of selfishness and, in a sense, he was masking against reality—the truth—instead of the virus itself (Zakaria, 2020b). Levy also explained that although the shelter-in-place initiatives were necessary, the unintended effect was to create a form of isolationist mentality. In other words, some individuals retreated from the world. Withdrawing from the world brought out the worst selfish traits in human beings. For him, it would have been better for humanity to show compassion and care about those who lacked health insurance and shelter. In the midst of the pandemic, the Trump administration filed a case to end the Affordable Care Act, which was, in the first place, meant to provide insurance for millions that had no coverage.

Propaganda

On June 30, 2020, President Donald Trump tweeted: "As I watch the pandemic spread its ugly face all across the world, including the tremendous damage it has done to the USA, I become more and more angry at China. People see it, and I can feel it!" Trump's remarks were echoed by his associates. The end of June 2020 marked a watershed period in American history. Over 100,000 Americans had lost their lives to COVID-19, more than any other country in the world. In response, the Trump administration blamed China.

The administration's top trade adviser, who had led the trade wars against China, found another data point in the propaganda arsenal against China when he, on an MSNBC show, advanced a conspiracy theory that China had deliberately permitted hundreds of thousands of infected people to leave the country and spread the virus abroad (Velshi, 2020). Navarro had made similar remarks on Fox News. Navarro also claimed that the virus was weaponized by the Chinese community party and its government. On its part, China's state-controlled newspapers blamed the Trump administration for incompetence in handling the COVID-19 pandemic. According to Chinese *Global Times*, lies from the Trump administration dominated the discourse. The July 3, 2020, editorial titled "Rampant US Epidemic to hurt the world" assessed the Trump administration as follows:

> *The US set another record for novel coronavirus cases on Thursday. Reuters reported that the country confirmed more than 55,000 new COVID-19 cases that day, which is "a new daily global record" for the pandemic. … The current epidemic in the US is completely out of control. However, the US federal government still insists that the rapid increase in infection figures is due to more testing. Data released by the US Labor Department on Thursday showed the country added 4.8 million jobs in June. The White House soon boasted about its economic achievements. But people doubt whether the US government is making utmost efforts to rein in the virus and save more American lives. Lies are dominating US society's recognition of the pandemic …. The US fight against the virus is paralyzed. There is no national strategy to alleviate the epidemic" (Global Times, 2020).*

China's offense against the United States was carried extensively in the *Global Times*, whose writers claimed the Trump administration was responsible for many of the missteps in the management of the COVID-19 pandemic.

Chen Qingqing and Leng Shumei (2020) wrote that a probe was necessary to unearth Trump's handling of the pandemic. The *Global Times* claimed that some American leaders had tested positive for COVID-19 antibodies in November 2019, which was way before the surge in China. And China was suspicious that the United States covered up the outbreak within its own borders. China also pointed a finger at the activities of various military biological labs in the United States, including the Pentagon-owned United States Army Medical Research Institute of Infectious Diseases (USAMRIID) at Fort Detrick, Maryland, which was shut down in 2019 (Chen & Leng, 2020). The closing of the lab was also reported: first in *The Frederick News Post* and then shortly after in *The New York Times* of August 5, 2019, and other credible outlets (Grady, 2019; Mongilio, 2019). The CDC had indicated that the lab had several violations in 2019 and was using a classified "select agent" which had the potential to severe public health and safety risks (DiGangi, 2020).

The CDC had reported that an individual at the lab performed procedures and interacted with animals without following procedures. The individual was likely exposed to select agent aerosols. Video reviews also showed that some of the primate used in experiments were housed in open caging and a laboratorian entered the lab without appropriate respiratory protection. The individual also disposed waste without wearing gloves. Overall, the "US Army Medical Research Institute of Infectious Diseases had systematically failed to ensure implementation of biosafety and containment procedures commensurate with the risks associated with working with select agents and toxins" (DiGangi, 2020).

China seized on the CDC reports concerning the failure of the US army to follow safety protocols in managing its biological weapons. Chinese defense experts noted that after the laps in safety protocols at the US lab, a wave of pulmonary pneumonia and similar diseases were observed in the United States. Consequently, the Chinese government demanded more accountability, especially because a few foreign athletes from western countries visited Wuhan for the Military World Games and were hospitalized due to an undisclosed infection.

There was widespread consensus among China's scientific community that the Trump administration failed to control COVID-19 thereby "dragging the globe into longer pandemic war" (Chen & Leng, 2020). Critics also claimed the administration travel ban involving flights from Europe came 5 weeks too late. Trump's own actions including ignoring early warnings about the diseases, silencing critics, ignoring scientific advice, and use of incorrect facts about the pandemic for political reasons became fodder for Chinese propaganda.

Russia also joined in the propaganda wars and demanded that the United States disclose the operations of its biological labs in Ukraine and Georgia, which were a direct threat to the Russian people. The United States rejected to disclose tails of its biological weapons capabilities. Many communities of nations had previously called for countries to be enjoined in the Protocols to the 1972 Convention on the Prohibition of the Development,

Production and Stockpiling of Bacteriological (Biological) and Toxin Weapons and also find ways to destroy such weapons.

The problem of coordination of efforts in governance networks

We learn from Luther Gulick (1937) and other classical writers to Guy Peters (2001) that coordination is an important administrative value. It mattered a lot during the COVID-19 management experience. For example, the state of New York, Connecticut, and New Jersey sought to coordinate efforts in their decisions to reopen their economies after the rates of infection had subsided. The governors of the three states worked together to prevent the spread of the disease.

The principle of coordination also featured prominently during a US Senate hearing on Global Health Security and Diplomacy meeting in which Sen. Jim Risch (R-Idaho), chair of the Foreign Relations Committee, and his colleagues lauded the role of coordination of roles of agencies such as the CDC, USAID, the State Department, and WHO in management of the COVID-19 pandemic. Senator Risch's Senate Bill S. 3829, cosponsored by Senators Chris Murphy (D-Conn.), Ben Cardin (D-Md.), and Rob Portman (R-Ohio), was meant to "detect, deter and contain infectious disease outbreaks overseas before they become global pandemics" (Risch, 2020). Among the key provisions of the Global Health Security and Diplomacy Act (GHSDA) was the provision:

Strengthening interagency coordination and diplomatic engagement:

- Establishes a coordinator for global health security and diplomacy at the Department of State, supported by a deputy at the United States Agency for International Development (USAID), to manage program coordination, lead diplomatic efforts, and ensure the efficient and effective use of U.S. foreign assistance for global health security.
- Encourages the president to appoint a senior director for global health security to the National Security Council, to coordinate the interagency process to ensure continuity of effort across the agencies engaged in international and domestic preparedness and response (Risch, 2020).

Although the US senate recognized the importance of the importance of bringing together the different agencies to work together to provide early warning systems, enhance preparedness, and build systems that would be effective at combatting future pandemics, there was much consensus that coordination was important on both domestic and international fronts.

This study shows that because some of the Indigenous communities are geographically dispersed in different jurisdictions, the expectation was that for meaningful management of the pandemic, coordination services would be available. Coordination of pandemic response services is in a large measure, about intergovernmental communications and relationships. In theory, the crisis demanded that first, all jurisdictions agree on a common definition and framing of the problem. After that, the different government agencies work in a collaborative way to solve the problem. The fact that the virus affected different communities unequally, and that the disease would inevitably affect all populations meant that a unified and well-coordinated mitigation approach was preferable. The strong evidence suggests that, for the Maasai, that was not the case. President Magufuli declined to join regional leaders on conference call to discuss COVID-19 matters, Tanzania, for the most part set its own agenda and at times ignored efforts to liaise with authorities from Kenya. As Peters (2001, pp. 184–5) notes, one approach on short-term basis for ensuring coordination is to create virtual organizations and the generation of task forces to comprehensively respond to problems facing governments and their citizens. The Maasai elders on both sides of the border were not provided with facilitative services to ensure that that happened.

It may also be noted that Kenyan authorities, especially corrupt officials, did not exude a level of professionalism that would have engendered a cooperative effort necessary for coordination. Social media clips captured truck drivers at border town of Namanga complaining that authorities in Kenya demanded bribes to offer COVID-19 clearance certificates. The authorities in Tanzania also complained that the tests carried on by authorities in Kenya were faulty. Thus, the two countries who have Maasai living across each other had different testing kits. Many Maasai were also disadvantaged by not being allowed to cross the borders freely.

Perhaps, from the point of view of the Navajo Nation, the issue of poor coordination can be linked directly to the question of sovereignty and federalism. Internally, going by the number of casualties of COVID-19, the coordination mechanism as established by the Navajo government was weak. The command center at the Navajo Department of Health lacked sufficient funding to build a robust monitoring command center. For mitigation, the reliance on CDC guidelines which were frequently distorted by the Trump administration proved a weak link in messaging. And at times, President Nez appeared to politicize the efforts to fight the virus by controlling messages through social media virtual town hall meetings. He had initially indicated his resolve to coordinate efforts of the Navajo Nation COVID-19 Preparedness Team (Nez, 2020). In essence, the wide variation in how the pandemic response ensued proved to be a disaster and the United States recorded the highest death rates in the world. Regardless of whether the high numbers were due to better record-keeping practices or not, for an advanced industrial society, clearly, the Trump administration's approach to pandemic response revealed that American federalism was under intense pressure to deliver on its positive promises.

From national signaling perspectives, COVID-19 exposed a major fault line in federalist practices in the United States. Above all, the pandemic exposed a lack of systemic model when it came to work with tribal authorities. We noted the earlier frustration among the

Sioux tribe in South Dakota which resisted the state governors attempt to disrupt tribal efforts to prevent spread of COVID-19 in their lands. By sending out confusing messages, the White House, under President Donald Trump, weakened the existing measures in enforcing national strategies to combat COVID-19 pandemic. Moreover, Congress too did not use its power of the purse to set uniform minimum standards in community mitigation efforts to reduce spread of the disease in the different states and Indigenous communities. Congress had power to use the interstate-commerce clause to regulate spread of COVID-19 from one state to another, but it did not act accordingly. Instead, the federalist impulse generated a cocktail of state reactions which worked against the larger national interests.

Enduring racism and survival strategies

It is well known that outside the framework on COVID-19, health disparities exist in America and throughout the world. In the United States research that demonstrates health and survival outcomes of the fortunes of white settlers and the enduring lag of Black persons is staggering. Starting from childhood, African Americans suffer from low birth weight and have higher rates of infant mortality (Cramer, 1995; Gortmaker, 1979; Hoyert et al., 2001; Ventura et al., 2000). With time, the cumulative differences extend into adulthood and African Americans have higher rates of homicides, hypertension, diabetes, kidney disease, septicemia, and other conditions (see for example, Murphy, 2000; National Center for Health Statistics, 1993). The obvious implication is that as a group, a higher percentage have underlying vulnerabilities and risks that impact their ability to deal with COVID-19.

In April 2020, US national political leaders, including Senators Cory Booker (D-NJ) Elizabeth Warren (D-MA), and Kamala Harris (D-CA), and Congresswoman Ayanna Pressley (D-MA), took the lead role in writing to the CDC to release data on ethnic and minority COVID-19 cases (Booker, 2020). Such data were meant to highlight the continued existence of health-care inequalities in the United States. Because of disproportionate casualties of COVID-19 within Indigenous societies, the low quality of health care in these communities was laid bare.

Race is a key fact in explaining why "Hispanic, Black, and Indigenous people are more likely to be hospitalized or die from Covid-19 than whites," wrote Northern Arizona University professor of medical Anthropology Lisa Hardy and her collaborators (Hardy et al., 2020). Similar remarks were widely shared among health-care experts, including Dr. Anthony Fauci, director of the National Institute of Allergy and Infectious Diseases, when he testified to a Congressional panel in June 2020 (Givens, 2020). Research published in the prestigious medical journal, *Nature*, also established that, in Europe, Black and South Asian COVID-19 patients died at higher rates than Whites, even after adjustment for other variables. The study traced 10,926 COVID-19-related deaths in the

United Kingdom (Williamson et al., 2020). The important part of this declaration is not that racial and ethnic identity somehow essentially created vulnerability, but that centuries of oppressive forces, dispossession, slavery, and genocide have not ended, and people still live with these ongoing forces today.

Yet, given the history and present of xenophobia related to contagion, this pandemic also sewed seeds among white supremacists to point to disparities and use them for racially motivated blame. The pandemic revealed that Indigenous communities also faced racist aggressions and violence during the COVID-19 pandemic. Dr. Hardy and her collaborators point that Daniel Franzen, a 34-year-old white person, called upon his Facebook followers to use lethal force against Indigenous people "because they are 100% infected." These were just not random acts but media all over the world reported incidence of emotional abuse meted toward Indigenous persons and communities.

A different dimension of the systems that marginalize the poor was evident in Kenya. COVID-19 was an equal opportunity disease and in developing countries, it affected both rich and poor. In political systems where the rich continuously evolve ways of growing richer amid abject poverty, as in most of Africa, affluent members of society took extra care to mitigate against the spread of COVID-19 within their families.

As an example of the last-mentioned mitigation endeavor, the rich in Kenya, knowing full well that Counties where the Maasai live lacked basic health care, did what in advanced countries would be politically unacceptable. They installed Intensive Care Units (ICUs) beds in their homes! (Wainaina, 2020). Wealthy Kenyans, including five county governors, installed high-end core ICU facilities in their private houses after realizing that flying overseas was not an available option since airports were closed. COVID-19 infections surged in June 2020 and most of the leading hospitals in Kenya were overwhelmed. This was done, against public outcry. The state-of-the art facilities were installed at high costs, and given the country's high rates of bureaucratic corruption, it is most likely that public resources were used. Many of the places where poor Kenyans, including the vast majority of Maasai people, lacked basic health-care infrastructure. Among the facilities installed included oxygen cylinders, dry air, ventilators, and patient monitors.

Overall, the enormity of the challenge of removing inequalities in health care was highlighted in public discourses. COVID-19 revitalized the consciousness of racial and ethnic disparities which had been entrenched through settler colonialism. However, throughout the pandemic, the presence of inequal virus responses did not trigger activism for the immediate change. For all its exertions, the WHO or even activist civil societies on global and national stages lacked the political and judicial activism to foster conditions to mitigate the rapid spread of COVID-19 in Indigenous societies. Editorials spoke about the need for nations to adhere to public health protocols, especially generic policies such as social distancing, masking, washing hands, vaccines, and avoidance of super-spreader events. Yet, after a year of the millions of deaths caused by various waves of COVID-19, examples of prototype legislations to promote health equalities and address disparities in Indigenous societies were remarkably absent.

Still, it is crucial to point toward the ability to advance and maintain public health in Indigenous communities throughout the world despite incredible oppression, lack of resources, and external control. Even when leaders outside of Indigenous nations fought to remove the ability for nations to close borders people fought to protect health and survival. This happened through immensely creative means through which people with little access to the Internet managed to send out health messages and create music. It also happened through the engagement of networks by foot and vehicle safely delivering supplies and fresh water. Understanding health disparities can only shed light on inequality when it is clear that they are not a unidimensional explanation for inequality but instead an outcome of ongoing oppression and one that does not go unchallenged. It is in this space where the texture and dynamism of Indigenous strategies for survival, sometimes successful and sometimes encumbered, become clear. This is also true of other oppressed groups who similarly have health struggles and descend from severely violent histories that continue on.

Public perceptions of the demographics of global leaders

In interviews during the pandemic conducted by one of the authors people in the United States sometimes referred to the nations with women at the helm as more successful that the United States in handling the pandemic. News media outlets published polls on performance of leaders (e.g., Euronews Now, 2020). Social media outlets, especially Facebook and WhatsApp, conveyed comments on the relatively commendable and remarkable performances of women leaders, at the global stage.

In comparative terms, policy outcomes are a function of responses which are also shaped by prevention and control measures implemented by different administrations. From studies on this subject, a number of countries that had lower per capita percentage death rates from COVID-19 were those in which women held the highest offices in the land (Table 7.2). The list of these countries included Germany, Taiwan, New Zealand, Iceland, Finland, Norway, and Denmark. Contrast this, for example, with leaders of countries such as the United States, Brazil, Russia, Spain, the United Kingdom, Italy, and France. There is an indication that women leaders had management styles that were more uplifting than their male counterparts (Table 7.3). The debate over nurturing versus strict parenting was rekindled as a possible explanation of the differential outcomes. Women, of course, being traditionally considered as more nurturing than male leaders.

Dr. Liji Thomas (2020), a medical physician (OB-GYN, based in India), summarized Luca Coscieme et al. (2020) and their colleagues well-designed study published in medRxiv, in

Table 7.2: Women leaders and COVID-19 policy management impact. Globally, leaders who focused on scientific solutions performed better.

Female leaders	Countries and COVID-19 qualitative score cards
Angela Merkel	Germany, earlier approval rating was strong. Her background as a scientist was touted as strength.
Tsai Yingwen	Taiwan, President Tsai was credited for her effective management of COVID-19 pandemic. Taiwan, with population of around 24 million, had only 451 cases and 7 deaths by mid-June. Taiwan had robust contact tracing and health declaration protocols. The country took the quarantine restrictions seriously (Farr & Gao, 2020). The country's Vice President Chen Chien-jen, an epidemiologist, was key to the success. Taiwan had also done well to prevent spread of SARS pandemic in 2003 (Ching-Tse, 2020; Hernández & Horton, 2020).
Jacinda Ardern	New Zealand, the country took an unprecedented early step to close the borders to all noncitizens or residents. This policy of go hard with rigorous alert systems, social distancing, and nationwide lockdowns led to no deaths by March 25. Only 102 cases were reported then. There was also effective communication by officials. By June 8, no community transmissions were reported (Jones, 2020).
Katrin Jakobsdòttir	Iceland had less than 2000 cases and 10 deaths by early July. Iceland beat coronavirus by employing effective surveillance techniques, quarantine techniques. Everybody knows everybody in Iceland approach worked! Prime Minister Jakobsdòttir announced foreigners arriving at Keflavik would have three options: Show certificate of recent negative COVID-19 test, be screened, or go into quarantine (Kolbert, 2020).
Sanna Marin	Finland, Prime Minister Marin, a millennial environmentalist, led the war against COVID-19. She was strong, decisive, and empowered. And was leader at enforcing test, trace, and treat mentality. She enjoyed 85% approval rating for handling COVID-19 pandemic in Finland. She invoked emergency powers and closed schools, museums, public gatherings, and borders (Mendoza, 2020).
Erna Solberg	Norway, Prime Minister Solberg construed the pandemic as a global problem and was leader in finding common solutions. Her government was generous at funding collaborative initiatives, including the WHO and the Oslo-based Coalition for Epidemic Preparedness Innovation (CEPI). Her approach toward preventing COVID-19 was over cautious and was controlled by fear and worse-case scenarios (Rolfe, 2020). Norway recorded 8455 cases and 237 deaths compared to Sweden's high numbers of 40,000 cases and 4500 deaths by the end of May (Rolfe, 2020).
Mette Frederiksen	Denmark, Prime Minister Frederiksen received high 88% (good work) rating for her management of the pandemic (Statista Research Department, 2020). She shut down schools early March and Parliament amended the 1915 and 1979 Epidemic Act which transferred power from regional Epidemic Commissions to the health minister. Subsequently ministerial orders were employed to give room for movement restrictions (Herrmann, 2020).

Table 7.3: Male leaders and COVID-19 policy management impact.

Male leaders	Countries and COVID-19 qualitative score cards
Donald Trump	United States; was the world's number 1 hot spot. By July 12, United States had reported 3,236,130 cases and 134,572 total deaths (CDC, 2020). 67% polled (ABC News/Ipsos) disapproved of how Trump handled COVID-19 crisis (Karson, 2020). Joe Biden, who replaced Trump, changed course and received 72% approval (Czachor, 2021, March 28).
Jair Bolsonaro	Brazil. The president was virus denier and was infected by COVID-19. Brazil was the world's number 2 hot spot. Brazil reported 1,864,681 cases and 72,469 deaths in early July 2020.
Vladimir Putin	Russia. Had 727,162 cases by July 12. Critics cast doubt over reported numbers (The Moscow Times, 2020).
Pedro Sanchez	Spain, Prime Minister Sanchez was criticized for treating the virus as an external threat. Over 12,000 Health-care workers had contracted the disease by early April (Minder, 2020). A case was filed against his cabinet alleging, "homicide due to grave negligence" (Harlan & Pitrelli, 2020).
Boris Johnson	United Kingdom. Prime minister was treated for COVID-19. By mid-May 2020, over 31,000 Britons had died from the disease.
Giuseppe Conte	Italy initially had the highest death tolls in Europe—over 27,967 at the end of April 2020. Prosecutors questioned Prime Minister Conte over his handling of crisis (Harlan & Pitrelli, 2020).
Emmanuel Macron	France had reported 170,752 cases and 30,004 deaths by July 12, among the hardest hit in Europe (Neuman, 2020). French public health agency had embraced the notion that the French had to live with the virus.

which leaders were assessed based on outcomes such as "the number of deaths, the number of days with reported deaths, the peak in daily deaths, deaths up to the first day of lockdown, and excess mortality" in different countries (Thomas, 2020). In all those indicators, women leaders outperformed male counterparts. As was found, in most instances, women leaders resorted to quick lockdowns and had less hesitation in saying yes to preserving life through preventing the spread of COVID-19 over economic growth (Thomas, 2020). Data show that populist and nationalist leaders mishandled the COVID-19 crisis. With time, the situation became more revealing after India's "second wave," which occurred during early 2021. Then, the country's COVID-19 management was placed on the global radar for recording the highest death rates of any nation on a single day. Numerous global news outlets criticized India's hypernationalist government led by Prime Minister Narendra Modi criticized for arrogance and bureaucratic incompetence that led to "COVID apocalypse" (Alluri, 2021). Modi's centralized and populist government had allowed Hindi festivals and massive political rallies to take place amid the ravaging pandemic. These were regarded as super-spreader events.

These are, of course, snapshots of decision-making practices in a few selected countries and only serve to stimulate debates over policy management under crisis situations. There are several variables that also count. For example, high degrees of centralization in different administrative systems can impact the problem framing and approaches to finding solutions. There is no magic bullet to solving a complex problem such as was presented by the coronavirus pandemic. Ideological divides and electioneering could have shaped some of the approaches to managing the pandemic. What we know, for now, is that presidential involvement in managing pandemics is a huge part of solving the problem because of the communication signaling and control of resources to combat the disease.

Moreover, having considerable international experience can help in building international consensus in managing global problems. Clearly, a president like Trump who campaigned as a "nationalist" with an America First agenda was not attentive to the international challenge that humankind experienced in most of 2020. He was widely seen in negative light in surveys from publics around the world. The Pew Research Center reported that 64% of those surveyed did not have confidence in Trump "to do the right thing in world affairs, while just 29% express confidence in the American leader." The Pew Research Center found anti-Trump sentiments to be particularly common in Western Europe and in Mexico (Wike et al., 2020). Besides, his indecisiveness left him unable to lead the world against the raging devastating consequences of the coronavirus pandemic. There were no clear signals from his administration that the pandemic would be managed jointly. Instead, the signals focused on blaming China and the WHO. Furthermore, the temporary solutions offered by members of his administration undermined the ability of global collective actions.

Competition over the rush to find a vaccine were undermined by the lack of trust in those who made the announcements about the possibility of a cure. Take for example, Russia's President Vladimir Putin's announcement that Russia had already started clinical trials on its COVID-19 vaccine. The Russian authorities claimed the vaccine, developed by the Gamaleya Research Institute of Epidemiology and Microbiology in Moscow, would confer 2 years of immunity to SARS-Cov-2, the virus that caused the disease, COVID-19. Because of Western media's perception of Russia as a rogue international actor, there was not any major excitement about the Russian solution. Instead, the Western European and American press treated the effort with contempt and skepticism. Writing for Science, Jon Cohen (2020) reported that the "Gamaleya vaccine has yet to complete a trial that convincingly shows it is safe and effective in a large group of people. Even some within Russia challenged the move .... it's ridiculous," says Svetlana Zavidova, a lawyer who heads the Association of Clinical Research Organizations in Russia. "I feel only shame for our country" (Cohen, 2020). COVID-19 amplified nationalistic and factional competitions instead of signaling global togetherness.

In the beginning of this work, we noted the inclination of governments, across the globe, to treat the pandemic as a war. This image further fueled the dominance of patriarchal discourses in the management of the crisis. Burrell and Ruxton (2020) offer that in the United Kingdom, for example, the government and its largely male cabinet created a war cabinet to combat COVID-19. The language applied in the war narratives was unmistakably, not only militaristic and toned with bureaucratic warnings but with heavy injection of machismo and sexism. This was quite clear in press briefings by governors and officials in the Kenyan and Tanzanian regions where Maasai people live.

And there is another perspective that would require further research. The Navajo Nation was previously matrilineal until the arrival of Spanish settlers and colonialists who used oppressive approaches to governance. Marcus Cleveland (2019) has argued that the shift to these policies through colonialism resulted in negative health and development outcomes. Thus, there is a link to the connection of oppressive rulership based on toxic masculinity and systems that do not effectively function for oppressed peoples. This is true now and has been throughout a colonial past of conquest during which masculine ideas of conquest and ownership of humans prevailed. Certainly, the populist machismo styles of neo-nationalists leaders in Brazil, India, and the United States were among the worst crisis managers of the crisis of 2020 and the signals they sent had negative impact on Indigenous communities under their care.

Conclusion

In the COVID-19 settings, such as those articulated here, numerous governments demonstrated some of the underlying deeply rooted societal divisions and misgivings about global public health infrastructures. In 2020, the world was clearly not prepared to solve future or present pandemics. By evaluating the roles of governance networks and key actors responsible for health care in Indigenous nations, we can see how communication mattered in management of the COVID-19 pandemic.

We note that leaders' problem-framing efforts matter. Inevitably, those who sent mixed messages fared very poorly and their failure to communicate effectively resulted in deaths of hundreds of thousands of people. The role of national leaders, whether as active or passive policy spokespersons, is a critical factor in management of pandemics. And while the availability of technological tools helps, specific uses in problem-solving are again a function of leadership. And for Indigenous societies, more generally, a call to investments in relevant ICT infrastructures. This work has shared a number of general assertions about the nature of public health policy judgments that have contributed to the current outcomes of COVID-19 on Indigenous communities. The task ahead is for policy experts to use the opportunity and be part of the solution to uphold the dignity of our fellow human beings in Indigenous societies. COVID-19 offers an opportunity to call to action new approaches that incorporate Indigenous solutions to our consciousness as members of the human family.

References

Abutaleb, Y., & Dawsey, J. (August 31, 2020). *New Trump pandemic adviser pushes controversial "herd immunity" strategy, worrying public health officials*. The Washington Post. https://www.boston.com/news/coronavirus/2020/08/31/scott-atlas-herd-immunity.

Alluri, A. (April 8, 2021). *India's Covid crisis delivers a blow to brand Modi*. BBC. https://www.bbc.com/news/world-asia-india-56970569.

Applebaum, A. (May 3, 2020). *The rest of the world is laughing at Trump*. The Atlantic. https://www.theatlantic.com.

Baker, D. (May 2, 2020). *National guard conducts flyover for Navajo Nation, other communities*. AZ family. KPHO broadcasting corporation. https://azfamily.com/news/.

BBC. (June 8, 2020). *Coronavirus: John Magufuli declares Tanzania free of COVID -19*. BBC News. https://www.bbc.com/news/world-africa-52966016.

Booker, C. (April 18, 2020). *Booker welcomes CDC's release of racial COVID-19 data*. U.S. Senate https://booker.senate.gov/news/press/booker-welcomes-cdcs-release-of-racial-covid-19-data.

Branson-Potts, H. (July 25, 2020). *A face mask is part of the 'scamdemic,' they say. But they'll be happy to sell you one*. Los Angeles Times. https://www.yahoo.com/news/face-mask-part-scamdemic-theyll-130015101.html.

Brulliard, K., & Wan, W. (September 3, 2020). *The pandemic is ruining our sleep. Experts say "coronasomnia" could imperil public health*. The Washington Post. https://www.msn.com/en-us/health/medical/the-pandemic-is-ruining-our-sleep-experts-say-coronasomnia-could-imperil-public-health/ar-BB18GxkA?li=BBnb7Kz&ocid=mailsignout.

Burrell, S., & Ruxton, S. (April 9, 2020). *Coronavirus reveals just how deep macho stereotypes run through society*. The Conversation. http://www.theconversation.com.

Carroll, R. (May 9, 2020). *Irish support for Native American Covid-19 relief highlights historic bond*. The Guardian. https://www.theguardian.com/us-news/.

Centers for Disease Control and Prevention (CDC). (July 12, 2020). *Cases in the U.S*. https://www.cdc.gov/coronavirus/2019-ncov/cases-updates/cases-in-us.html.

Cereceda, R. (April 4, 2020). *New coronavirus not real killer: The patient's immune system damaging vital organs*. Euronews. https://www.euronews.com/2020/04/09/understanding-covid-19-the-unknown-disease-with-multiple-faces.

Chakravorti, B. (July 10, 2020). *Digital contact tracing's mixed record abroad spells trouble for U.S. Efforts to rein in COVID 19*. Government executive. https://www.govexec.com.

Chen, Q., & Shumei, L. (May 6, 2020). *US Owes world an explanation on COVID-19*. Global Times. https://www.globaltimes.cn.

Ching-Tse, C. (May 15, 2020). *Chen Chien-jen becomes Taiwan's first vice president to give up pension*. Taiwan Times. https://www.taiwannews.com.tw/en/news/3934346.

Citizen, T. V. (July 9, 2020). *Man charged in court for preparing false COVID-19 results*. https://www.youtube.com/watch?v=UKTPRPpy-Ac.

Clark, J. (April 24, 2020). *What's the safest way to shop? Are certain blood types at higher risk? Answering your questions about COVID-19*. Quantum nutrition. https://qnlabs.com/blog/covid-19-pt-3.html?.

Cleveland, M. (2019). Patriarchy and Paternalism: The navajo nation and dine social, political and cultural emergence. PhD dissertation, Northern Arizona University, Flagstaff.

CNN. (June 30, 2020). *COVID-19 drug to cost K sh. 41K for one vial*. CNN for Citizen Digital. https.www.citizentv.co.ke.

Cohen, J. (August 11, 2020). Russia's approval of a COVID -19 vaccine is less than meets the press release. *Science, 369*(6509). https://www.sciencemag.org/news/2020/08/russia-s-approval-covid-19-vaccine-less-meets-press-release.

Coscieme, L., et al. (2020). Women in power: Female leadership and public health outcomes during the COVID-19 pandemic. *medRxiv*. https://doi.org/10.1101/2020.07.13.20152397

Couzin-Frankel, J. (2020, July 31). From "brain fog" to heart damage, COVID-19's lingering problems alarm scientists. *Science*. https://www.sciencemag.org/news/2020/07/brain-fog-heart-damage-covid-19-s-lingering-problems-alarm-scientists.

Cramer, J. C. (1995). Racial and ethnic differences in birth weights: The role of income and financial assistance. *Demography, 32*, 231–247.

Curnow, R. (July 4, 2020). *CNN newsroom with robyn curnow*. https://www.cnn.com.

Curtis, C. (May 6, 2020). *Arizona air national guard to fly over Navajo nation in honor of frontline workers*. Arizona republic. http://azcentral.com/story/news/.

Czachor, E. (March 28, 2021). *72 % of Americans approve of Joe Biden's handling of the coronavirus pandemic: Poll*. https://www.msn.com/en-us/news/politics/72percent-of-americans-approve-of-joe-bidens-handling-of-the-coronavirus-pandemic-poll/ar-BB1f3qJT?ocid=mailsignout&li=BBnb7Kz.

Davis, W. (August 6, 2020). *The Unraveling of America: Anthropologist Wade Davis on how COVID-19 signals the end of the American era*. Rolling Stone. https://www.rollingstone.com/politics/political-commentary/covid-19-end-of-american-era-wade-davis-1038206/?fbclid=IwAR0AKpUHGnzxNvadNpkwowQIJ88ci82c-Lk-VqBWlvu5LwBkbJDeUorsj-Y.

Dettmer, J. (May 6, 2020). *Hungary: The first dictatorship in the EU?* VOA. http://wwwvoanews.com.

Diamond, L. (April 30, 2020). *COVID-19 and democracy around the world*. Hoover Institution. Briefing Series. https://www.hoover.org/events/larry-diamond-covid-19-and-democracy-around-world.

DiGangi, D. (January 22, 2020). *Army germ lab shut down by CDC in 2019 had several 'serious' protocol violations that year*. WJLA. https://wjla.com/news/local/cdc-shut-down-army-germ-lab-health-concerns.

Elkind, E. (May 5, 2020). *Pediatrician explains why she's 'not surprised' the coronavirus is being linked to a rare disease in children*. CBS. https://www.cbsnews.com/news/coronavirus-kawasaki-disease-rare-children/.

Estrin, D. (July 7, 2020). *Amid new surge in virus cases, Israel's top public health official resigns*. National Public Radio, NPR. https://npr.org/sections/coronavirus-live-updates/2020/07/07/888148317/amid-new-surge-in-virus-cases-israels-top-public-health-official-resigns.

Euronews. (May 28, 2020). *EU leadership poll: Merkel and conte lauded, Macron Panned for response*. Euronews Now. https://www.youtube.com/watch?v=xcRyx95kf60.

Farr, C., & Gao, M. (July 15, 2020). *How Taiwan beat the coronavirus*. CNBC. https://www.cnbc.com/2020/07/15/how-taiwan-beat-the-coronavirus.html.

FreedomHouse. (2021). *Freedom in the World-India Country Report*. https://freedomhouse.org/country/india.

Frenk, J. (July 5, 2020). *Interview on Fredricka whitfield, CNN newsroom, 11.25 Pm*. https://tv.youtube.com/browse/cnn-newsroom-with-fredricka-whitfield.

Garari, K. (June 5, 2020a). *Here's how doctors are being put to risk in Telangana. I'm one of 48 corona+ docs*. Deccan Chronicle. https://www.deccanchronicle.com/nation/in-other-news/050620/.

Garari, K. (July 2, 2020b). *New symptoms delay Covid care*. Deccan Chronicle Hyderabad. https://www.deccanchronicle.com/.

Ghabreyesus, T. (2020). Lack of unity is a bigger threat than Coronavirus, WHO Chief says in emotional speech. In Chappell, B. NPR Transcript. https://www.npr.org/sections/coronavirus-live-updates/2020/07/09/889411047/lack-of-unity-is-a-bigger-threat-than-coronavirus-who-chief-says-in-emotional-sp

Givens, D. (June 23, 2020). *Dr. Fauci says institutional racism played a huge part in disproportionate impact of COVID-19*. Black Enterprise. https://www.blackenterprise.com/dr-fauci-says-institutional-racism-played-a-huge-part-in-disproportionate-impact-of-covid-19/.

Global Times. (July 3, 2020). *Rampant US epidemic to hurt the world*. Editorial. https://www.globaltimes.cn.

Gortmaker, S. L. (1979). Poverty and infant mortality in the United States. *American Sociological Review, 44*, 280–297.

Grady, D. (August 5, 2019). *Deadly Germ Research is shut down at Army Lab over safety concerns*. New York Times. https://www.nytimes.com.

Gross, A. (May 29, 2020). *Rights restrictions and securitization of health in Israel during COVID-19*. Harvard Law School. Petrie Flom Center. https://blog.petrieflom.law.harvard.edu/2020/05/26/denmark-global-responses-covid19/.

Grover, A. (May 18, 2020). COVID-19 in India: Lockdown, legal challenges, and disparate impacts. In *The Petrie –Flom Center, Harvard University Blog symposium*. https://blog.petrieflom.law.harvard.edu/2020/05/18/india-global-responses-covid19/.

Gulick, L. H. (1937). Notes on the theory of organization. In L. Gulick, & L. Urwick (Eds.), *Papers on the science of administration* (pp. 1–46). Institute of Public Administration.

Haglage, A. (August 5, 2020). *CDC: Drinking methanol hand sanitizers poisoned 15, killed 4*. https://www.yahoo.com/lifestyle/cdc-methanol-hand-sanitizers-poisoned-15-killed-4-172919134.html.

Harari, Y. N. (March 19, 2020). *Yuval NOoah Harari: The world after coronavirus*. Financial Times. https://www.ft.com.

Hardy, L. J. (2020). Connection, contagion, and COVID-19. *Medical Anthropology, 39*(8), 655–659. https://doi.org/10.1080/01459740.2020.1814773

Hardy, L. J., Saul, G., & Smith, S. M. (June 30, 2020). *As Arizona coronavirus cases surge from early reopening, Indigenous nations suffer not only more COVID-19 but also the blame*. The Conversation. https://theconversation.com/.

Harlan, C., & Pitrelli, S. (June 12, 2020). *Italian prosecutors question Conte over handling of coronavirus outbreak*. Washington Post. https://washingtonpost.com.

Hatcher, S. M., Agnew-Brune, C., Anderson, M., et al. (January 31–July 3, 2020). COVID-19 among American Indian and Alaska native persons — 23 states. *Morbidity and Mortality Weekly Report 2020, 69*, 1166–1169. https://doi.org/10.15585/mmwr.mm6934e1external icon

Hedgpeth, D. (May 13, 2020). *The Irish are repaying a favor from 173 years ago in Native Americans' fight against coronavirus*. The Washington Post. http://www.washingtonpost.com/history/2020/05/13.

Hernández, J. C., & Horton, C. (May 9, 2020). *Taiwan's weapon against coronavirus: Epidemiologist as vice president*. New York times. https://www.nytimes.com/2020/05/09/world/asia/taiwan-vice-president-coronavirus.html.

Herrmann, J. R. (May 26, 2020). *How Denmark's epidemic act was amended to respond to COVID-19. The Petrie-Flom Center staff. University of Copenhagen Centre for advanced studies in biomedical innovation law*. Blog. https://blog.petrieflom.law.harvard.edu/2020/05/26/denmark-global-responses-covid19/.

Hoyert, D., Arias, E., Smith, B., Murphy, S., & Kochanek, K. (2001). *Deaths: Final data for 1999. National vital statistics reports* (Vol. 49). U.S. Department of Health and Human Services, National Center for Health Statistics (8).

Janis, I. (1971). Groupthink: The desperate drive for consensus at any cost, 185–92. In J. Shafritz, S. Ott, & Y. S. Jang (Eds.), *Classics of organizational theory* (6th ed.). Thomson Wadsworth, 2005.

Jensen, A. (July 17, 2020). *4 dead, 26 hospitalized in Arizona after ingesting hand sanitizer with methanol*. Arizona Republic. https://www.azcentral.com/story/news/local/arizona-health/2020/07/17/people-dead-hospitalized-drinking-hand-sanitizer-methanol-arizona/5448890002/?fbclid=IwAR1LvKv1jounmE4WftuPK0_kcMm6YswGMMiOK7zpp-mtq6jvIiAzwMfW8OE.

Jones, A. (July 10, 2020). *Coronavirus: How New Zealand went 'hard and early' to beat COVID-19*. BBC. https://www.bc.com/news/world-asia-53274085.

Karson, K. (July 10, 2020). *Broad disapproval for Trump's handling of coronavirus, race relations: POLL*. ABC News. https://abcnews.go.com/Politics/broad-disapproval-trumps-handling-coronavirus-race-relations-poll/story?id=71704889.

Kelly, G. (June 9, 2020). *Asymptomatic cases not infectious*. Newsmax TV. Greg Kelly Reports https://www.facebook.com/newsmax/videos/doctor-responds-to-whos-latest-covid-19-statements/262667375071822/.

Kimuyu, H. (July 9, 2020). *Lang'ata Hospital hit by COVID-19 tests scandal.* Daily Nation. https://nairobinews.nation.co.ke/news/langata-hospital-hit-by-covid-19-tests-scandal.

Kolbert, E. (June 1, 2020). *How Iceland beat the Coronavirus: The country didn't just manage to flatten the curve; it virtually eliminated it.* The New Yorker. https://www.newyorker.com/magazine/2020/06/08/how-iceland-beat-the-coronavirus.

Krisst, R. (July 10, 2020). *Line-item vetoes 'send wrong message': Nez nixes CARES Act funds for chapters, Hataalii group.* Navajo Times. https://ww.navajotimes.com.

Lasswell, H. (1951). The policy orientation. In D. l Lerner, & H. D. Lasswell (Eds.), *The policy sciences: Recent developments in scope and method.* Stanford University Press.

Lauer, S. A., Grantz, K. H., Bi, Q., et al. (May 11, 2020). *Estimated incubation period of COVID-19.* American College of Cardiology. https://www.acc.org.

Lavezzo, E., Franchin, E., Ciavarella, C., et al. (June 30, 2020). Suppression of a SARS-CoV-2 outbreak in the Italian municipality of Vo'. *Nature, 584,* 425–429. https://doi.org/10.1038/s41586-020-2488-1

Levy, B.-H. (2020). *The virus in the age of madness.* Yale University Press.

Logical Indian. (June 9, 2020). *Fact check: Did WHO say that asymptomatic COVID-19 cass 'rarely' spread the infection.* The Logical Indian. https://thelogicalindian.com/fact-check/who-asymptomatic-covid-19-rarely-spread-21705.

Lovelace, B., Jr. (June 10, 2020). *Dr. Anthony Fauci says WHO's remark on asymptomatic coronavirus spread 'was not correct'.* CNBC. https://www.cnbc.com/2020/06/10/dr-anthony-fauci-says-whos-remark-on-asymptomatic-coronavirus-spread-was-not-correct.html.

Lusaka Times. (June 28, 2020). *Malawi opposition leader Lazarus Chakwera wins presidential vote Re-run.* https://www.lusakatimes.com/2020/06/28/malawi-opposition-leader-lazarus-chakwera.

Majumder, K. (April 23, 2020). *Mission journal: Journalists in India's Uttar Pradesh say threat of attack or prosecution looms large.* Committee to Protect Journalists. (CPJ). https://cpj.org/2020/04/journalists-in-indias-uttar-pradesh-say-threat-of/.

Masina, L. (June 28, 2020). *Malawi opposition leader wins fresh elections with landslide victory.* VOA. http://www.voanews.com/africa/malawi-opposition-leader-wins-fresh-elections-landslide-victory.

Mendoza, K. A. (June 23, 2020). *Meet Sanna Marin-Finland's female millennial prime minister, from an LGBT family, leading the war on coronavirus.* Style. https://www.scmp.com/magazines/style/news-trends/.

Merrill, R. (May 14, 2020). *UMAss Dartmouth biology professor gains international attention for COVID-19 blog.* Umass News. https://www.umassd.edu/news/2020/umassd-biology-professor-bromage-covid-19-blog.html.

Minder, R. (April 7, 2020). *Spain's coronavirus crisis accelerated as warnings went unheeded.* New York Times. https://www.nytimes.com.

Mongilio, H. (August 2, 2019). *Fort Detrick lab shut down after failed safety in inspection; all research halted immediately.* The Frederick News-Post. https://www.fredericknewspost.com/news/health/fort-detrick-lab-shut-down-after-failed-safety-inspection-all-research-halted-indefinitely/article_767f3459-59c2-510f-9067-bb215db4396d.html.

Moscow Times. (July 12, 2020). *Russia's coronavirus cases near 730K as restrictions ease.* https://www.themoscowtimes.com/2020/07/12.

Munguti, R. (July 12, 2020). *KNH Nurse charged over fake COVID-19 certificates.* Daily Nation. https://www.nation.co.ke/kenya/news/knh-nurse-charged-over-fake-covid-19-certificates–1447730.

Murphy, S. (2000). *Deaths: Final data for 1998. National vital statistics reports* (Vol. 48). U.S. Department of Health and Human Services, National Center for Health Statistics (11).

National Center for Health Statistics. (1993). *Health, United States, 1992.* Hyattsville, MD: U.S. Public Health Service.

Navajo Times. (June 25, 2020). *'We must return to the teachings of Hozho': Hataalii Assoc. poised to receive $ 1M for coronavirus response.* https://www.navajotimes.com.

Neuman, S. (June 15, 2020). *France announces further reopening amid declining number of coronavirus cases.* National Public radio. https://www.npr.org/.

New York Times. (June 1, 2020). *After 6 months, important mysteries about coronavirus endure.* New York Times. https://www.nytimes.com/2020/06/01/health/coronavirus-mysteries.html.

Nezt, J. (February 28, 2020). *Navajo Nation COVID-19 preparedness team established to address coronavirus precautionary efforts.* Press Release. Navajo Nation Office of President and Vice President Communications. http://www.opvp.navajo-nsn.gov/.

O'Toole, F. (April 25, 2020). *Donald Trump has destroyed the country he promised to make great again.* The Irish Times. https://www.irishtimes.com/opinion/fintan-o-toole-donald-trump-has-destroyed-the-country-he-promised-to-make-great-again-1.4235928?.

Peters, G. B. (2001). *The future of governing.* University Press of Kansas.

Pfeiffer, S. (July 4, 2020). *The ethics of coronavirus vaccine trials in developing countries. All things considered. 5.05 PM ET. Transcript.* National public radio. NPR. https://www.npr.org/2020/07/04/887325575/the-ethics-of-coronavirus-vaccine-trials-in-developing-countries?utm_medium=RSS&utm_campaign=allthingsconsidered.

Puntmann, V. O., Ludovica, M. C., Wieters, I., et al. (July 27, 2020). Outcomes of cardiovascular magnetic resonance imaging in patients recently recovered from coronavirus disease 2019 (COVID-19). *JAMA Cardiology.* https://doi.org/10.1001/jamacardio.2020.3557

Ramsay, S. (July 13, 2020). *Coronavirus warning from Ital: Effects of COVID-19 could be worse than first thought.* Sky News. https://news.sky.com/story/coronavirus-warning-from-italy-effects-of-covid-19-could-be-worse-than-first-thought-12027348.

Rezaian, J. (June 1, 2020). *In India, the pandemic is cover for Modi's war on journalists.* Washington Post: Global Opinions Writer. https://www.washingtonpost.com/opinions/2020/06/01/india-pandemic-is-cover-modis-war-journalists/.

Ricard, P. (May 4, 2020). *Coronavirus sets balancing act between science, politics.* AFP. https://www.barrons.com/news/coronavirus-sets-balancing-act-between-science-politics-01588609507.

Rim, C. (May 20, 2020). *Why colleges may be able to order students to get COVID -19 vaccinations.* Forbes Magazine. http://www.forbes.com.

Risch, J. (June 5, 2020). *Chairman Risch touts global health pandemic solutions bill on senate floor. US. Senate Committee on foreign relations. 116th congress (2019–2020).* Chairman Press. https://foreign.senate.gov/press/chaor/release/chairman-risch-touts-global-pandemic-solutions-bill-on-senate-floor.

Rogers, K., Hanser, C., Yuhas, A., & Haberman, M. (April 24, 2020). *Trump's suggestion that disinfectant could be used to treat coronavirus prompts aggressive pushback.* New York Times. htts://www.nytimes.com.

Rogo, K. (March 11, 2020). *Health care in Kenya. Keynote address Prof. Khama. Kenya health convention, YouTube new dawn AV.* https://www.youtube.com/watch?v=u0nXcB_BwH4.

Rolfe, B. (June 3, 2020). *'Was it necessary'? The PM who regrets taking tough coronavirus lockdown measures.* Yahoo News. https://au.news.yahoo.com.

Safi, M. (April 3, 2018). *India backs down over plan to ban journalists for 'fake news'.* The Guardian. https://www.theguardian.com/world/2018/apr/03.

Sanchez, Y. W., & Oxford, A. (July 2, 2020). *Vice President Mike Pence announces 500 medical personnel on their way to help Arizona with COVID -19 crisis.* Arizona Republic. https://www.azcentral.com.

Scoop, U. (June 5, 2020). *South Korea sends 10K masks to Navajo Nation to honor service as 'code talkers' in Korean war.* https://scoop.upworthy.com.

Shivji, S. (July 1, 2020). *U.S. buys global supply of COVID-19 drug remdesivir.* CBC News. https://www.facebook.com/cbcnews/videos/us-buys-global-supply-of-covid-19-drug-remdesivir/2237789216366265/.

Simon, H. (1976). *Administrative behavior* (3rd ed.). The Free Press.

Sky News. (May 25, 2020). *Coronavirus 'disappearing' so fast Oxford vaccine has 'only 50% chance of working.'*. UK: Sky News. https://news.sky.com/story/amp/coronavirus-disappearin-so-fast-oxford-vaccine-has-only-50-chance-of-working-11993739.

Slotkin, J. (April 25, 2020). *NYC Poison Control sees uptick in calls after Trump's disinfectant comment.* KNAU, NPR. https://www.npr.org.

Smith, T. (August 2, 2020). *A supercomputer analyzed COVID-19 and an interesting new theory has emerged. A closer look at the Bradykinin hypothesis.* Elemental Medium.com. https://elemental.medium.com/a-supercomputer-analyzed-covid-19-and-an-interesting-new-theory-has-emerged-31cb8eba9d63.

Statista. (June 29, 2020). *Opinions on the prime minister handling COVID-19 in Denmark 2020? How well do you think the Prime Minister, Mette Frederiksen, has handled the coronavirus situation in Denmark?* Statista Research Department. https://www.statista.com/.

Steenhuysen, J. (June 26, 2020). *Scientists just beginning to understand the many health problems caused by COVID-19.* Reuters. https://www.reuters.com/article/us-health-coronavirus-effects/scientists-just-beginning-to-understand-the-many-health-problems-caused-by-covid-19-idUSKBN23X1BZ?fbclid=IwAR3FV93k7Q3fcjCIbtan884tizfxHdHo6Ihoyz9IfdfeMnLvAwvDz49jQGU.

Stevens, M., & Kaplan, T. (June 30, 2020). *Biden says Trump 'surrendered' to coronavirus in Blistering speech.* https://www.nytimes.com/2020/06/30/us/politics/biden-speech-trump-coronavirus.html.

Thomas, L. (July 17, 2020). *COVID -19 outcomes better in countries with female leaders.* News Medical Life Sciences. https://www.news-medical.net/news/20200717/COVID-19-outcomes-better-in-countries-with-female-leaders.aspx.

Varmus, H., & Shah, R. (August 31, 2020). *It has come to this: Ignore the C.D.C.* New York times. https://www.nytimes.com/2020/08/31/opinion/cdc-testing-coronavirus.html.

Velshi, A. (July 3, 2020). *Peter navarro on MSNBC. Breaking news. Top Trump official speaks as new coronavirus cases surge, 1:32 PM PT.* MSNBC TV.

Ventura, S., Martin, J., Curtin, S., Mathews, T. J., & Park, M. (2000). *Births: Final data for 1998. National vital statistics reports* (Vol. 48). Washington, DC: U.S. Department of Health and Human Services, National Center for Health Statistics (3).

Wainaina, E. (2020). *Covid-19 Race against time as rich Kenyans install ICU beds in homes. PD Online.* https://www.pd.co.ke/news/covid-19-race-against-time-as-rich-kenyans-install-icu-beds-in-homes-43404/.

Walton, A. G. (June 15, 2020). *Face masks may be the key determinant of the COVID-19 curve, study suggests.* Forbes Media, LLC. https://wwww.forbes.com.

Weiss, Y. (May 21, 2020). *How fear, groupthink drove unnecessary global lockdowns.* Real clear politics. https://www.realclearpolitics.com/articles/2020/05/21/how_fear_groupthink_drove_unnecessary_global_lockdowns_143253.html.

Werman, M. (July 2, 2020). *Fauci: We've got to do something about the coronavirus. Podcast. The World.* National Public Radio (NPR). https://www.npr.org/podcasts/381444246/pri-s-the-world.

Wike, R., Poushter, J., Fetterolf, J., & Schumacher, S. (January 8, 2020). *Trump ratings remain low around globe, while views of U.S. Stay mostly favorable.* Pew Research Center. https://www.pewresearch.org/global/2020/01/08/trump-ratings-remain-low-around-globe-while-views-of-u-s-stay-mostly-favorable/.

Wilkie, C. (June 30, 2020). *Joe Biden rips Trump as coronavirus surges: the wartime president has surrendered.* CNBC. https://www.cnbc.com/2020/06/30/joe-biden-rips-trump-on-coronavirus-the-wartime-president-has-surrendered.html.

Williamson, E. J., Walker, A. J., Bhaskaran, K., Goldacre, B., et al. (2020). Factors associated with COVID-19 death using OpenSAFELY. *Nature, 584*, 430–436. https://doi.org/10.1038/s41586-020-2521-4

Woodyatt, A., & Howard, J. (April 27, 2020). *Coronavirus could be tied to a rare but serious illness in children, UK doctors say.* CNN. https://www.cnn.com/2020/04/27/health/children-covid-19-illness-intl-scli-gbr/index.html.

World Bank. (February 28, 2012). *Improving health services for tribal populations*. https://www.worldbank.org/en/news/feature/2012/02/28/improving-health-services-for-tribal-populations.

Yong, E. (September, 2020). *How the pandemic defeated America*. The Atlantic. https://www.theatlantic.com/magazine/archive/2020/09/coronavirus-american-failure/614191/.

Yurth, C. (June 25, 2020). *Lizer defends attendance at Trump event on lobbying*. Navajo Times. https://navajotimes.com/reznews/lizer-defends-attendance-at-trump-event-as-lobbying/.

Zakaria, F. (June 28, 2020a). *On GPS: UN secretary general Antonio Gueterres*. CNN. GPS. https://www.cnn.com/shows/fareed-zakaria-gps.

Zakaria, F. (July 19, 2020b). *On GPS: What has COVID done to humanity? GPS program*. https://www.cnn.com/videos/tv/2020/07/19/exp-gps-0719-bernard-henri-levy-on-covid-and-humanity.cnn.

CHAPTER 8

Unequal virus beyond 2020: paths forward

Social sciences learning and pandemics

Recent government and societal COVID-19 mitigation actions provide an indication of what the immediate future policies will learn from in case other pandemics arise. During uncertain times of the novel coronavirus pandemic, it behooved social scientists to work collaboratively to find solutions to the public health emergency. Perhaps, more importantly, it was the need for equality in the scholarly attention to impacts of the scourge on the economically and politically marginalized societies. The outlook for more aggressive scholarship on COVID-19 is quite promising given that most major academic journals in the social sciences have already released special issue editions that speak to COVID-19 academics and practitioner concerns. From our point of view, the emphasis on finding solutions must be undergirded through reflection and recommendations on the ways forward. The energies of policymakers and health-care practitioners will also be increasingly directed toward preparation for future pandemics. Initially, as far as managing COVID-19, the performance of many governments was disjointed and wanting. There is benefit in reflection for self-correction and this last chapter intends to offer practical and theoretical points to consider. The desirability of changes in how future pandemics will be managed is reflected in much of this volume. The path leading to future solutions will again be hinged on not only the scientific community, which did a terrific job at vaccine development but also on behavioral and social sciences whose insights inform much of what governance prescriptions are designed to do.

Social science research has affirmed that inequalities in health care are a function disparity enabled through among other things, forces of marginalization, and economic depravity. And when coronavirus appeared in Wuhan, China, and rapidly spread first through advanced economies in Europe and North America, there was fear that given the stronger health-care systems in those countries, the pandemic would be quickly vanished. Perennial arguments about state capacity to solve challenges in health offered little hope for developing countries and marginalized Indigenous societies. China, which has transitioned to the ranks of countries with immense economic and technological capabilities, was among the list of few countries to lower the number of fatalities associated with

The Inequality of COVID-19. https://doi.org/10.1016/B978-0-323-99867-3.00002-8

coronavirus in a relatively shorter time, yet it was the original epicenter. Its rapid turnaround signified that robust government action was essential in managing the pandemic. Perceived lack of capacity in Indigenous societies caused much anxiety within the United Nations and other multilateral organizations.

There was consensus that COVID-19 would be a virus of grave consequences in those communities (Hardy, 2020). The unequal capacity of these communities was manifest in lower human power, lack of ICU facilities, lowly paid doctors, absence of personal protection equipment (PPE), and more significantly data that showed that Indigenous societies have been prone to more significant loads of debilitating diseases which made it easier for coronavirus and other respiratory pandemics to overwhelm already weak health-care systems.

Weak health-care systems and Indigenous populations

Of course, the overlap between weak health-care systems and colonially induced violence unmistakable. And there is also continued violence inflicted through persistent lack of political power and enduring violation of rights of Indigenous people. It matters that the probabilities of Indigenous communities to gain power and leverage health policies in their favor are heavily curtailed by a biased political system. Thus, health outcomes experienced in marginalized socioeconomic classes and ethnicities generally reflect both prevailing policy vulnerabilities and inequalities in political resources. In the United States, in particular, the virus did more harm to the Tribal nations because the federal government mishandled its containment of nationwide community spread, which, in turn, trickled down to Navajo people, among others. Thus, from the beginning, the virus was unequal in its target and outcomes. The differential resources base including unequal distribution of information on pandemic mitigation played a part in the COVID-19 management efforts.

The raw numbers from Johns Hopkins (2021) show that by end of February 2021, for example, 511,787 deaths occurred in the United States. India reported 156,938 deaths, compared to Kenya's 1853, Tanzania's 21, and Australia's 909. Of course, the numbers from African countries are underreported and worse still, Tanzania declined to release data until February 2021. These numbers do not tell the whole story of the impact of COVID-19 on Indigenous societies. Significantly, the unequal virus's manifestation is directly related to government actions and a reflection of existing hierarchies that place Indigenous people at lower ranks of society.

Oxfam International (2021) researchers led by Esme Berkhout, Nick Galasso, Max Lawson, Pablo Morales, Angela Tanjela, and Diego Pimentel considered the inequality virus as a drawback to global economic recovery. Oxfam articulated coronavirus pandemic's potential to lead to increased poverty especially in low-income communities.

Its primary message is that government workers who manage virus matters such as the "healthcare professionals and other essential workers are overwhelmingly people from ethnic minorities and are the most overworked, underpaid, undervalued and often not protected, even in the midst of deadly pandemic" (Oxfam International, 2021). This inequality is also manifest in the existence of "chronically neglected health-care systems" especially those that serve communities such as Native Americans and Aboriginal people (Yashadhana et al., 2020). Likewise, the Siddi people, who live mostly in economically marginalized and densely populated zones of Western Indian states of Gujarat and Maharashtra, dealt with the highest burden of coronavirus disease (Rath et al., 2020). By virtue of being among the poorest communities, Indigenous people are likely to experience more devastating effects of coronavirus. In general, across populations the likelihood of premature death from COVID-19 increases with lower incomes statuses (Brown & Ravallion, 2020; Palmer, 2020). Further, the inequality virus is about the increase in wealth for the fortunes of the top billionaires during the first 9 months of the COVID-19 crisis. Their wealth, Oxfam argues, is sufficient to "prevent anyone on Earth from falling into poverty" and to pay for vaccines for all (Oxfam International, 2021).

One of the reasons for writing this book was to reflect on the prevailing responses to COVID-19 within selected countries that have significant Indigenous populations. This book described what governments and citizens chose to do or not do about the COVID-19. In other words, governments are complicit in the outcomes of the COVID-19 pandemic. Political science scholars assume that the policy actions implemented to mitigate the pandemic are in accordance with government goals and professional judgment of public officials, especially those in public health organizations and/or administrative departments. From a democratic governance standpoint, the decisions taken are meant to reflect the will of the people governed. Yet, critical analytical lenses that include postcolonial approaches regard the prevailing health-care conditions in Indigenous communities as a continuation of systems that abrogated the existing moral and legal treaties. These treaties shape and structure the intergovernmental relationships between dependent marginalized communities and national governments.

Before we looked at the individual country situations, we made a point that governments across the globe maintained close and multifaceted ties with Indigenous communities. Through these networks, some aspects of the COVID-19 mitigation actions were handled through fragmented authorities. Each of the communities considered in this book had its own interests and community organizations which performed with limited means and capacities in efforts to curb the spread of COVID-19. Given this situation, we have a clear picture of how the hierarchical relation between the Indigenous societies and the national governments permitted or directed these societies to respond to the pandemic. In other words, the dominant national governments activated governance networks in bid to address the pandemic producing the varied results. Yet, we also know that the governmental bodies and leaders involved should have arrived at joint approaches to establish structured interactions and communications geared toward tackling the global pandemic. Ideally these agreements would include leadership by Indigenous people themselves exercising sometimes political and sometimes social sovereignty. This did not happen, perhaps because "although everyone is connected, no one is in control" (Zakaria, 2020. p.13).

Essentially, we assumed that tribal and Indigenous sovereignty and autonomy in managing an existential such as the COVID-19 crisis would be guided by some moral campus, particularly dominant powers were expected to recognize the limited resources available in Indigenous communities and move forward collaboratively to address the existential threat posed by the coronavirus. This would also mean tangibly honoring the magnitude of social and cultural loss that was and continues to be at stake for Indigenous peoples and nations. This discussion reveals that the responses by national governments impacted these marginalized communities in varied ways that did not successfully lead toward public health outcomes, and instead Indigenous people negotiated and developed successful strategies in spite of external circumstances in some cases.

However, for the most part, the response in the different regions was immediate but often disjoined, considering the global nature of the pandemic. And within the different countries selected for this study, there were certainly unequal choices made at every level of government. Perhaps that was because the different regions experienced the pandemic and the outcomes of the virus in uneven and unequal ways. The outcomes were a product of unequal health-care conditions that undergird the inequalities in resources to combat challenges that exit its health-care systems across the globe. The choices made in regions inhabited by Indigenous communities were, for the most part, a reflection of the deployment of national strategies or lack thereof in the respectful regions where these communities are found.

General descriptions of actions and responses taken by Indigenous community leaders are somewhat understudied and often ignored in early appraisals on how COVID-19 pandemic was managed. However, generic policy prescriptions prescribed by health authorities across the globe found unique meanings within tribal and Indigenous communities. The abstract policy prescriptions conveyed different sets of realities for tribal societies, especially the great impact they had on people's own sense of survival and tribal identity existence. The examples of these realities are myriad. COVID-19 had enormous consequences for the ability of elders to carry on their role of custodian of Indigenous cultures. So too, did the pandemic take heavy toll on health-care workers within these regions.

Governments and clustered public policy issues

Public health-care policies that provide for the needs of these communities and beyond were brought under scrutiny. In short, the governments concerned should be aware of at least five different clustered public policy issues for the benefit of future management of pandemics. In no particular order, these issues must be factored in future pandemic mitigation policy actions.

The first issue going forward is that of the digital divide and its attendant negative externalities. Perhaps, more than any other variable, lower institutionalization of new technological innovations in marginalized communities weakened their capacity for COVID-19 relief, remedy, and mitigation. The second area of attention is the lingering effects of low investments in public health-care systems in marginalized communities. And to add on to that, the World Health Organization (WHO) "was not structured or funded for independence or agility," to contain the disease, especially in the developing countries (Tufekci, 2021). Initially, WHO's communication and messaging strategies were also counterproductive. The announcements that coronavirus spread by aerosols came way after thousands had already been infected.

Third, how should global public health institutions evaluate risk in marginalized societies? Risk evaluation is about responding to the question, "what challenges did these communities face that undermined the virus mitigation strategies?" Was the risk amplified through specific cultural behaviors? And was it understood by a majority of tribal members? In order to communicate about risks and to monitor and create action plans against risks, the vibrancy of information and communication technologies (ICTs) again comes into play. Risk also cannot be understood well without attention to agency and efforts to survive.

Fourth, concerning our humanity, just who are we as humanity and what did the COVID-19 experience teach us about our common home and ourselves as a human family? These two moral questions should be discussed in tandem with the assertion in the previous chapter that richer nations, especially those that decidedly adhered to nationalistic agendas, appeared to act in narrow interests to the detriment of the human family. The global COVID-19 vaccine distribution apparatus seemed to point at perpetuation of colonially established barriers in providing equitable solutions to pandemics (Richardson, 2020). Even after the COVAX Facility begun to supply vaccines to developing countries, the in-country mechanisms did not prevent ruling elites from short-changing the marginalized communities. And even within developed nations, some states failed to prioritize providing mitigation measures to marginalized and homeless groups (Ness, 2021).

Fifth, knowing what we know from the scientific community, moving forward, will governments put in place policy frameworks that inspire confidence in solutions prescribed to mitigate pandemic attacks? This is the arena in which pandemic mitigation policy prescriptions, including those that seemingly infringe on personal liberties like mandating mask wearing and abuse of freedoms of speech through disinformation are thoroughly interrogated. It is also the arena where people must explore why there are fears of vaccines and provide clear communication about rapidly emerging health research. New ways of public health messaging solutions to instill confidence in scientific solutions are necessary. We suppose, moving forward, there must be some shared values, especially adherence to scientific knowledge. Crucial for the success of such efforts is better collaboration and listening to tribal nationals and societies about what is at stake and how to best respond.

We identify and discuss these issues and provide social science perspectives on the importance of paying attention to concerns that may be a natural part of successful managing pandemic crisis. These are areas that demand immediate attention by stakeholders in the health-care sectors concerned. In other words, we respond to the question, what should governments and Indigenous societies do to mitigate against the unequal devastation of viruses like COVID-19 in the coming years?

The inequality of COVID-19 and ICT regimes

The year 2020 can be easily understood to be the first year of pandemic mitigation efforts in the age of the internet and social media. As such, many proposals have been suggested, especially the need to tackle the perennial challenge of unequal access to ICTs. These technologies are vital for multiple reasons, ranging from information about prevention methods, e-learning, telemedicine, to scheduling vaccine appointments. Thus, the pandemic's silver lining may be an increased investment in new technologies to combat broadband deserts and enhance capacities of Indigenous societies ability to build educational content for combatting epidemics.

Perhaps, more than ever, new technologies must be harnessed for promoting fact-based public health messages, especially concerning the symptoms of the diseases, treatments and importance of vaccines, social distancing, and other mitigation policies. Freedom House (2020) called for increased training in use of online resources as a way of promoting freedom in times of pandemics like COVID-19. In addition, availability of online resources in Indigenous societies serves to increase Native-led efforts and full participation in decision-making.

Products such as AVA Robotics, which connected families to elderly patients with mobile video robot named AVA, have emerged as relevant for conditions in Indigenous areas. There is more, virtual care created in response to the pandemic will be part of the cloud collaboration platforms telehealth options available moving forward. These technologies will be included in strategies that increasingly rely on data-driven care delivery systems that, hopefully, will enhance clinical communication and collaboration at a global level. Such innovations will, hopefully, make management of the next pandemic much more transformative and effective in Indigenous societies.

COVID-19 increased awareness of broadband limitations in marginalized societies. As such, further investments will have to be made in platforms that support e-learning, particularly directed toward health literacy and future COVID-19 and pandemics communication strategies. This is an important point. Otenyo and Lind's (2011, pp. 200–201) earlier observations had pointed to the perennial poor execution of ICT projects on the Navajo Nation. COVID-19 is a tragic reminder of the urgency for improved delivery of ICT services in all tribal lands. The WHO (2020) not only advises that all countries "expand capacity for communication and proactively manage media" relations to ensure that uniform and truthful messages govern the mitigation processes. In addition, modern ICTs will continue to provide forums for coordination of programs designed for mitigation of future pandemics.

Investments in public health

Incremental and piecemeal additions to health-care budgets will not work in pandemic situations of the future. Instead, bold policy strategies with clear crystal balls and pandemic policy planning are required. Most importantly, COVID-19 has taught us that investments in strategic public health infrastructure are essential. Moving forward, some critical changes will have to be made including establishment of strategic reserves of certain equipment, disease data and surveillance systems, and clinical care facilities in all public health services. The concept of strategic reserves already exists in the supply of grains and other critical food supplies in developing countries. It is appropriate that such reserves be available for critical public health facilities including emergency hospital beds, ventilators, syringes, and durable Personal Protection Equipment (PPE).

The uncertainties which were manifest in ICU and other hospital settings are a wake-up call. Moreover, the low investments in laboratories and virology facilities in developing countries will require greater collaborations among developing countries or states that have limited capacity to build their own state-of-the-art facilities. One way to understand the future is to look at the lessons of COVID-19 management systems. From a scientific perspective, some templates to managing future pandemics are now in place. For one, there are already numerous other viruses circulating which means investments in science and technology are a must for all governments. And it will help for the international community to cooperate and use data gathered for the good of humanity. Research must be interdisciplinary, collaborative, and rapid to combat future pandemics. An approach that ignores Indigenous communities in finding solutions to combat new viruses of concern is bound to fail. Indigenous leaders must be at the table when decisions are made.

Evaluation of risk

Some countries like the United States have often shown a bias toward market solutions to managing health inequalities. The pronounced unequal virus spread in the Navajo Nation, for example, was perhaps a reflection of the tacit bias against communities that are left out of market calculations (Dyer, 2020). This kind of neglect, even if it is couched in racial or ethnic terms, fuels the calculation that "their virus" won't touch "us." The other version of this argument is the Not-In-My-Back Yard (NIMBY) attitude. In February 2020, President Trump had, for example, publicly stated that the "China virus" was not going to affect the United States and that the virus would disappear (Olorunnipa et al., 2020). His Vice President Mike Pence also repeated the remarks noting that the risk of the pandemic affecting Americans was low. Trump went on, on numerous occasions to insinuate that the virus would not pose any major risks to Americans. And wearing masks and social distancing and other precautionary measures were a burden! He was wrong. His assessment of the risk of the global pandemic was utterly misguided and, perhaps, based on his administration's bias against some societies. The bias and mistrust trickled down to President Trump's followers whose risky behaviors became subject of condemnation from health experts. In fact, after President Trump and members of his family became infected, the White House was described by the liberal press as "as a coronavirus super spreader" (Cummings, 2020; Haberman, 2020).

Likewise, there were several government officials who regarded many Indigenous communities as prone to disease because those communities were already vulnerable to other diseases anyways. In many ways, the politicization of disease is not new. It comes with the blame games, scapegoating, xenophobia, and social stigmas associated with communicable diseases and contagion. Instances of members of a society placing blame on others are frequently a product of social inequality and structural biases associated with variables such as gender, race, ethnicity, ageism, and class, among other markers of difference.

This exaggeration of richer communities' own sense of infallibility in terms of their ability to manage pandemics turned out to be short sighted. The risks of mismanaging the pandemic clearly disrupted the markets and economies of the most powerful societies in the world. Refusing to manage COVID-19 and taking risks of opening up the economies even when the community spread of the virus was ill-advised was one of the major stories of the pandemic management failure in many states. The various waves of the pandemic are attributed to both risky decision-making and societal behaviors. The fixation on short-term economic goals provided to be a disaster for the long-term public health goals, resulting in worse outcomes for many communities.

Another element of risk evaluation, from a social science and government behavioral aspect, is the larger question of how governments address the root sources of viruses that transcend national borders. We have in mind, the preliminary results from the WHO sponsored examination of the roots of COVID-19. The existing consensus is, "this pandemic was due to a natural transmission event" and "we can … expect multiple further transmissions of similar, related coronaviruses from the same animal host to human over coming years" (Mallapaty et al., 2021). If the evidence from immunologists is that the virus originated from animals, then policy must assess the nexus with climate change and its impact on animal habitation. The climate change context must be part of the insight and context of developing mitigation strategies across the globe.

There is, however, an element of positive externality in the manner in which business sector made a calculated risk to join hands with the governments to produce vaccines. Operation Warp Speed, which was initiated during the Trump administration, was a public–private partnership (PPP) enterprise. Under this model, vaccines, therapeutics, and diagnostics for COVID-19 were produced quickly. Students of public management will in future use the operation as a model for various reasons. For example, pharmaceuticals like Moderna and Pfizer applied to the Food and Drug Administration (FDA) for emergency use authorization of their mRNA vaccines in record time. These vaccines were based on technology that had been studied before. The whole effort bringing together the National Institutes for Health, universities, FDA, and manufacturing corporations was a remarkable learning experience. Moderna, for example, took a risk and recalibrated its local manufacturing facilities to invest in production of vaccines. The response to the pandemic was a risk in more than just financial. It was also an innovation and technology risk. The FDA, on its part, also without sacrificing drug efficacy and public safety, worked around the clock to approve the use of vaccinations to mitigate the spread of coronavirus.

Thus, from an administrative point of view, collaborative networks matter in emergency situations. In retrospect, one may suggest here that the pandemic underscores the nexus between emergency management and immediate communication and technology systems that enabled rapid transmission of information and policy solutions to urgently end human suffering and death. The COVID-19 vaccine rollout under the Joe Biden administration proved to be a successful operation as measured by the number of vaccine dosages produced and used for purposes of combatting the virus. The government worked amicably with private producers like Johnson and Johnson, Moderna, and Pfizer to spread the risk and manage vaccine prices and distribution and political barriers. Vaccine plans in the Navajo Nation worked remarkably well and tribal nations across the United States rolled out vaccination plans more successfully than the rest of the country. In fact, the Navajo Nation outpaced the states in vaccinations.

According to data from the Indian Health Service and tribal health-care systems, the Navajo Nation administered over 100,000 COVID-19 shots by end of February 2021 (Krisst, 2021). This was 92% of the total 146,980 doses received, a much higher percentage than most states in the union. This meant that 26% of 327,000 enrolled members of the Navajo community were vaccinated (Krisst, 2021). Just to illustrate, by comparison, during the same time frame, in the state of New Mexico about 22.4% of the population had received at least one shot. Arizona had 17.1% and Utah 12.2%, respectively (Krisst, 2021). This endeavor again illustrates the prominence of PPPs. In addition, the leadership of the Navajo Nation and the work of Navajo front-line health-care workers, success in vaccination drives, indicate that sovereign Indigenous societies contribute to knowledge on governance in emergency networks. The resilience shows that investments in Indigenous societies help to improve performance of governments managing new virus disasters.

In the immediate period of the pandemic, successful mitigation initiatives were also reported in Australia. Although as journalist Julie Power (2020) reported citing distinguished Australian epidemiologist Professor Fiona Stanley declared that Australia had the "best result for any Indigenous population in the world," there is still hard work to be done for future pandemics communication and trust building strategies.

In Australia, vaccine skepticism was reported, primarily because of the distrust of government policy (Hoyle, 2021). The mistrust comes from the unfinished job of healing the trauma of settler colonialism and genocide. While we laud Indigenous efforts to reduce deaths from COVID-19, in Australia, the report that an estimated 50–75 of Aborigine people in parts of Australia would not take vaccines represents a challenge in communications and trust building for the future management of pandemics. The same skepticism was also reported in parts of India, which, ironically, is the largest producer of COVID-19 and other vaccines in the world. Therefore, given that some of the areas where Aboriginal and Siddi people live are remote and lack basic infrastructure, the absence of basic infrastructures is an indictment to the global consciousness and a blot on capitalism's ability to foster human dignity. It is an inequality that the pandemic has highlighted.

A Kantian ideal: why we should be humans first

The International Monetary Fund (IMF) has maintained a catalog of the various economic measures initiated in response to managing the COVID-19 economy (IMF, 2021). Countries like India instituted multiple Lockdown and Unlock down measures (Unlock 3.0 was announced September 2020). An important part of the unequal virus is the outreach and communication strategy. COVID-19 has provided tragic lessons on how not to frame a dangerous disease in crisis management situations. Because the marginalized communities suffered disproportionately as a result of the pandemic, the reluctance among some members of the marginalized groups to take vaccines can be attributed to previous experiences of cruelty and abuse. This calls attention to the value of Native-led efforts that involve community members leading and communicating with one another as trusted insiders. Support for these efforts is much more important for crisis response than those that originate outside of Native nations and are imposed externally in ways that do not consider the multiplicity of lived experiences and concerns within multiple Indigenous societies.

Various mitigation measures and the initial disjointed management of the vaccines and therapeutics reflected weak coordinative capacity within the institutional framework governing global public health ecosystem. Furthermore, the WHO's leadership role was immensely diminished, as was the American Centers for Disease Control and Prevention (CDC) especially under the Trump administration. Trump's administration gave notice to leave the WHO in July 2020. This action weakened the agency in more than financial terms. The United States provided about 20% of its budget and also vital public health expertise. In 2021, President Joseph Biden rejoined the WHO, a move that was hailed not just for global solidarity and symbolic terms, but as a step toward enhancing scientific collaboration and norm building in managing pandemics (Weintraub, 2021).

The failure of the international system to contain the spread of COVID-19 in a unified manner has heightened the need for a return to debates that informed global responses to major human crisis such as the World Wars and the 1918 pandemic. Witness that leaders of several nations who framed COVID-19 as a war. As such lessons from history are beneficial, at least in framing responses to a calamity with such horrible outcomes. During major calamities, humans have set aside social and economic differences to combat a common enemy. In those times, the decisions reaffirm our common humanity as opposed to narrow and selfish factional interests.

Scholarly historical debates frequently shed some light on moral and bioethical imperatives for collective action against forces of darkness. Philosophers teach about rights granted to human beings. Humans have rights to life, and in many nations, access to health through combatting pandemics is a responsibility of the respective government. The moral argument for collective action against the spread of COVID-19 was manifest in WHO's proclamations calling for joint efforts and transparency in implementation of mitigation plans. International solidarity in combatting the pandemic was recognition that no society is alone in its struggles to achieve healthy societies. The commitment to Kantian ideals of preserving lives was indeed required in the management of the pandemic primarily because of the unequal virus effects on Indigenous and marginalized societies. Kantian thinking on ruling elites providing more resources to prevent threats to life in communities that we looked at was, therefore, a useful frame to guide future pandemic responses.

Moving forward, the primary questions should be, did powerful institutions act to further the lives of those already marginalized by forces beyond their control? Did they move public health measures forward or provide further challenges to these communities by failing to support and respect agency within? Commitment to the preservation of life in Indigenous societies must be a goal not to be undermined by the emergent vaccine nationalism and other forms of individualistic set of actions that were likely to limit some possibilities of a more comprehensive approach to combating the effects of the pandemic.

Strategies applied for pandemic management can be of benefit to humanity only if policymakers are influenced by this Kantian thinking. For practical purposes, opposition to health-care systems that provide access to vulnerable members of society must be seen as a direct challenge to our interpretation of material equity, which in our view would be in direct support to utilitarian goals championed by ruling elites in advanced Western societies. In other words, supporting public health infrastructures including Native-led and designed pandemic literacy and improved sanitation in tribal and marginalized societies is a win—win proposition. It is not about handouts. A long-term strategy is also one that raises strong generations of Indigenous researchers, scientists, medical professionals, policymakers, and global leaders to join those who have already become prominent global leaders and participants.

COVID-19 demonstrated examples of Native-led initiatives that were better than those in border towns like the management and response of the Navajo Nation. Their call for federal support, which had been delayed, was therefore, to ensure that they too as members of the American polity received fair treatment from federal institutions. Indeed, Kantian arguments present opportunities for the federal and national governments to treat all people fairly and to prevent death and illness in areas of the country that have the highest vulnerabilities. Indigenous people all over the world are particularly vulnerable to injustice and policies that perpetuate inequalities in quality of life. This necessitates that future responses to pandemics must be more proactive in such communities. Ignoring conditions in marginalized communities is a reflection of injustices and moral inadequacy. And future management policies must recognize that social justice through health care can only work if it is not localized but universalized. Protecting vulnerable communities, therefore, is a global enterprise that requires solutions by all countries, rich and poor.

Science and global shared values

To deal with misinformation and disinformation during pandemics, should the nations put a hold on freedoms of speech? Or should each government impose heavy penalties on groups or persons who deliberately misinform members of the public on dangers and dynamics of pandemics of the future? That is, should global institutions like the WHO be empowered to override national government communications on matters of global public health? Particular international norms and local statutory or regulatory provisions can be adopted to help curb the effects of disinformation and misinformation and the impacts on especially marginalized society. This point has also been reinforced in Freedom House (2020) prescriptions for enlargement of democratic spaces and freedoms.

Most important, policy conflicts such as the well-documented difference between Maasai living in two different countries of Kenya and Tanzania made it increasingly difficult for generic COVID-19 mitigation plans to succeed. The different views were not just about roles of government in regulating behavior but about the nature of regional politics. Pandemics management requires international cooperation and not competition about which solutions boost political credentials of leaders seeking reelection. Inevitably, policy-making involves choices about global health as social value for international peace and security. In the heat of debates over who caused the virus to spread, there will be national interest groups seeking to control the narrative in their favor. Who will arbitrate these differences in opinion for the sake of global peaceful coexistence? Moving forward, international organizations must forge mechanisms for consensus building rather than promote narrow interests because deep conflicts during a pandemic do not serve the interest of any single nation.

The noted devaluation of sciences and undermining of scientific ways of mitigating the scourge of COVID-19 will remain one of the worst manifestations of the antiscience movement that has taken root in some societies globally. The antiscience movement is not unique to developing countries and has frequently been associated with religious measures. For example, many religious communities in Maasai regions placed enormous pressure on local governments to allow churches to reopen and allow congregations to participate in religious ceremonies. And framing religious meetings as super spreader evens often conjured the impression that the secular institutions were at war with faith-based organizations. That is not the case. Almost certainly, weak enforcement of COVID-19 mitigation protocols in regions that were unwilling to accept scientific prescriptions was by many measures a failure to respond to negative effects of fanaticism. The scientific mitigation plans are humanity's shared values. And that is how all governments may frame their policy responses. Again, this is an area where policy decisions will have to be bold enough to communicate effectively the dangers of unreasonable risks to public health imposed by religiosity.

Existential threat and enduring structural barriers to equitable public health

Perhaps, the closing remarks in this book should be to appraise the learning experience in management of the tragedy of COVID-19. The pandemic provided huge lessons for humanity. It was the most serious existential threat to global livelihoods and the targets of mitigations policy continued to shift by the day. Public health professionals learned new methods of treating patients and understood the symptoms and effects of vaccines developed to combat the spread of the virus. The community of health-care workers readily learned from each other new information about the novel virus, regardless of the source of the emerging clinical ideas.

Thus, due to the huge loss of lives and grief, the pandemic was also an opportunity for pursuance of global solidarity and striving for common humanity interests. In which case, the emerging policy recommendations on pandemic mitigation by health experts should be shared as part of the global commons. Emerging literature highlights the need for data-driven strategies. This includes collection of data on immigrants and Indigenous citizens in developed countries to address the disparities in the spread of COVID-19. Health professionals note that some communities faced structural disadvantages that leave them vulnerable to communicable diseases (Langellier, 2020). For instance, poor housing conditions make it much harder to control diseases. This implies a renewed attention to global housing and sanitation conditions (Hardy et al., 2021). We noted in the case of the Navajo Nation and among the Maasai communities that the coronavirus pandemic exposed the risks caused by lack of access to water and sanitation. Governments must, therefore, pay close attention to enduring structural barriers to poor housing and resolve contradictions in neoliberal economic policies which sustain social and economic inequalities in low-income communities.

Conclusion

This work has sketched out some of the factors and forces that influenced the ineffectiveness of national government COVID-19 management and especially in pandemic communication strategies. While the international institutions initiated generic COVID-19 mitigation strategies, these too did not achieve the desired outcomes. Inequalities in capacity to mitigate against the ravages of the coronavirus hampered effectiveness in all cases discussed here. Of course, diligent national strategies in Australia resulted in relatively better outcomes for the Indigenous people of that country. Australia's leaders provided clear national signals to warn against lack of adherence to science and evidence-based remedial measures. In addition, their whole-society approach and leadership that instilled public trust seemed to work well.

On the other hand, the Indian states where a vast number of Siddi people reside, initially, had high incidence of COVID-19. And since many Siddi people were among the migrant workers performing domestic work in highly populated cities they were among those with disproportionately high incidence of COVID-19. These laborers (shramik) serve the large cities like Mumbai in Maharashtra State and Surat, Gujarat, where their housing conditions are, for the most part, crowded, squalid, and lacking proper sanitation facilities. Although relatively few, Siddi people constitute part of the fairly large number of the Scheduled Tribes, many of whom also suffer from social isolation and marginalization. Initially, many of these groups only knew about coronavirus after people around them started dying. This signified the low level of health literacy in these communities. Although the Indian government under President Narendra Modi at policy proclamations level rejected the place of caste in management of COVID-19, the underprivileged castes were initially not included in mitigation management (Gettleman et al., 2020). The major lesson from the Indian experience is that migrant workers from cities who traveled to their rural homes were a significant factor in the spread of COVID-19. This has significant policy implications, for future pandemic responses. At the very minimum, controlling movements within the country is as important as the globally instituted ban on international flights and other international border control measures.

At the same time, the disjointed and lack of national strategy in the United States under Donald Trump's administrative apparatus produced disastrous side effects and high death rates within the Navajo and other tribal societies. By March 2021, 1233 deaths and close to 30,007 cases of COVID-19 had been reported in the Navajo Nation. Through amateur administration practices, the CDC, White House, and lead agencies caused the 50 states to develop uncoordinated mitigation policies with disastrous results.

Among Maasai people, those in rural areas fared better than their counterparts in urban settings. Worse still, different outcomes appeared to mirror the different national strategies in the two countries where Maasai people reside. Certainly, the COVID-19 denial and carefree style of President John Magufuli's administration in Tanzania proved to be a disaster as Maasai in Tanzania lost lives. The government refused to report numbers of casualties or cases. President Magufuli, who lost several members of his cabinet from ailments that were associated with COVID-19, died in March of 2021 from heart ailments. As an avid denier, the press relentlessly associated his demise to coronavirus pandemic (Kaiser Health News (KHN), 2021).

Magufuli's disregard for interventions promoted by the WHO caused President Joseph Biden's US State Department to issue travel bans to Tanzania. In February 2021, the CDC issued a Level 4 Travel Health Notice for Tanzania, which included the heavily trafficked Maasai tourism regions. This is significant because it is the highest level reserved mostly to countries that are experiencing natural disasters, political and social unrest, and war. Tanzania had not even bothered to be CDC compliant testing for COVID-19. Thus, at a minimum, future public health diplomacy will have to be a part of the pandemic mitigation planning and implementation in all countries. Tourism and other international travel enterprises will require much more investments in efforts to monitor incidences of communicable diseases and virus outbreaks.

Moving forward, the challenge for citizens in marginalized communities is to consider their cumulative knowledge of how to manage pandemics and to preserve what worked while discarding old methods that showed little promise. If Indigenous societies hold to their effective and proactive measures of managing pandemics, they must preserve those while addressing the concerns that were experienced from the COVID-19 pandemic.

This book confirms our observations that Indigenous initiatives buttressed in local cultures proved to be an important part of the framing of the pandemic and preventing further devastation to these communities. American Indian and other Indigenous societies proved to be remarkably strong, bouncing back when waves of the pandemic disrupted their social order. Navajo responses, for example, added value to the COVID-19 mitigation efforts. It would appear reasonable to acknowledge that the tribal structural relationships helped members to help one another with a deeply personal knowledge of what was at stake. And many Indigenous communities shared the power of partnerships with nonprofit organizations and even corporations to tackle the pandemic-related problems at hand.

What is more redeeming is that local responses among groups like Maasai people and Australia's Aboriginal communities were based on self-reflective community efforts. These groups agreed to both traditional and government designed COVID-19 mitigation policies. These communities stood out as examples of what can be accomplished when people gather their own collective identity. These societies suspended a vast majority of their collective ceremonial events and practiced mitigation protocols such asocial distancing at higher levels than most communities. They also adapted new ways of interacting at marketplaces where many interactions with outsiders occur.

Finally, this book is a small part of the conversation about the experience of Indigenous societies with COVID-19 management. It focused on these communities to highlight ways to prevent future disappointment in effectiveness of pandemic management. It is imperative that these conversations continue to take place among all stakeholders who ultimately have to provide answers to resolving issues of inequality in health care and pandemic response. This book contends that the unequal virus impact on Indigenous people is rooted within the context of entrenched political and socio-economic marginalization of these communities. Our hope is that policy adjustments are implemented to end unequal virus loads on economically marginalized societies across the globe.

In closing, we pause to remember the countless lives lost to COVID-19 including those of many tribal elders carrying the culture, language, history, and present of global Indigenous Nations. There are no words for the depth of pain and grief of the fallout of COVID-19. We offer this work in the hopes that we, as a global community, will respond more effectively, respectfully, and swiftly in future global crises in ways that honor sovereignty, Native-led organizations, and survivance across this precious world.

References

Brown, C., & Ravallion, M. (2020). *Poverty, inequality, and COVID-19 in the US.* https://voxeu.org/article/poverty-inequality-and-covid-19-us.

Cummings, W. (October 9, 2020). *The data speak for themselves': Dr. Anthony Fauci says White House held a 'superspreader event' for coronavirus.* USA Today. https://www.usatoday.com/story/news/politics/2020/10/09/anthony-fauci-white-house-held-covid-superspreader-event/5940483002/.

Dyer, O. (2020). Covid-19: Black people and other minorities are hardest hit in U.S. *British Medical Journal, 369*, m1483.

Freedom House. (2020). *Policy recommendations: COVID-19 and the global struggle for freedom.* https://freedomhouse.org/report/report-sub-page/2020/policy-recommendations-covid-19-and-global-struggle-freedom.

Gettleman, J., Raj, S., Yasir, S., & Singh, K. D. (December 15, 2020). *The virus trains: How lockdown chaos spread COVID-19 across India.* New York Times. htps://www.nytimes.com.

Haberman, M. (October 5, 2020). *Trump's physician says the president's oxygen levels dropped and he took steroid treatment.* New York Times. https://www.nytimes.com/live/2020/10/02/world/covid-19-coronavirus.

Hardy, L. J. (2020). Negotiating inequality: Disruption and COVID-19 in the United States. *City & Society.* https://doi.org/10.1111/ciso.12312. American Anthropological Association.

Hardy, L. J., Thompson, K. F., & O'Connell, M. C. (March 15, 2021). *Fixing indoor air pollution problems that are raising native Americans' COVID-19 risk.* The Conversation. https://theconversation.com/fixing-indoor-air-pollution-problems-that-are-raising-native-americans-covid-19-risk-156727.

Hoyle, R. (March 27, 2021). Indigenous Australians are skeptical about COVID -19 vaccines, and they're hard to reach. *The Wall Street Journal.* https://www.wsj.com/articles/indigenous-australians-are-skeptical-about-covid-19-vaccines-and-theyre-hard-to-reach-11616846401.

International Monetary Fund (IMF). (2021). *Policy responses to COVID-19.* Washington, D.C: IMF. https://www.imf.org/en/Topics/imf-and-covid19/Policy-Responses-to-COVID-19.

Johns Hopkins University. (2021). *COVID-19 dashboard by the Center for Systems Science and Engineering (CSSE) at Johns Hopkins University (JHU).* Coronavirus Resource Center. https://coronavirus.jhu.edu/map.html.

Kaiser Health News, KHS. (March 18, 2021). *President of Tanzania, a COVID denier, dies amid rumor he had coronavirus.* Kaiser Family Foundation. https://khn.org/morning-breakout/president-of-tanzania-a-covid-denier-dies-amid-rumors-he-had-coronavirus/.

Krisst, R. (March 9, 2021). *Navajo outpacing states in rate of vaccinations.* Navajo Times. https://navajotimes.com/reznews/navajo-outpacing-states-in-rate-of-vaccinations/.

Langellier, B. A. (2020). Policy recommendations to address high risk of COVID-19 among immigrants. *American Journal of Public Health, 110*(8), 1137−1139. https://doi.org/10.2105/AJPH.2020.305792

Mallapaty, S., Mazman, A., & Callaway, E. (2021). February 10). Major stones unturned: COVID-origin search must continue after WHO report, say scientists. *Nature, 590*, 371−372. https://doi.org/10.1038/d41586-021-00375-7.

Ness, L. Van (March 1, 2021). *States fail to prioritize homeless people for vaccines.* Route Fifty. https://www.route-fifty.com/health-human-services/2021/02/five-pandemic-mistakes-we-keep-repeating/172345/.

Olorunnipa, T., Dawsey, J., & Abutaleb, Y. (February 20, 2020). *Trump downplays risk, places Pence in charge of coronavirus outbreak response.* Washington Post. https://www.washingtonpost.com/politics/trump-downplays-risk-places-pence-in-charge-of-coronavirus-outbreak-response/2020/02/26/ab246e94-58b1-11ea-9000-f3cffee23036_story.html.

Otenyo, E. E., & Lind, N. S. (2011). *e-Government: The use of information and communication technologies in administration.* Teneo Press.

Oxfam International. (January 2021). *The inequality virus: Bringing together a world torn apart by coronavirus through a fair, just and sustainable economy.* Oxford, UK: Oxfam Briefing Paper. Oxfam House. https://oxfamlibrary.openrepository.com/.

Palmer, B. (June 5, 2020). *Chart of the week: Covid-19 kills people in the most deprived areas at double the rate of those in the most affluent.* Nuffield Trust. https://www.nuffieldtrust.org.uk/resource/chart-of-the-week-covid-19-kills-the-most-deprived-at-double-the-rate-of-affluent-people-like-other-conditions.

Power, J. (November 13, 2020). *When it came to COVID-19, Indigenous Australians sent it packing.* The Sydney Morning Herald. https://www.smh.com.au/national/when-it-came-to-covid-19-indigenous-australians-sent-it-packing-20201112-p56e5u.html.

Rath, R. S., Dixit, A. M., Koparkar, A. R., Kharya, P., & Joshi, H. S. (2020). COVID-19 pandemic in India: A Comparison of pandemic pattern in Selected States. *Nepal journal of epidemiology, 10*(2), 856−864. https://doi.org/10.3126/nje.v10i2.28960.

Richardson, E. T. (2020). *Epidemic illusions: On the coloniality of global public health.* M.I.T. Press.

Tufekci, Z. (February 28, 2021). Five pandemic mistakes we keep repeating. *Commentary*. Route Fifty. https://www.route-fifty.com/health-human-services/2021/02/five-pandemic-mistakes-we-keep-repeating/172345/

Weintraub, K. (January 22, 2021). *Biden administration renewed support for World Health Organization is 'good news for America and the world,' scientists say*. USA Today. https://www.usatoday.com/story/news/health/2021/01/22/scientists-applaud-biden-decision-rejoin-world-health-organization/4243377001/.

World Health Organization (WHO). (April 6, 2020). *Policy, guidance and recommendations for COVID-19 decision making*. Geneva: WHO, Regional Office Europe. https://healthmanagement.org/c/hospital/news/policy-guidance-and-recommendations-for-covid-19-decision-making.

Yashadhana, A., et al. (July 24, 2020). Indigenous Australians at increased risk of COVID-19 due to existing health and socioeconomic inequalities. *The Lancet Regional Health-Western Pacific, 1*, 100007. https://doi.org/10.1016/j.lanwpc.2020.100007

Zakaria, F. (2020). *Ten lessons for a post-pandemic world*. W.W. Norton.

Index

Note: 'Page numbers followed by "t" indicate tables.'

R

S

T

9780323998673